AF581639

Abiotic Stresses, Biostimulants and Plant Activity

Abiotic Stresses, Biostimulants and Plant Activity

Editors

Daniele Del Buono
Primo Proietti
Luca Regni

MDPI • Basel • Beijing • Wuhan • Barcelona • Belgrade • Manchester • Tokyo • Cluj • Tianjin

Editors
Daniele Del Buono
University of Perugia
Perugia
Italy

Primo Proietti
University of Perugia
Perugia
Italy

Luca Regni
University of Perugia
Perugia
Italy

Editorial Office
MDPI
St. Alban-Anlage 66
4052 Basel, Switzerland

This is a reprint of articles from the Special Issue published online in the open access journal *Agriculture* (ISSN 2077-0472) (available at: https://www.mdpi.com/journal/agriculture/special_issues/Abiotic_Stresses_Biostimulant_Plant).

For citation purposes, cite each article independently as indicated on the article page online and as indicated below:

LastName, A.A.; LastName, B.B.; LastName, C.C. Article Title. *Journal Name* **Year**, *Volume Number*, Page Range.

ISBN 978-3-0365-6964-2 (Hbk)
ISBN 978-3-0365-6965-9 (PDF)

Contents

About the Editors . . . . **vii**

Daniele Del Buono, Luca Regni and Primo Proietti
Abiotic Stresses, Biostimulants and Plant Activity
Reprinted from: *Agriculture* **2023**, *13*, 191, doi:10.3390/agriculture13010191 . . . . **1**

İsmail Sezer, Hasan Akay, Zeki Mut, Hakan Arslan, Elif Öztürk, Özge Doğanay Erbaş Köse and Mehmet Sait Kiremit
Effects of Different Water Table Depth and Salinity Levels on Quality Traits of Bread Wheat
Reprinted from: *Agriculture* **2021**, *11*, 969, doi:10.3390/agriculture11100969 . . . . **7**

Talaat Ahmed, Ahmed Abou Elezz and Muhammad Fasih Khalid
Hydropriming with Moringa Leaf Extract Mitigates Salt Stress in Wheat Seedlings
Reprinted from: *Agriculture* **2021**, *11*, 1254, doi:10.3390/agriculture11121254 . . . . **21**

Luca Regni, Daniele Del Buono, Begoña Miras-Moreno, Biancamaria Senizza, Luigi Lucini, Marco Trevisan, et al.
Biostimulant Effects of an Aqueous Extract of Duckweed (*Lemna minor* L.) on Physiological and Biochemical Traits in the Olive Tree
Reprinted from: *Agriculture* **2021**, *11*, 1299, doi:10.3390/agriculture11121299 . . . . **35**

Tzu-Ya Weng, Taiken Nakashima, Antonio Villanueva-Morales, J. Ryan Stewart, Erik J. Sacks and Toshihiko Yamada
Assessment of Drought Tolerance of *Miscanthus* Genotypes through Dry-Down Treatment and Fixed-Soil-Moisture-Content Techniques
Reprinted from: *Agriculture* **2022**, *12*, 6, doi:10.3390/agriculture12010006 . . . . **49**

Muhammad Azeem, Muhammad Zulqurnain Haider, Sadia Javed, Muhammad Hamzah Saleem and Aishah Alatawi
Drought Stress Amelioration in Maize (*Zea mays* L.) by Inoculation of *Bacillus* spp. Strains under Sterile Soil Conditions
Reprinted from: *Agriculture* **2022**, *12*, 50, doi:10.3390/agriculture12010050 . . . . **67**

Akanksha Sehgal, Kambham Raja Reddy, Charles Hunt Walne, T. Casey Barickman, Skyler Brazel, Daryl Chastain and Wei Gao
Individual and Interactive Effects of Multiple Abiotic Stress Treatments on Early-Season Growth and Development of Two *Brassica* Species
Reprinted from: *Agriculture* **2022**, *12*, 453, doi:10.3390/agriculture12040453 . . . . **89**

Tessia Rakgotho, Nzumbululo Ndou, Takalani Mulaudzi, Emmanuel Iwuoha, Noluthando Mayedwa and Rachel Fanelwa Ajayi
Green-Synthesized Zinc Oxide Nanoparticles Mitigate Salt Stress in *Sorghum bicolor*
Reprinted from: *Agriculture* **2022**, *12*, 597, doi:10.3390/agriculture12050597 . . . . **111**

Alberto Marco Del Pino, Luca Regni, Alessandro Di Michele, Alessandra Gentile, Daniele Del Buono, Primo Proietti and Carlo Alberto Palmerini
Effects of Selenium-Methionine against Heat Stress in Ca^{2+}-Cytosolic and Germination of Olive Pollen Performance
Reprinted from: *Agriculture* **2022**, *12*, 826, doi:10.3390/agriculture12060826 . . . . **127**

Qi Xin, Bangdi Liu, Jing Sun, Xinguang Fan, Xiangxin Li, Lihua Jiang, et al.
Heat Shock Treatment Promoted Callus Formation on Postharvest Sweet Potato by Adjusting Active Oxygen and Phenylpropanoid Metabolism
Reprinted from: *Agriculture* **2022**, *12*, 1351, doi:10.3390/agriculture12091351 . . . . **139**

Nawroz Abdul-razzak Tahir, Djshwar Dhahir Lateef, Kamil Mahmud Mustafa and Kamaran Salh Rasul
Under Natural Field Conditions, Exogenous Application of Moringa Organ Water Extract Enhanced the Growth- and Yield-Related Traits of Barley Accessions
Reprinted from: *Agriculture* **2022**, *12*, 1502, doi:10.3390/agriculture12091502 . . . . **155**

Yan Ma, Caihong Bai, Xincheng Zhang and Yanfeng Ding
Nitrate Increases Aluminum Toxicity and Accumulation in Root of Wheat
Reprinted from: *Agriculture* **2022**, *12*, 1946, doi:10.3390/agriculture12111946 . . . . **179**

Kanagaraj Muthu-Pandian Chanthini, Sengottayan Senthil-Nathan, Ganesh-Subbaraja Pavithra, Pauldurai Malarvizhi, Ponnusamy Murugan, Arulsoosairaj Deva-Andrews, et al.
Aqueous Seaweed Extract Alleviates Salinity-Induced Toxicities in Rice Plants (*Oryza sativa* L.) by Modulating Their Physiology and Biochemistry
Reprinted from: *Agriculture* **2022**, *12*, 2049, doi:10.3390/agriculture12122049 . . . . **189**

Nawroz Abdul-razzak Tahir, Kamaran Salh Rasul, Djshwar Dhahir Lateef and Florian M. W. Grundler
Effects of Oak Leaf Extract, Biofertilizer, and Soil Containing Oak Leaf Powder on Tomato Growth and Biochemical Characteristics under Water Stress Conditions
Reprinted from: *Agriculture* **2022**, *12*, 2082, doi:10.3390/agriculture12122082 . . . . **205**

About the Editors

Daniele Del Buono

Daniele Del Buono (Prof), who graduated in Chemistry, holds a PhD in Environmental and Agricultural Sciences. Since 2015, he has been an Associate Professor in Agricultural Chemistry at the Department of Agricultural, Food and Environmental Sciences of the University of Perugia. He is the author of numerous publications in international journals. He is included in the World's Top 2% of Scientists List of 2021.

Primo Proietti

Primo Proietti (Prof) is a Full Professor for Tree Cultivation at the University of Perugia (UNIPG), Italy. At UNIPG, he is the Coordinator of the "Tree science" Research Unit, Coordinator of one Degree Course and of one Master's Degree Course, and a Member of four PhD Committees. He is included in the World's Top 2% of Scientists List of 2021 and 2022.

Luca Regni

Luca Regni (Dr) is Researcher for the scientific sector Tree Cultivation at the University of Perugia, Italy. He obtained a PhD in Agricultural, Food and Environmental Sciences and Biotechnology in 2017. He is the author of numerous publications in international and national journals.

Editorial

Abiotic Stresses, Biostimulants and Plant Activity

Daniele Del Buono *,†, Luca Regni *,† and Primo Proietti *,†

Department of Agricultural, Food and Environmental Sciences, University of Perugia, Borgo XX Giugno, 06121 Perugia, Italy

* Correspondence: daniele.delbuono@unipg.it (D.D.B.); luca.regni@unipg.it (L.R.); primo.proietti@unipg.it (P.P.)

† These authors contributed equally to this work.

1. Introduction

Contemporary agriculture is characterized by a highly intensive nature and productivity. Furthermore, this activity is denoted by the substantial impact on natural ecosystems caused by water consumption and the use of fertilizers, plant growth promoters and herbicides/pesticides. In addition, it should also be considered that along the entire supply chain, such activity produces large quantities of waste and causes the emission of significant amounts of greenhouse gases (GHG) [1]. This places agriculture among the anthropogenic activities that contribute significantly to climate change. At the same time, agriculture is seriously affected by these processes, suffering the impact of abiotic stresses such as salinity, drought and high temperatures.

Abiotic stresses can significantly decrease plant growth and development, as well as negatively influence crop quality and productivity. As a result of climate change, abiotic stresses will be characterized by increasing intensity and frequency in the coming years and will put more intense pressure on agricultural systems. Consequently, crop production could dramatically decline, an especially worrisome prospect considering that agricultural systems must also cope with the food needs of the world ever-growing population. For these reasons, new or effective low or no climate-impacting measures need to be considered and developed to maintain/increase crop production and the resilience of agricultural systems, while working to minimize the impact of abiotic stresses.

An agronomic tool of increasing interest is the use of different formulations of certain organic materials and microorganisms, defined by the term biostimulants. Biostimulants are usually grouped into different families based on the raw materials used for their production: humic substances, complex organic materials, beneficial chemical elements (e.g., silicon), inorganic salts, algae and plant extracts, protein hydrolysates, chitin and chitosan derivatives, antiperspirants (e.g., kaolin), amino acids and other compounds [2]. These substances can improve plant stress tolerance, crop nutrient use efficiency, the bioavailability of nutrients in the soil or rhizosphere and quality traits. For the abovementioned reasons, biostimulants can benefit crops when applied under optimal environmental conditions and in states of abiotic and biotic stress.

In this context, the aim of this Special Issue *Agriculture*, entitled "Abiotic Stresses, Biostimulants and Plant Activity", was to advance knowledge on the effect of biostimulants, both in crops grown under normal conditions and in the presence of abiotic stress conditions. We focused, in particular, on heat, salt, drought stress, potentially toxic metals, multiple stresses, and improving plant tolerance. Therefore, this Special Issue paid attention to scientific contributions regarding the stimulatory and protective effects of different biostimulants on crops, their mechanism of action, and their qualitative, economic, and environmental benefits.

2. Special Issue Overview

Related to the use of plant extracts capable of promoting beneficial effects in crops grown in non-stressful conditions, interesting experimental evidence has emerged regard-

Citation: Del Buono, D.; Regni, L.; Proietti, P. Abiotic Stresses, Biostimulants and Plant Activity. *Agriculture* **2023**, *13*, 191. https://doi.org/10.3390/agriculture13010191

Received: 9 January 2023
Accepted: 11 January 2023
Published: 12 January 2023

ing two plant extracts obtained from Moringa and *Lemna minor*. In particular, a foliar extract of Moringa (*Moringa oleifera* L.) was tested on different barley accessions [3]. The leaf extract positively influenced the crop growth and yield, albeit with different strengths in the different accessions considered. The greatest effect of Moringa extract was ascertained on total crop yield, followed by increases recorded in straw weight and number of tillers per plant. This research, therefore, showed that the foliar application of Moringa extract could be an effective solution to stimulate growth and yield in barley.

Another study proposed the use of an extract from duckweed (*Lemna minor* L.), a free-floating aquatic species, to promote beneficial effects in olive plants. The foliar application of the extract stimulated plant growth and improved physiological and biochemical traits in the treated samples [4]. Indeed, the extract positively influenced leaf net photosynthesis, stomatal conductance, sub-stomatal CO_2 concentration and chlorophyll content. Furthermore, the duckweed extract increased the uptake of nutrients like nitrogen (N), potassium (K), calcium (Ca), magnesium (Mg), iron (Fe) and zinc (Zn). The broad diversity of bioactive compounds (including phytohormones, phenolics and glutathione) found in the extract, through untargeted metabolomic profiling, explained the stimulatory effect observed on the olive plants.

Heat, found among the abiotic stresses exacerbated by climate change, is expected to affect plant growth, crop yields, and product quality. The negative effects of this environmental stress can also impact the characteristics of olive pollen. In this context, the potential of selenium–methionine (Se-met) to mitigate the negative effects of heat stress on olive pollen germinability, morphology and cytosolic Ca^{2+} content was investigated [5]. A temperature of 40 °C caused a marked reduction in the olive pollen germination rate, changes in the morphology of the external pollen wall, and a decreased response to Ca^{2+}-agonist agents. The adverse effects of heat stress were counteracted by Se-met, which improved the germination rate, and Ca^{2+}-cytosolic homeostasis, containing the hydrogen peroxide toxicity.

Since heat stress can influence the metabolic processes that enable callus formation, a study investigated the effect of rapid high-temperature (RHT) treatment at 50, 65, and 80 °C on sweet potato tubers [6]. The results showed that appropriate RHT treatment at 65 °C can stimulate the metabolism of reactive oxygen species (ROS) at the injury site of sweet potato on the first day. However, significant ROS formation and scavenging activity were maintained within five days after RHT treatments. Consequently, ROS induced the phenylalanine ammonia-lyase (PAL), 4-coumarate-CoA ligase and cinnamate-4-hydroxylase activities of the phenylpropane metabolic pathway and promoted the rapid synthesis of chlorogenic acid, p-coumaric acid, rutin, and caffeic acid at the injury site, which stacked to induce callus formation. These results evidenced that appropriate high-temperature rapid treatment can promote sweet potato callus formation through ROS modulation and phenylpropane metabolism. Moreover, antioxidant enzymes, PAL, and chlorogenic acid appeared to be key factors in promoting the metabolic pathways involved in the sweet potato callus formation.

Salt stress is one of the abiotic factors that cause significant problems for crops, and it has been estimated to be responsible for more than 50% of crop losses worldwide. Moreover, its impact is increasing, especially in arid and semi-arid regions or coastal areas, a phenomenon also due to climate change. In recent years, the application of nanotechnologies in agriculture has gained particular attraction since specific nanomaterials can show stimulatory effects on crops and increase their capacity to cope with environmental stress. In particular, green synthesized nanoparticles (NPs) can be used as eco-friendly and cost-effective methods to counteract salt stress, thanks to their potential to exert biostimulant effects. The efficacy of zinc oxide nanoparticles (ZnO NPs), synthesized using *Agathosma betulina*, to mitigate severe salt stress was investigated in *Sorghum bicolor* [7]. Salt negatively affected *S. bicolor* growth, causing severe deformation on the epidermis and vascular bundle tissue anatomical structure, while increasing the Na^+/K^+ ratio, oxidative stress (ROS and malondialdehyde content), the activity of some antioxidant enzymes and the content of proline and soluble sugars. Seed priming with ZnO NPs counteracted the negative effects of the salt stress, promoting plant growth, allowing plants to recover a well-organized anatomical structure, decreasing the Na^+/K^+ ratio and reducing the cellular oxidative stress.

Since salinity can affect seed germination and the early stages of plant development that shape the entire life of the crop, the use of biostimulants or plant extracts containing bioactive compounds is a strategic way to counteract the adversity caused by this stress. In this vein, a study published in this Special Issue sheds light on the negative effect exerted by salt stress on wheat in terms of seed germination, plant growth and yield. An interesting method, proposed by the authors of this research, is the priming of wheat seeds with Moringa leaf extract [8]. This approach positively influenced the seed development and germination parameters, seedling growth and nutrient uptake in salt stress conditions. This effect resulted from the Moringa extract capacity to control and reduce the concentrations of Na and ROS, stimulating, at the same time, nutrient uptake.

Again, related to salinity stress, the effects of different water table depths and groundwater salinity levels under irrigated and rainfed conditions were investigated in wheat, studying some quality parameters (hectoliter weight, fat ratio, starch ratio, protein content, Zeleny sedimentation, wet gluten content, ash ratio, acid detergent fiber (ADF), and neutral detergent fiber (NDF)) [9]. Water table depths positively influenced the quality traits of the above crop, while increased salinity levels resulted in decreases in hectoliter weight, fat ratio, starch ratio, and NDF values and increases in protein ratio, sedimentation value, wet gluten content, ash ratio, and ADF values.

The use of a *Chaetomorpha antennina* aqueous extracts was investigated to decrease the detrimental effects of salt stress on rice [10]. As a result, the seaweed extract promoted rice seed germination, plant growth, leaf water, and photosynthetic pigments, while also reversing the negative impact on protein and phenol content, as well as superoxide dismutase (SOD) activity. Moreover, the extract improved grain protein content and enhanced rice nutritional profile and marketability.

Biostimulants can be efficiently used also against the detrimental effects of drought stress. Indigenously isolated *Bacillus* spp. strains were used in *Zea mays* L., which had been subjected to drought stress, in order to promote plant growth characteristics, including mineral uptake and phytohormone profile [11]. The obtained results, especially those regarding biochemical properties, lipid peroxidation and antioxidant responses, suggested that the amelioration of plant capacity to cope with stress depended on a specific plant–strain interaction. Furthermore, oak leaf extract, biofertilizer, and soil containing oak leaf powder were successfully used to enhance growth and biochemical traits in four tomato genotypes under water stress [12]. The use of oak leaf powder and the extract is particularly interesting since they are low-cost substances, simple to use and represent an environmentally sustainable technique for enhancing tomato resistance to drought. Two methods were evaluated for assessing *Miscanthus* (a high-yielding, warm-season C_4 grass) tolerance to drought stress: a dry-down treatment and a system where soil moisture content (SMC) was maintained at fixed levels using an automatic irrigation system [13]. Since the dry-down treatment simulates the water-stress conditions in the field, it appears to be the more suitable method for selecting drought-tolerant genotypes. On the other hand, the SMC can be used to understand the physiological responses of plants to a certain level of drought stress.

Another highly relevant and complex topic concerns the presence of pollutants in the environment that, if present in cultivated areas, can also be absorbed in high amounts by crops. In this context, the effect of nitrate (N) on Al toxicity and accumulation in the roots of two wheat genotypes, Shengxuan 6 hao (SX6, Al-tolerant genotype) and Zhenmai 168 (ZM168, Al-sensitive genotype), was investigated in a hydroponic experiment with four treatments (control without N or Al, N, Al, and Al+N, respectively) [14]. The results showed that N increased the inhibition of root elongation and aluminium accumulation in roots. The Al-sensitive genotype suffered more serious Al toxicity than its Al-tolerant counterpart. Histochemical observation clearly showed that Al prefers binding on the root apex 7–10 mm zones, and that the Al-sensitive genotype accumulated more Al in these zones. Compared with other treatments, the Al+N treatment had significantly higher O_2^-, superoxides dismutase (SOD), catalase (CAT), peroxidase (POD) activities, H_2O_2, Evans blue uptake, malondialdehyde (MDA), ascorbic acid (AsA), pectin, and hemicellulose 1 (HC1) contents in both genotypes. Under Al+N treatment, O_2^- activity, Evans blue uptake, MDA, and HC1 contents of SX6 were significantly lower than those of ZM168, but SOD, CAT, and POD activities and AsA content exhibited an opposite trend. Therefore, aluminum toxicity and accumulation were aggravated in the roots of wheat seedlings by nitrate.

Climate change can also result in a combination of different abiotic stresses detrimental to plant growth. Two *Brassica* species: *B. oleracea* L. and *B. juncea* were exposed to different abiotic stresses: CO_2, UV-B, temperature (T), CO_2+UV-B, CO_2+T, and CO_2+UV-B+T [15]. Plant growth and development were significantly decreased by each of these stresses and their combinations in both species, except in the case of elevated CO_2 concentrations. On the contrary, increasing CO_2 concentrations alleviated some deleterious impacts of high temperature and UV-B stresses.

3. Conclusions

This Special Issue of *Agriculture*, entitled "Abiotic Stresses, Biostimulants and Plant Activity", includes studies conducted on the impact of some abiotic stresses on widely grown crops and how the use of biostimulants could represent an effective and environmentally friendly means of counteracting the aforementioned adversities. The Academic Editors of this Special Issue hope that this collection of research articles will significantly increase knowledge and further stimulate research in this key area for future agriculture, especially in view of ongoing climate change that will increasingly exacerbate the effects of abiotic stresses on crops.

Author Contributions: Conceptualization, D.D.B., P.P. and L.R.; writing—original draft preparation, D.D.B., P.P. and L.R.; writing—review and editing, D.D.B., P.P. and L.R. All authors have read and agreed to the published version of the manuscript.

Conflicts of Interest: The authors declare no conflict of interest.

References

1. Del Buono, D. Can Biostimulants Be Used to Mitigate the Effect of Anthropogenic Climate Change on Agriculture? It Is Time to Respond. *Sci. Total Environ.* **2021**, *751*, 141763. [CrossRef] [PubMed]
2. Rouphael, Y.; Colla, G. Editorial: Biostimulants in Agriculture. *Front. Plant Sci.* **2020**, *11*, 40. [CrossRef] [PubMed]
3. Tahir, N.A.; Lateef, D.D.; Mustafa, K.M.; Rasul, K.S. Under Natural Field Conditions, Exogenous Application of Moringa Organ Water Extract Enhanced the Growth-and Yield-Related Traits of Barley Accessions. *Agriculture* **2022**, *12*, 1502. [CrossRef]
4. Regni, L.; Del Buono, D.; Miras-Moreno, B.; Senizza, B.; Lucini, L.; Trevisan, M.; Morelli Venturi, D.; Costantino, F.; Proietti, P. Biostimulant Effects of an Aqueous Extract of Duckweed (*Lemna minor* L.) on Physiological and Biochemical Traits in the Olive Tree. *Agriculture* **2021**, *11*, 1299. [CrossRef]
5. Del Pino, A.M.; Regni, L.; Di Michele, A.; Gentile, A.; Del Buono, D.; Proietti, P.; Palmerini, C.A. Effects of Selenium-Methionine against Heat Stress in Ca^{2+}-Cytosolic and Germination of Olive Pollen Performance. *Agriculture* **2022**, *12*, 826. [CrossRef]
6. Xin, Q.; Liu, B.; Sun, J.; Fan, X.; Li, X.; Jiang, L.; Hao, G.; Pei, H.; Zhou, X. Heat Shock Treatment Promoted Callus Formation on Postharvest Sweet Potato by Adjusting Active Oxygen and Phenylpropanoid Metabolism. *Agriculture* **2022**, *12*, 1351. [CrossRef]
7. Rakgotho, T.; Ndou, N.; Mulaudzi, T.; Iwuoha, E.; Mayedwa, N.; Ajayi, R.F. Green-Synthesized Zinc Oxide Nanoparticles Mitigate Salt Stress in Sorghum Bicolor. *Agriculture* **2022**, *12*, 597. [CrossRef]
8. Ahmed, T.; Abou Elezz, A.; Khalid, M.F. Hydropriming with Moringa Leaf Extract Mitigates Salt Stress in Wheat Seedlings. *Agriculture* **2021**, *11*, 1254. [CrossRef]
9. Sezer, İ.; Akay, H.; Mut, Z.; Arslan, H.; Öztürk, E.; Erbaş Köse, Ö.D.; Kiremit, M.S. Effects of Different Water Table Depth and Salinity Levels on Quality Traits of Bread Wheat. *Agriculture* **2021**, *11*, 969. [CrossRef]
10. Chanthini, K.M.-P.; Senthil-Nathan, S.; Pavithra, G.-S.; Malarvizhi, P.; Murugan, P.; Deva-Andrews, A.; Janaki, M.; Sivanesh, H.; Ramasubramanian, R.; Stanley-Raja, V. Aqueous Seaweed Extract Alleviates Salinity-Induced Toxicities in Rice Plants (*Oryza sativa* L.) by Modulating Their Physiology and Biochemistry. *Agriculture* **2022**, *12*, 2049. [CrossRef]
11. Azeem, M.; Haider, M.Z.; Javed, S.; Saleem, M.H.; Alatawi, A. Drought Stress Amelioration in Maize (*Zea mays* L.) by Inoculation of *Bacillus* Spp. Strains under Sterile Soil Conditions. *Agriculture* **2022**, *12*, 50. [CrossRef]

12. Tahir, N.A.; Rasul, K.S.; Lateef, D.D.; Grundler, F.M. Effects of Oak Leaf Extract, Biofertilizer, and Soil Containing Oak Leaf Powder on Tomato Growth and Biochemical Characteristics under Water Stress Conditions. *Agriculture* **2022**, *12*, 2082. [CrossRef]
13. Weng, T.-Y.; Nakashima, T.; Villanueva-Morales, A.; Stewart, J.R.; Sacks, E.J.; Yamada, T. Assessment of Drought Tolerance of Miscanthus Genotypes through Dry-Down Treatment and Fixed-Soil-Moisture-Content Techniques. *Agriculture* **2021**, *12*, 6. [CrossRef]
14. Ma, Y.; Bai, C.; Zhang, X.; Ding, Y. Nitrate Increases Aluminum Toxicity and Accumulation in Root of Wheat. *Agriculture* **2022**, *12*, 1946. [CrossRef]
15. Sehgal, A.; Reddy, K.R.; Walne, C.H.; Barickman, T.C.; Brazel, S.; Chastain, D.; Gao, W. Individual and Interactive Effects of Multiple Abiotic Stress Treatments on Early-Season Growth and Development of Two Brassica Species. *Agriculture* **2022**, *12*, 453. [CrossRef]

 agriculture

Article

Effects of Different Water Table Depth and Salinity Levels on Quality Traits of Bread Wheat

İsmail Sezer [1], Hasan Akay [1,*], Zeki Mut [2], Hakan Arslan [3], Elif Öztürk [1], Özge Doğanay Erbaş Köse [2] and Mehmet Sait Kiremit [3]

[1] Department of Field Crops, Faculty of Agriculture, Ondokuz Mayıs University, Samsun 55270, Turkey; isezer@omu.edu.tr (İ.S.); elif.ozturk@omu.edu.tr (E.Ö.)
[2] Department of Field Crops, Faculty of Agriculture and Natural Sciences, Bilecik Şeyh Edebali University, Bilecik 11230, Turkey; zeki.mut@bilecik.edu.tr (Z.M.); ozgedoganay.erbas@bilecik.edu.tr (Ö.D.E.K.)
[3] Department of Agricultural Structures and Irrigation, Faculty of Agriculture, Ondokuz Mayıs University, Samsun 55270, Turkey; hakan.arslan@omu.edu.tr (H.A.); mehmet.kiremit@omu.edu.tr (M.S.K.)
* Correspondence: hasan.akay@omu.edu.tr

Abstract: Abiotic stress factors encountered in production lands influence both the yield and the quality traits of bread wheat. This study investigated the effects of three different water table depths (30, 55, and 80 cm) and four different groundwater salinity levels (0.38, 2.0, 4.0, and 8.0 dSm^{-1}) on some quality traits of bread wheat under irrigated and unirrigated conditions. The experiments were conducted in the 2018 and 2019 growing seasons in randomized blocks—factorial (three factors) experimental design with three replications under controlled conditions. The hectoliter weight, fat ratio, starch ratio, protein content, Zeleny sedimentation, wet gluten content, ash ratio, acid detergent fiber (ADF), and neutral detergent fiber (NDF) values were investigated. The hectoliter weights varied between 66.1 and 77.8 kg, fat ratios between 1.49% and 1.70%, starch ratios between 61.9% and 67.8%, protein contents between 11.9% and 13.8%, Zeleny sedimentation values between 23.5 and 28.0 mL, wet gluten contents between 25.0% and 28.8%, ash ratios between 1.43% and 1.75%, and ADF values between 2.85% and 4.12%. The quality traits were positively influenced by increasing the water table depths. With increasing the groundwater salinity levels, the hectoliter weight, fat ratio, starch ratio, and NDF values decreased, while the protein ratio, sedimentation value, wet gluten content, ash ratio, and ADF values increased.

Keywords: bread wheat; water table; salinity; gluten; sedimentation

Citation: Sezer, İ.; Akay, H.; Mut, Z.; Arslan, H.; Öztürk, E.; Erbaş Köse, Ö.D.; Kiremit, M.S. Effects of Different Water Table Depth and Salinity Levels on Quality Traits of Bread Wheat. *Agriculture* **2021**, *11*, 969. https://doi.org/10.3390/agriculture11100969

Academic Editors: Daniele Del Buono, Primo Proietti and Luca Regni

Received: 31 August 2021
Accepted: 2 October 2021
Published: 6 October 2021

1. Introduction

Wheat is among the most widely cultivated agricultural crop worldwide. It constitutes the primary calorie source in human nutrition [1,2]. Annually, 766 million tons of wheat are produced every year globally, and 19 million tons of wheat are produced in Turkey [3]. Rain-fed farming is practiced in the wheat cultivation of arid and semi-arid regions, and the yields are decreasing significantly because of insufficient water resources [4].

In the Mediterranean climate zone, producers generally practice one or two supplementary irrigations in a year (except for dry years) in wheat fields using the surface irrigation method. In these regions, the March and May months coincide with the flowering and milk dough stages of wheat, which are the sensitive growth periods. Insufficient precipitations in these months may result in serious yield losses [5]. The initiation of irrigations in arid and semi-arid regions subsequently brought about drainage problems, and such problems then resulted in the rise of the water table and salinity problems [6]. Global warming is the most challenging environmental problem that humanity should deal with. Global warming alters the seasonal normal and increases soil salinity through insufficient precipitations and high evaporations. The water table and salinity control in these regions are largely dependent on a well arrangement of the water balance.

Salinity problems could be overcome with a well water balance [7]. Salinity is among the most significant problems encountered in agricultural fields worldwide. Salinity-induced yield decreases are experienced in various parts of the world, and salinity ultimately terminates agricultural practices. High irrigation water salinity or soil salinity raises the osmotic pressure of the soil solution, then reduces the water uptake of roots from the soil and, consequently, decreases the crop yield and quality [8]. Therefore, crop and soil-based water management strategies should be developed to sustain water and soil resources.

The wheat yield was decreased by approximately 17% with increasing the irrigation water salinity from 0.6 to 10 dS/m [9]. Besides the protein content, the wet and dry gluten content values of the wheat crops were increased with the increasing salinity and drought stresses [10]. Extreme droughts can cause significant decreases in the protein content, wet gluten content, and sedimentation volume content of wheat [11].

Several factors designate the wheat quality, and the quality criteria vary significantly based on the producer, industry, and consumer demands [12,13]. The protein ratio is the most important quality criterion in wheat [14], and the protein ratios of different wheat varieties under different environmental conditions vary between 6 and 22% [15]. In bread wheat, the sedimentation value and wet gluten ratio designate the protein quality. Therefore, besides the protein ratio, the protein quality also plays a great role in the quality of bread wheat [16]. The protein ratio and quality may change with the growing conditions and climate factors [17]. In Turkey, the hectoliter weight of wheat varies between 70 and 84 kg and is mainly dependent on the cultivars and climate conditions. The starch content of wheat grain constitutes about 65–70% of the grain dry weight [13]. Acid detergent fiber (ADF) is an indicator of the cellulose, lignin, and insoluble protein contents of the cell wall and reveals information about the digestibility of the products [18]. Neutral detergent fiber (NDF) expresses the indigestible substances like cellulose, hemicellulose, lignin, cutine, and insoluble protein of the cell wall. High NDF values negatively influence the feed quality [19].

Wheat production, which is one of the most basic foods globally, is severely affected by drainage and salinity problems. Moreover, the quality parameters are as important as the yield in wheat.

In the literature, many studies have been carried out on the effects of the water table depth and salinity on the quality parameters of wheat. However, there is no detailed study on the combined effects of these stress factors on the quality parameters of wheat. Therefore, the present study was conducted to investigate the effects of different water table depths and salinity levels on the quality parameters of bread wheat plants under with and without irrigation conditions.

2. Materials and Methods

2.1. Experimental Site Description

This study was conducted at the Agricultural Research and Implementation Center (41°21′ N and 36°11′ E, elevation 192 m above sea level), which belongs to the Faculty of Agricultural, University of Ondokuz Mayıs, Northern Turkey. The experiment was laid out over a two-year growing season, from December to June 2018 to 2019. The lysimeters were conducted on a four-sides open land area (120 m^2) with a plastic cover above to protect from precipitation. The daily temperature and relative humidity values were measured with a datalogger located at the middle of the research area from 2 m above the ground for two growing seasons. The mean monthly temperature and relative humidity data are illustrated in Table 1.

Table 1. The average monthly temperature and relative humidity values in the first and second growing seasons.

	November	December	October	February	March	April	May	June
				Min Temperature (°C)				
2017 to 2018	3.4	2.1	0.5	2.5	0	4.6	7.7	12.3
2018 to 2019	4.5	0.8	−0.2	0.6	0.2	3.2	8.1	15.8
				Average Temperature (°C)				
2017 to 2018	9.7	11.8	7.7	9.3	11.6	16.6	19.2	24.4
2018 to 2019	11.6	8.8	8.1	8.1	8.9	12.7	18.2	23.9
				Max. Temperature (°C)				
2017 to 2018	16.3	22.3	23.6	25.4	28.4	28.8	34.9	38.4
2018 to 2019	23.7	22.3	21	22	25.9	32.6	30.6	33.7
				Average Relative Humidity (%)				
2017 to 2018	73.1	64.8	76	80	77	68.2	79.7	66
2018 to 2019	77.5	78.2	70.9	79.8	74.3	78.6	81.7	83.8

The experimental soil was obtained from the top 30-cm layer, and its texture was loam with 25.3% clay, 31.3% silt, and 43.4% sand. Additionally, its chemical properties were 2.4% organic matter, 7.1-mg kg^{-1} phosphorus, 0.33-me 100 g^{-1} potassium, 7.99 pH, and 0.27-dSm^{-1} electrical conductivity. Phosphorous (P) was determined with a UV–Visible spectrophotometer according to Reference [20]. Potassium was measured using flame photometers.

2.2. Lysimeter Set-Up

A 5-cm layer of gravel and a 5-cm layer of sand were placed at the bottom of each lysimeter to provide a continuous water supply from Mariotte bottles to lysimeters. Then, each lysimeter was filled with 330 kg of soil sieved through a 4-mm sieve, and the soil in the lysimeter was compacted layer by layer (10 cm) to reach a soil bulk density of 1.297 gr/cm^3. Schematic views of the lysimeters used in the present experiment are presented in Figure 1 [21]. The groundwater depths in the lysimeters were controlled at the constant levels of 30, 55, and 80 cm from below the soil surface. The groundwater was checked daily by keeping the water in the Mariotte bottles at a constant level. The daily amount of water moving into each lysimeter was calculated by water loss from the Mariotte bottle. The drainage pipe was placed above the groundwater depth into each lysimeter to drain out excess water automatically.

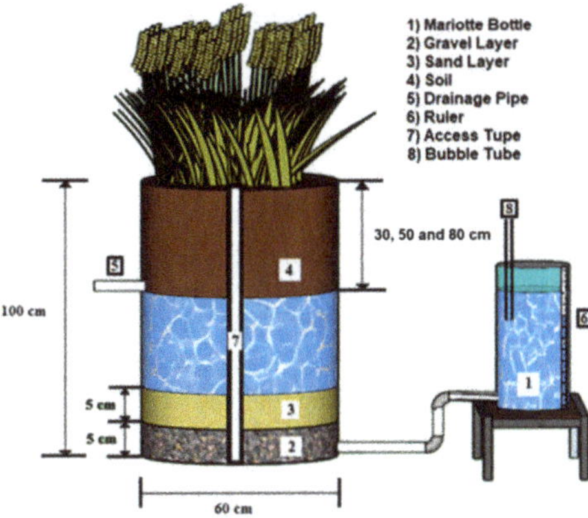

Figure 1. Schematic view of lysimeter and Mariotte bottle used in the study.

2.3. Experimental Design and Treatments

The experimental traits were conducted in 72 lysimeters, which were 100 cm deep with 60-cm inner diameters. The experimental design was an arrangement in a randomized complete block with an irrigation treatment as the main plot, groundwater depth as the subplot, and groundwater salinity as the sub-subplot with three replicates. The experimental traits contained two irrigation treatments of I_1 (with irrigation) and I_2 (without irrigation) and three groundwater depths of 30 cm, 55 cm, and 80 cm and four groundwater salinities of 0.38 dSm^{-1}, 2.0 dSm^{-1}, 4.0 dSm^{-1}, and 8.0 dSm^{-1} (Table 2).

Table 2. Experimental treatments of the irrigation treatments, groundwater depth, and groundwater salinity.

Irrigation Treatments	Groundwater Depth (cm)	Groundwater Salinity (dSm^{-1})
I_1 (with irrigation)	D_1 = 30 cm	S_1 = 0.38 dSm^{-1}
I_2 (without irrigation)	D_2 = 55 cm	S_2 = 2.0 dSm^{-1}
	D_3 = 80 cm	S_3 = 4.0 dSm^{-1}
		S_4 = 8.0 dSm^{-1}

At the end of the tillering period, the lysimeters were saturated from the bottom with different water salinities of 0.38, 2.0, 4.0, and 8.0 dSm^{-1} at up to 30, 55, and 80-cm groundwater depths for two weeks. The saline waters were prepared with the use of highly soluble $MgSO_4$ (99% purity), $CaCI_2$ (99% purity), and NaCl (99.5% purity) salts. The amount of salt to be added to prepare relevant salt concentrations (EC values) was calculated with the use of QBASIC software to achieve a sodium adsorption ratio (SAR) of <5 and a Ca/Mg ratio of 1:3.

In both growing years, all lysimeters were supplemented with 10-mm irrigation water until the end of the tillering period. The first irrigation was applied after establishing the constant groundwater depths. For this, each lysimeter's volumetric soil moisture content above the groundwater depth was measured with a neutron scattering method (CPN 503 Dr Hydro probe). The polyvinyl chloride (PVC) pipes used for the neutron meter measurements were placed in the middle of the lysimeters. Under S_1 (with irrigation) conditions, the soil moisture content of each lysimeter was measured, and irrigation water was applied to bring the available soil moisture to field capacity. However, under I_2 (without irrigation) conditions, the soil moisture content of each lysimeter was measured, but irrigation water was not applied from the end of the tillering period to harvesting.

After the first irrigation, the volumetric soil moisture content above the groundwater depth in all the lysimeters was monitored every seven days. When 50% of the available soil water above the groundwater depth was depleted, irrigation water was added to fill up to field capacity in the $_{I1}$ treatments.

2.4. Crop Management

A Pandas wheat cultivar was used as the seed material of the study. In the first year, wheat seeds were sown on 11 November 2017, and at 14 November 2018 in the second year, to have a sowing density of 500 seeds per m^2. Fertilization was applied according to the soil analysis. With this aim, 100-kg ha^{-1} pure nitrogen (N) and 60-kg ha^{-1} P_2O_5 were applied to each lysimeter. All phosphorus was applied in diammonium phosphate form prior to sowing. Additionality, nitrogen was applied in the form of urea (46% N) at two different times; half of the nitrogen was applied at sowing, and the other half was applied just before the bolting period. Weed control was practiced manually. Wheat crops from each lysimeter were harvested at full maturity on 8 June 2018 and 19 June 2019.

2.5. Grain Quality Parameters

The quality indicators included the hectoliter weight (kg), ash content (%), fat content (%), protein content (%), Zeleny sedimentation (mL), and wet gluten values (%). These parameters were determined following Reference [22]. The acid detergent fiber (ADF) (%)

and neutral detergent fiber (NDF) contents (%) were determined according to Reference [23] and the starch ratio was determined with the use of the Ewers Polarimetric method [24].

2.6. Data Analysis

Firstly, all data were subjected to the homogeneity test, and they were shown normal distribution. The experimental data were subjected to a variance analysis using JMP statistical software (SAS Institute, Cary, NC, USA) [25]. All treatment means were compared using Tukey's test. Biplot and Pearson correlation analyses were conducted to assess the relations among the investigated parameters.

3. Results

The variance analysis results of the effects of the irrigation level, groundwater depth, salinity, year, and their interactions on the quality parameters of wheat are given in Table 3.

Table 3. Variance analysis table of the analyzed parameters *.

VS.	DF	Average of Squares								
		HW	SC	PC	ZSV	WG	FC	AC	ADF	NDF
Year	1	534.15 **	179.23 **	292.75 **	4289.27 **	4112.02 **	0.38 **	0.043 **	1.41 **	10.79 **
Irrigation	1	147.42 **	0.09	0.20	22.18 **	6.18 **	0.01	0.054 **	0.84 **	0.05
D	2	222.73 **	14.51 **	6.36 **	39.40 **	18.40 **	0.16 **	0.071 **	1.57 **	3.86 **
S	3	196.38 **	34.05**	7.18 **	32.29 **	22.41 **	0.07 *	0.304 **	4.25 **	2.70 **
Block	4	0.09	0.27	0.14	0.04	0.08	0.03	0.002	0.01	0.02
Y×I. Int.	1	6.63 *	4.87 **	0.23	6.17 **	0.10	0.09 *	0.001	0.01	1.70 **
Y×D. Int	2	14.63 **	44.04 **	0.47 *	0.61 **	0.30 *	0.23 **	0.001	0.01	0.68 **
Y×S. Int.	3	0.09	1.39 **	0.30 *	5.01 **	1.45 **	0.002	0.003 *	0.03 **	0.18 **
D×I. Int.	2	14.29 **	8.77 **	1.13 **	13.53 **	0.07	0.04	0.002	0.18 **	0.27 **
S×I Int.	3	0.25	0.30	0.06	1.01 **	1.08 **	0.01	0.001	0.21 **	0.04 *
S×D Int.	6	3.41 **	0.38	0.10	0.61 **	0.13	0.01	0.002 *	0.02 *	0.03
Y×I×D Int.	2	2.80	5.46 **	0.10	1.21 **	0.25	0.02	0.001	0.01	0.53 **
Y×I×S Int.	3	0.10	2.19 **	0.11	0.32 **	0.42 **	0.02	0.001	0.01	0.02
Y×D×S Int.	6	3.70 **	1.41 **	0.07	0.60 **	0.13	0.02	0.001	0.01	0.04 *
S×D×I nt.	6	0.58	0.75 **	0.03	0.32 **	0.17	0.01	0.0002	0.04 **	0.15 **
Y×I×D×S Int.	6	0.85	0.82 **	0.10	0.14 **	0.09	0.01	0.0003	0.01	0.14 **
Error	**92**	0.99	0.19	0.11	0.07	0.10	0.02	0.0009	0.01	0.01
% CV		**1.37**	**6.81**	**2.58**	**1.03**	**1.19**	**8.73**	**1.87**	**2.07**	**6.50**

* The $p < 0.05$ and ** $p < 0.01$ levels are significant. HW = Hectoliter weight (kg), FC: Fat Content (%), NC: Starch Content (%), PC: Protein Content (%), ZSV: Zeleny Sedimentation Value (mL), WG: Wet gluten (%), AC: Ash Content (%), ADF: Acid Detergent Insoluble Fiber (%), NDF: Neutral Detergent Insoluble Fiber (%), DF: Degree of Freedom, Y: Year, I: Irrigation, D: Depth of Ground Water, and S: Ground Water Salinity.

3.1. Hectoliter Weight

The hectoliter weights at different water table depths were significantly different ($p \leq 0.01$) (Table 3). The highest hectoliter weight (73.9 kg) was obtained from the D_3 level, followed, respectively, by the D_2 (73.3 kg) and D_1 (69.9 kg) levels (Table 4). Significant differences were also observed in the hectoliter weights at different groundwater salinity levels ($p \leq 0.01$). The greatest hectoliter weight (75.2 kg) was obtained from the S_1 level, followed, respectively, by the S_2 (73.1 kg), S_3 (71.5 kg), and S_4 (69.7 kg) levels. In terms of the irrigation treatments, the greatest hectoliter weight (73.4 kg) was obtained from the I_1 treatment and the lowest (71.4 kg) from the I_2 treatment. The differences in the hectoliter weights of the irrigation treatments were found to be significant ($p \leq 0.01$). In terms of the salinity x water table depth interactions ($S \times D$), the greatest hectoliter weight (76.2 kg) was obtained from the $S_1 \times D_3$ combination and the lowest (66.9 kg) from the $S_4 \times D_1$ combination (Table 5). In terms of the $D \times I$ interactions, the highest hectoliter weight (75.6 kg) was obtained from the D_3 x I_1 combination and the lowest (69.3 kg) from the $D_1 \times I_2$ combination (Table 5). In the present study, the greatest hectoliter weight (77.8 kg) was obtained from the $S_1 \times D_3 \times I_1$ combination and the lowest (66.1 kg) from the $S_4 \times D_1 \times I_2$ combination (Table 6).

Table 4. The mean of the two years for the effects of the four salinity levels, three water table depths, and two irrigation levels on the wheat parameters *.

Source of Variance	HW	SC	PC	ZSV	WG	FC	AC	ADF	NDF
				Salinity (S)					
S_1	75.2 a	65.1 a	12.3 d	24.8 d	25.8 d	1.64 a	1.49 d	3.04 d	15.7 a
S_2	73.1 b	64.3 b	12.7 c	25.4 c	26.2 c	1.62 a	1.57 c	3.33 c	15.5 b
S_3	71.5 c	63.6 c	13.1 b	26.3 b	26.9 b	1.63 a	1.65 b	3.59 b	15.3 c
S_4	69.7 d	62.8 d	13.4 a	26.9 a	27.6 a	1.59 b	1.73 a	3.85 a	15.1 d
				Water table depth (D)					
D_1	69.9 c	63.3 b	12.5 c	25.2 c	26.0 c	1.56 B	1.57 c	3.27 c	15.7 a
D_2	73.3 b	64.3 a	12.9 b	25.5 b	26.7 b	1.67 A	1.60 b	3.45 b	15.3 b
D_3	73.9 a	64.3 a	13.2 a	26.9 a	27.2 a	1.63 A	1.65 a	3.63 a	15.1 c
				Irrigation (I)					
I_1	73.4	63.9	12.9	26.2 a	26.8 a	1.61	1.62	3.37	15.3
I_2	71.4	64.0	12.8	25.4 b	26.4 b	1.62	1.58	3.53	15.4

* There is no difference in the significance level of 0.01 between the averages shown with the same letter in each column. HW = Hectoliter weight (kg), FC: Fat Content (%), SC: Starch Content (%), PC: Protein Content (%), ZSV: Zeleny Sedimentation Value (mL), WG: Wet gluten (%), AC: Ash Content (%), ADF: Acid Detergent Fiber (%), and NDF: Neutral Detergent Fiber (%).

Table 5. The mean of the two years for the interaction effects of the salinity level × water table depth, salinity × irrigation, and water table depth × irrigation on the wheat parameters *.

Source of Variance	HW	PC	ZSV	WG	AC	ADF	NDF
			Salinity (S) × Water Table Depth (D)				
$S_1 \times D_1$	73.5 bc	11.9	23.9 g	25.2	1.45 g	2.91 h	16.0
$S_2 \times D_1$	70.1 fg	12.3	24.5 f	25.6	1.53 ef	3.15 g	15.8
$S_3 \times D_1$	69.1 g	12.8	25.9 d	26.3	1.63 d	3.40 e	15.6
$S_4 \times D_1$	66.9 h	13.1	26.3 c	26.8	1.66 cd	3.63 d	15.3
$S_1 \times D_2$	75.8 a	12.4	24.5 f	25.9	1.49 fg	3.00 h	15.6
$S_2 \times D_2$	74.3 bc	12.8	25.2 e	26.2	1.56 e	3.30 f	15.5
$S_3 \times D_2$	72.2 de	13.1	25.8 d	26.9	1.64 cd	3.63 d	15.3
$S_4 \times D_2$	70.8 f	13.3	26.4 c	27.7	1.71 ab	3.87 b	15.1
$S_1 \times D_3$	76.2 a	12.8	25.8 d	26.5	1.54 e	3.22 fg	15.5
$S_2 \times D_3$	74.9 ab	13.1	26.5 c	26.6	1.62 d	3.55 d	15.2
$S_3 \times D_3$	73.3 cd	13.4	27.3 b	27.5	1.68 bc	3.74 c	14.9
$S_4 \times D_3$	71.4 ef	13.7	27.9 a	28.3	1.73 a	4.03 a	14.7
			Salinity (S) × Irrigation (I)				
$S_1 \times I_1$	76.2	12.3	24.9 e	26.0 de	1.51	3.03 e	15.7 a
$S_2 \times I_1$	74.0	12.8	25.8 d	26.2 d	1.59	3.31 d	15.4 b
$S_3 \times I_1$	72.6	13.2	26.8 b	27.1 b	1.67	3.49 bc	15.3 c
$S_4 \times I_1$	70.8	13.4	27.4 a	28.1 a	1.72	3.67 b	15.0 d
$S_1 \times I_2$	74.2	12.3	24.6 f	25.7 e	1.47	3.05 e	15.7 a
$S_2 \times I_2$	72.2	12.7	25.0 e	26.1 d	1.55	3.35 d	15.5 b
$S_3 \times I_2$	70.5	13.0	25.8 d	26.7 c	1.63	3.68 b	15.2 c
$S_4 \times I_2$	68.6	13.3	26.3 c	27.1 b	1.68	4.02 a	15.1 d
			Water Table Depth (D) × Irrigation (I)				
$D_1 \times I_1$	70.5 d	12.7 d	25.7 d	26.2	1.58	3.14 d	15.7 a
$D_1 \times I_2$	69.3 e	12.4 e	24.6 e	25.8	1.55	3.40 c	15.6 b
$D_2 \times I_1$	74.1 b	13.0 bc	26.3 c	26.9	1.62	3.36 c	15.3 d
$D_2 \times I_2$	72.4 c	12.8 cd	24.6 e	26.4	1.57	3.53 b	15.4 c
$D_3 \times I_1$	75.6 a	13.1 ab	26.7 b	27.4	1.67	3.62 a	15.0 f
$D_3 \times I_2$	72.3 c	13.4 a	27.0 a	27.0	1.62	3.64 a	15.2 e

* There is no difference in the significance level of 0.01 between the averages shown with the same letter in each column. HW = Hectoliter weight (kg), PC: Protein Content (%), ZSV: Zeleny Sedimentation Value (mL), WG: Wet gluten (%), AC: Ash Content (%), ADF: Acid Detergent Insoluble Fiber (%), and NDF: Neutral Detergent Insoluble Fiber (%).

Table 6. The mean of the two years for the interaction effects of the salinity level × water table depth × irrigation on the wheat parameters *.

Source of Variance	SC	ZSV	ADF	NDF
Salinity (S) × Water Table Depth (D) × Irrigation (I)				
$S_1 \times D_1 \times I_1$	64.5 bc	24.3 de	2.85 f	16.2 a
$S_2 \times D_1 \times I_1$	63.8 c	25.0 d	3.13 de	15.9 ab
$S_3 \times D_1 \times I_1$	62.9 d	26.4 bc	3.25 d	15.6 ab
$S_4 \times D_1 \times I_1$	61.9 e	27.0 b	3.31 cd	15.2 bc
$S_1 \times D_2 \times I_1$	65.2 b	24.9 d	2.94 ef	15.5 ab
$S_2 \times D_2 \times I_1$	64.5 bc	26.0 c	3.24 d	15.4 b
$S_3 \times D_2 \times I_1$	63.0 d	26.8 b	3.53 c	15.3 bc
$S_4 \times D_2 \times I_1$	63.0 d	27.6 ab	3.74 b	15.0 c
$S_1 \times D_3 \times I_1$	65.5 b	25.5 c	3.29 cd	15.4 b
$S_2 \times D_3 \times I_1$	64.8 bc	26.3 bc	3.57 bc	15.1 bc
$S_3 \times D_3 \times I_1$	64.5 bc	27.0 b	3.69 bc	14.9 c
$S_4 \times D_3 \times I_1$	64.1 c	27.8 a	3.94 ab	14.7 c
$S_1 \times D_1 \times I_2$	67.8 a	23.5 e	2.96 ef	15.8 ab
$S_2 \times D_1 \times I_2$	63.5 cd	24.0 e	3.17 d	15.6 ab
$S_3 \times D_1 \times I_2$	62.9 d	25.4 cd	3.54 c	15.5 ab
$S_4 \times D_1 \times I_2$	62.3 dd	25.7 c	3.95 ab	15.4 b
$S_1 \times D_2 \times I_2$	65.7 b	24.2 de	3.06 e	15.7 ab
$S_2 \times D_2 \times I_2$	65.0 b	24.4 de	3.35 cd	15.5 ab
$S_3 \times D_2 \times I_2$	64.6 bc	24.7 d	3.73 b	15.3 bc
$S_4 \times D_2 \times I_2$	63.4 cd	25.3 cd	3.99 ab	15.1 bc
$S_1 \times D_3 \times I_2$	65.0 b	26.1 bc	3.14 de	15.6 ab
$S_2 \times D_3 \times I_2$	64.2 bc	26.7 b	3.52 c	15.4 b
$S_3 \times D_3 \times I_2$	63.8 c	27.5 ab	3.78 b	14.9 c
$S_4 \times D_3 \times I_2$	62.6 d	28.0 a	4.12 a	14.7 c

* There is no difference in the significance level of 0.01 between the averages shown with the same letter in each column. SC: Starch Content (%), ZSV: Zeleny Sedimentation Value (mL), ADF: Acid Detergent Insoluble Fiber (%), and NDF: Neutral Detergent Insoluble Fiber (%).

3.2. Starch Ratio and Protein Content

For the starch ratios, the water table depths and groundwater salinity levels were found to be highly significant ($p \leq 0.01$). The effects of the irrigation treatments on the starch ratios were not found to be significant (Table 3). In terms of the water table depths, the greatest starch content (64.3%) was obtained from the D_3 and D_2 levels and the lowest (63.3%) from the D_1 level (Table 4). Groundwater salinity negatively influenced the starch ratios. The greatest starch content (65.1%) was obtained from the S_1 level and the lowest (62.8%) from the S_4 level.

The protein contents of wheat were considerably affected by the water table depths and salinity (Table 3). In terms of the water table depths, the protein ratios varied between 12.5% (D_1) and 13.2% (D_3) (Table 4). Decreasing protein ratios were observed with the rising water table levels. In terms of the groundwater salinity levels, the greatest protein ratio was obtained from the S_4 (13.4%) salinity level and the lowest from the S_1 (12.3%) salinity level. The effects of the irrigation treatments on the protein ratios were not found to be significant. In terms of the D × I interactions, the greatest protein ratio (13.1%) was obtained from the D_3 x I_1 combination and the lowest from the $D_1 \times I_2$ combination (Table 5).

3.3. Zeleny Sedimentation Value, Wet Gluten, and Fat Content

Significant differences were observed in the Zeleny sedimentation values at the different irrigation, water table depth, and groundwater salinity treatments ($p \leq 0.01$) (Table 3). The sedimentation value was measured as 26.2 mL in the irrigated treatments (I_1) and as 25.4 mL in the non-irrigated treatments (I_2) (Table 4). In terms of the water table depths, the greatest sedimentation value (26.9 mL) was obtained from the D_3 treatment, respectively, followed by the D_2 (25.5 mL) and D_1 (25.2 mL) treatments. In terms of the groundwater salinity, the greatest sedimentation value (26.9 mL) was obtained from the S_4 salinity level and the lowest (24.8 mL) from the S_1 salinity level. Increasing sedimentation values were observed with the increasing groundwater salinity levels. In terms of the S × D interactions, the greatest sedimentation value (27.9 mL) was obtained from the $S_4 \times D_3$ combination and the lowest (23.9 mL) from the $S_1 \times D_1$ combination (Table 5). In terms of the D × I interactions, the greatest sedimentation value was obtained from the $D_3 \times I_1$ combination. The greatest sedimentation value was obtained from the $S_4 \times D_3 \times I_2$ combination (Table 6).

For the wet gluten contents, the experimental treatments were significant ($p \leq 0.01$) (Table 3). In terms of the salinity levels, the greatest wet gluten content (27.6%) was obtained from the S_4 level, respectively, followed by the S_3 (26.9%), S_2 (26.2%), and S_1 (25.8%) salinity levels (Table 4). Increasing gluten contents were observed with the increasing salinity levels. The wet gluten contents decreased with the decreasing water tables (respectively, D_3, D_2, and D_1). The wet gluten content was measured as 26.8% in the irrigated treatments (I_1) and 26.4% in the non-irrigated treatments (I_2). In terms of the salinity × irrigation interactions, the greatest wet gluten content (28.1%) was obtained from the $S_4 \times I_1$ combination and the lowest (25.7%) from the $S_1 \times I_2$ combination (Table 5).

The water table depth and salinity had significant effects on the fat content, while the effects of the irrigation treatments on the fat content were not significant. Salinity decreased the fat content by 1.22%, 0.61%, and 3.05% in the S2, S3, and S4 treatments, respectively, compared to the S_1 treatment. The highest plant fat content value (1.67) was determined in D_2, while the lowest (1.56 cm) was observed from the D_1 treatment.

3.4. *Ash Content, Acid Detergent Fiber, and Neutral Detergent Fiber*

There were significant differences in the ash contents at the different irrigation, water table depth, and salinity treatments (Table 3). The ash contents increased with the increasing water table depths (D_1, D_2, and D_3, respectively, with 1.57, 1.60, and 1.65%) (Table 4). The ash contents also increased with the increasing salinity levels (S_1, S_2, S_3, and S_4, respectively, with 1.49, 1.57, 1.65, and 1.73%). The ash content was measured as 1.62% in the irrigated plots (I_1) and as 1.58 in the non-irrigated plots (I_2) (Table 5).

There were highly significant differences in the acid detergent fiber (ADF) and neutral detergent fiber (NDF) values of the experimental treatments ($p \leq 0.01$) (Table 3). In terms of the salinity levels, the greatest ADF and NDF (1.73 and 3.85%, respectively) values were obtained from the S_4 salinity level. In terms of the water table depth, the lowest ADF and NDF (1.57 and 3.27%, respectively) values were obtained from the D_1 level (Table 4).

3.5. *Biplot Analysis*

A biplot analysis allows researchers to visually assess the relationships between the experimental treatments and investigated parameters and offers some advantages over the correlation analysis revealing relationships only between two traits [26]. The classification of the investigated traits based on the experimental treatments and changes in the investigated parameters of the experimental treatments are presented in Figure 2. In the biplot analysis, two principal components explained 77.0% of the total variation (PC1 57.1% and PC2 19.9%) (Figure 1). As shown in the biplot graph, the hectoliter weight, wet gluten, sedimentation, and protein ratio were positioned in the upper-right section of the graph. Since the vector angles of these traits were less than 90°, there were significant positive relationships between these parameters. There were significant positive relationships between the NDF and starch ratios. The starch content, NDF values, S_1 and S_2 salinity levels, and D_1 and D_2 water table levels were placed in the upper-left section of the graph. Therefore, it was thought that the S_1, S_2, D_1, and D_2 treatments were prominent for starch and NDF. There was a significant positive relationship between the ash content and ADF. The ash content, ADF, S_3 and S_4 salinity levels, and D_3 water table level were placed into the lower-right section of the biplot graph. According to this result, there was a significant positive relationship between the ash content and ADF. The fat content was placed into the lower-left section of the graph. Approaching the center of the graph, the D_2, S_2, I_1, and I_2 treatments were prominent for more than one trait (Figure 2).

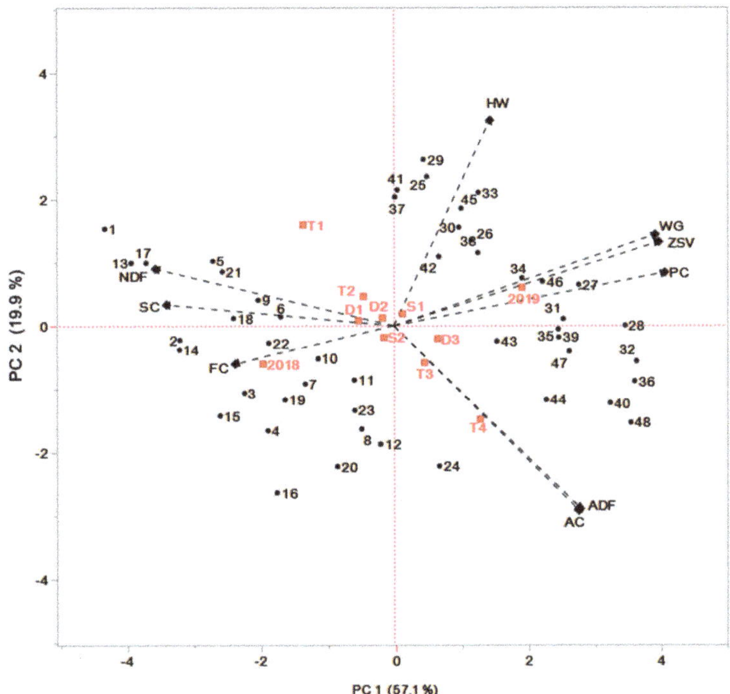

Figure 2. Categorization of the examined features by the biplot analysis method and the relationships of the traits examined.

3.6. Correlation Analysis

The correlations between the investigated parameters are depicted in Table 7. The hectoliter weight had significant positive correlations with the protein ratio, Zeleny sedimentation and wet gluten content. Additionally, the protein ratio was positively correlated with the Zeleny sedimentation, wet gluten, ash content, and ADF values while negatively associated with the fat content, starch ratio, and NDF values. The wet gluten content had significant positive correlations with the ash ratio and ADF values and significant negative correlations with the oil ratio, starch ratio, and NDF value.

Table 7. Correlation coefficients between the features and significance levels*.

	HW	PC	ZSV	WGC	ASH	FC	SC	ADF
PC	0.44 **							
ZSV	0.50 **	0.98 **						
WGC	0.52 **	0.96 **	0.99 **					
ASH	−0.17	0.48 **	0.38 **	0.34 *				
FC	−0.06	−0.42 **	−0.46 **	−0.45 **	−0.18			
SC	0.01	−0.65 **	−0.67 **	−0.65 **	−0.49 **	0.78 **		
ADF	−0.18	0.51 **	0.40 **	0.39 **	0.86 **	−0.12	−0.46 **	
NDF	−0.27	−0.72 **	−0.67 **	−0.66 **	−0.69 **	0.24	0.49 **	−0.71 **

* The $p < 0.05$ and ** $p < 0.01$ levels are significant. HW = Hectoliter Weight (kg), PC = Protein Content (%), ZSV =Zeleny Sedimentation Value (mL), WGC = Wet Gluten Content (%), ASH = Ash (%), FC = Fat Content (%), SC = Starch Content (%), ADF = Acid Detergent Insoluble Fiber (%), and NDF = Neutral Detergent Insoluble Fiber (%).

4. Discussion

The present study found that the hectoliter weights were affected by the abiotic stress factors and decreased with the increasing stress conditions. Salinity, drought, waterlogging, and other abiotic stresses adversely affect the biochemical and physiological processes in plants and cause deterioration of the grain quality [21]. The change in the grain quality varies according to the amount of stress. As a result of the accumulation of salts in the root zone of the plant, the water and mineral intakes of the plants decrease, and this causes the grain quality to deteriorate. The hectoliter weight is an important physical quality criterion designating, especially, the flour yield. The hectoliter weights were negatively influenced as the water table depths approached the soil surface and decreased with the increasing groundwater salinity levels. Wheat plants had greater hectoliter weights under irrigated conditions than under non-irrigated conditions. It was reported in previous studies that hectoliter weights were mostly influenced by cultural practices and biotic and abiotic stressors [13,27–29]. The hectoliter weights of bread wheat cultivars change between 77.90 and 79.86 kg [30].

The starch and protein contents have significant effects on the bread quality. In our study, the starch content values of the grain quality were considerably affected by the groundwater depths and salinities. Previous researchers indicated that the starch contents greatly varied with the growing conditions [13,31]. The starch contents were increased with the increasing water table depths (1.0, 1.4, 1.8, 2.2, 2.6, and 3.0 m), and also, researchers reported that the starch contents varied between 70.2 and 77.8% in the first year and between 71.2 and 77.2% in the second year [32]. Especially in bread wheat, the protein quantity and quality are among the most important quality traits [14,33]. The protein content of bread wheat should be $\geq$11% [30]. Differences in the grain protein contents were mostly attributed to climate factors and cultural practices [34–36]. The protein ratios decrease with the increasing abiotic salt concentrations [37–39]. It was reported that the protein ratios increased with the increasing water table depths [32] and salinity [40].

The bread volume increases with the increasing sedimentation values; thus, bread wheat is desired to have high sedimentation values [41]. Different groundwater depths and salinities under with or without irrigation conditions considerably influenced the sedimentation values. According to this result, the sedimentation values were increased with the increasing groundwater depths and salinities. Additionally, the I_1 (with irrigation) conditions had a higher sedimentation value in comparison to the I_2 conditions. This situation could be attributed to the salt accumulation and soil moisture variations in the root zone according to the different groundwater depths, salinities, and irrigation conditions. Previous researchers also indicated that the sedimentation values varied with the environmental conditions and cultivars [13,35].

Wet gluten is the most important quality characteristic of bread wheat and refers to the bread quality of wheat. [13]. In our study, the gluten content values increased with the increasing salinity and depth levels of the groundwater. Additionally, the influence of the I_1 condition was greater than the I_2 condition on wet gluten. The gluten protein gives rising and elasticity attributes to wheat flour [42]. Increased gluten contents were reported with the increasing irrigation water and soil salinity levels [43–45]. The ash content in wheat is closely related to the flour yield, and the climatic factors can change the ash content. Increasing abiotic stress conditions such as salinity and drought increase the ash contents [46]. The ash contents of wheat species may vary with the climate and soil conditions [16,42].

Significant variations were reported in the quality traits of wheat farming practiced under abiotic stress factors induced especially by high temperatures and insufficient precipitation [13]. There were positive correlations between the ash content and ADF values and highly significant negative correlations between the starch ratio and NDF values. Previous researchers indicated that the ADF and NDF values of bread wheat genotypes generally varied with the genotypes and environmental conditions [28,47]. There were highly significant positive correlations between the fat ratio and starch ratio. Similar correlations were also reported in previous studies [7,10,29,30]. There were significant negative correlations between the ADF and NDF values.

5. Conclusions

The initiation of irrigations in bread wheat farming fields, unconscious irrigations, and excessive fertilizations generated both drainage and salinity problems in these fields. Such problems reduced not only the yields but, also, some quality attributes of bread wheat. In the present study, the water table depth and groundwater salinity levels had highly significant correlations with the irrigation treatments. The investigated quality parameters were positively influenced by increasing the water table depths from 30 cm to 80 cm. The greatest values for the quality traits were obtained from the 80-cm (D3) water table depth treatments. With the increasing groundwater salinity levels from the S_1 to S_4 levels, the NDF, hectoliter weight, fat ratio, and starch ratios decreased and the protein ratio, sedimentation value, wet gluten content, ash ratio, and ADF values increased. In bread wheat, the protein ratio, sedimentation, and wet gluten contents are important quality traits for the milling industry. Increasing the salinity levels positively influenced these traits. Groundwater salinity may increase the accumulation of salts in the soil and can generate persistent damages in the soil structure. However, the present findings revealed the better quality of bread wheat cultivated in saline lands. Bread wheat cultivation is recommended to be done in places with deep water table levels and irrigation opportunities.

Author Contributions: Conceptualization, H.A. (Hasan Akay), H.A. (Hakan Arslan), Z.M. and İ.S.; methodology, H.A. (Hasan Akay), H.A. (Hakan Arslan), Z.M. and İ.S.; software, M.S.K.; validation, H.A. (Hasan Akay), H.A. (Hakan Arslan), Z.M. and İ.S.; formal analysis, H.A. (Hasan Akay); investigation, Ö.D.E.K.; resources, E.Ö., Ö.D.E.K. and M.S.K.; data curation, H.A. (Hasan Akay), H.A. (Hakan Arslan), Z.M., Ö.D.E.K. and İ.S.; writing—original draft preparation, İ.S., H.A. (Hasan Akay), Z.M., H.A. (Hakan Arslan), Ö.D.E.K., E.Ö. and M.S.K.; writing—review and editing, İ.S., H.A. (Hasan Akay), Z.M., H.A. (Hakan Arslan), Ö.D.E.K., E.Ö. and M.S.K.; visualization, M.S.K. and H.A. (Hasan Akay); supervision, H.A. (Hasan Akay), H.A. (Hakan Arslan), Z.M. and İ.S.; project administration, H.A. (Hakan Arslan) and İ.S.; funding acquisition, H.A. (Hasan Akay) All authors have read and agreed to the published version of the manuscript.

Funding: This study was supported by the Scientific and Technological Research Council of Turkey (TÜBİTAK), project number 116O492.

Institutional Review Board Statement: Not applicable.

Informed Consent Statement: Not applicable.

Data Availability Statement: The data presented in this study are available on request from the corresponding author.

Conflicts of Interest: The authors declare no conflict of interest.

References

1. Kün, E. *Cereals-I (Cool Climate Cereals)*; Ankara Univ. Faculty of Agriculture, Ankara University: Ankara, Turkey, 1996; Publication No: 1451.
2. Çakmak, I. Enrichment of cereal grains with zinc: Agronomic or genetic biofortification? *Plant. Soil.* **2008**, *302*, 1–17. [CrossRef]
3. TUİK. Turkish Statistical Institute. 2019. Available online: http://tuik.gov.tr/PreTablo.do?alt_id=1001 (accessed on 15 May 2019).
4. Wakchaure, G.C.; Minhas, P.S.; Ratnakumar, P.; Choudhary, R.L. Optimizing supplemental irrigation for wheat (*Triticum aestivum* L) and the impact of plant bio-regulators in a semi-arid region of deccan plateau in India. *Agric. Water Manag.* **2016**, *172*, 9–17. [CrossRef]
5. Alghory, A.; Yazar, A. Evaluation of crop water stress index and leaf water potential for deficit irrigation management of sprinkler-irrigated Wheat. *Irrig. Sci.* **2019**, *37*, 61–77. [CrossRef]
6. Middleton, W.; Burnett, P.; Raphael, B.; Martinek, N. The bereavement response: A cluster analysis. *Br. J. Psychiatry* **1996**, *169*, 167–171. [CrossRef] [PubMed]
7. Smedema, L.K. Natural Salinity Hazards of Irrigation Development in (Semi) Arid Regions. Symposium on Land Drainage for Salinity Control. In *Arid and Semi-Arid Regions, Feb/March 1990*; Drainage Research Institute, Ministry of Public Works: Cairo, Egypt, 1990.

8. Arslan, H.; Guler, M.; Cemek, B.; Demir, Y. Assessment of groundwater quality in Bafra plain for irrigation. *J. Tekirdag Agric. Fac.* **2007**, *4*, 213–217. Available online: https://dergipark.org.tr/tr/download/article-file/178536 (accessed on 20 May 2019).
9. Mosaffa, H.R.; Sepaskhah, A.R. Performance of irrigation regimes and water salinity on winter wheat as influenced by planting methods. *Agric. Water Manag.* **2019**, *216*, 444–456. [CrossRef]
10. Houshmand, S.; Arzani, A.; Mirmohammadi-Maibody, S.A.M. Effects of Salinity and Drought Stress on Grain Quality of Durum Wheat. *Commun. Soil Sci. Plant. Anal.* **2014**, *45*, 297–308. [CrossRef]
11. Ozturk, A.; Erdem, E.; Aydin, M. The effects of drought after anthesis on the grain quality of bread wheat depend on drought severity and drought resistance of the variety. *Cereal Res. Commun.* **2021**. [CrossRef]
12. Konak, C.; Yılmaz, R.; Arabacı, O. Salt Tolerance in Eagean Region's Wheats. *Tr. J. Agric. For.* **1999**, *23*, 1223–1229.
13. Mut, Z.; Erbaş Köse, Ö.D.; Akay, H. Determination of grain yield and quality characteristics of some bread wheat (*Triticum aestivum* L.) varieties. *Anatol. J. Agric. Sci.* **2017**, *32*, 85–95. [CrossRef]
14. Sade, B. *Grain breeding (Wheat and Corn)*; Selcuk University Faculty of Agriculture Publications: Konya, Turkey, 1997; No: 31.
15. Ünal, S.S. Importance of wheat quality and methods in wheat quality determination. In *Cereal Products Technology Congress and Exhibition*; Hububat Ürünleri Teknolojisi Kongre ve Sergisi: Gaziantep, Turkey, 2002; pp. 25–37.
16. Mut, Z.; Aydın, N.; Bayramoğlu, N.O.; Özcan, H. Determination of yield and main quality characteristics of some bread wheat (*Triticum aestivum* L.) genotypes. *J. OMÜ Fac. Agric.* **2007**, *22*, 193–201. Available online: https://dergipark.org.tr/tr/pub/omuanajas/issue/20227/214339 (accessed on 20 October 2019).
17. Linina, A.; Ruţa, A. Weather conditions effect on fresh and stored winter wheat grain gluten quantity and quality. In Nordic View to Sustainable Rural Development. In Proceedings of the 25th NJF Congress, Riga, Latvia, 16–18 June 2015; pp. 148–153. Available online: https://llufb.llu.lv/conference/NJF/NJF_2015_Proceedings_Latvia-148-153.pdf (accessed on 20 July 2019).
18. Van Soest, P.J. *Nutritional Ecology of the Ruminant*, 2nd ed.; Cornell University Press: Ithaca, NY, USA, 1994.
19. Hansey, N.C.; Lorenz, J.A.; DeLeon, N. Cell wall composition and ruminant digestibility of various maize tissues across development. *Bioenerg. Res.* **2010**, *3*, 28–37. [CrossRef]
20. Olsen, S.R.; Cole, C.V.; Watanabe, F.S.; Dean, L.A. Estimation of available phosphorus in soils by extraction with sodium bicarbonate. In *Circular*; US Department of Agriculture: Washington, DC, USA, 1954; Volume 939, p. 19.
21. Kiremit, M.; Arslan, H.; Saıdy, A. Seedling growth characteristics of wheat seeds grown at different groundwater depth under non-ırrigation conditions. *Mustafa Kemal Univ. J. Agric. Sci.* **2019**, *24*, 241–248. Available online: https://dergipark.org.tr/tr/pub/mkutbd/issue/51091/651501 (accessed on 15 October 2020).
22. Elgün, A.; Türker, S.; Bilgiçli, N. Analytical Quality Control. In *Grain and Products*; Konya Commodity Exchange Publication: Konya, Turkey, 2001; No. 2.
23. Van Soest, P.J.; Robertson, J.B.; Lewis, B.A. Methods for dietary fiber, neutral detergent fiber, and nonstarch polysaccharides in relati onto animal nutrition. *J. Dairy Sci.* **1991**, *74*, 3583–3597. [CrossRef]
24. AACC. *American Association of Cereal Chemists. Approved Methods of the AACC*, 11th ed.; AACC: St. Paul, MN, USA, 2005; Available online: https://methods.aaccnet.org/about.aspx (accessed on 20 May 2020).
25. JMP. *JMP Users Guide. Version 13.0.0*; SAS Institute Inc.: Cary, NC, USA, 2020.
26. Yan, W.; Reid, J.F. Breeding line selection based on multiple traits. *Crop. Sci.* **2008**, *48*, 417–423. [CrossRef]
27. Özkaya, H.; Kahveci, B. *Tahıl ve Ürünleri Analiz Yöntemleri*; Gıda Teknol Dernei Yayınları: Ankara, Turkey, 1990; pp. 146–148.
28. Camphell, L.D.; Boila, R.J.; Stother, S.C. Variation in the chemical composition and test weight of barley and wheat grain grown at selected locations throughout Manitoba. *Can. J. Anim. Sci.* **1995**, *31*, 2–75.
29. Türköz, M.; Mut, Z. Determination of Yield and Quality Traits of Some Durum Wheat Genotypes in Konya Ecology. *Selcuk J. Agric. Food Sci.* **2019**, *31*, 27–36. [CrossRef]
30. Erbaş Köse, Ö.D.; Mut, Z. Grain yield and some quality traits of bread wheat cultivars. In Proceedings of the 3. Internationa Conference on Agriculture, Food, Veterinary and Pharmacy Sciences, Trabzon, Turke, 16–18 April 2019; pp. 250–256.
31. Koca, H.; Bor, M.; Özdemir, F.; Türkan, İ. The effect of salt stress on lipid peroxidation, antioxidative enzymes and proline content of sesame cultivars. *Environ. Exp. Bot.* **2007**, *60*, 344–351. [CrossRef]
32. He, F.; Pasam, R.; Shi, F.; Kant, S.; Keeble-Gagnere, G.; Kay, P.; Forrest, K.; Fritz, A.; Hucl, P.; Wiebe, K.; et al. Exome sequencing highlights the role of wild-relative introgression in shaping the adaptive landscape of the wheat genome. *Nat. Genet.* **2019**, *51*, 896–904. [CrossRef] [PubMed]
33. Erekul, O.; Yiğit, A.; Koca, Y.O.; Ellmer, F.; Weib, K. Quality potential of some bread wheat (*Triticum aestivum* L.) varieties and their ımportance in nutrition physiology. *J. Field Crop. Cent. Res. Inst.* **2016**, *25*, 31–36. [CrossRef]
34. Akkaya, A. *Buğday Yetiştiriciliği*; Sütçü İmam Üni. Ziraat Fakültesi Genel Yayın No:1; Ders Kitapları Yayın: Kahramanmaraş, Turkey, 1994.
35. Özen, S.; Akman, Z. Determination of yield and quality characteristics of some bread wheat varieties in Yozgat ecological conditions. *Süleyman Demirel Univ. J. Fac. Agric.* **2015**, *10*, 35–43.
36. Anjum, F.; Wahid, A.; Javed, F.; Arshad, M. Influence of foliar applied thiourea on flag leaf gas exchange and yield parameters of bread wheat (*Triticum aestivum*) cultivars under salinity and heat stresses. *Int. J. Agric. Biol.* **2014**, *10*, 619–626.
37. Çağlar, Ö. Investigation of Yield, Plant and Grain Protein Relationships in Some Winter Bread Wheat (*Triticum aestivum* L.) Cultivars and Lines. Master's Thesis, Department of Field Crops, Ataturk University, Erzurum, Turkey, 1990.

38. Atli, A.; Köksel, H.; Kocak, N.; Ercan, E. The qualities of domestic and foreign wheat cultivars grown in Turkey. In Proceedings of the 3rd Technical Congress of Agricultural Engineering, Ankara, Turkey, 8–12 January 1990; Chamber of Agricultural Engineers: Ankara, Turkey; pp. 272–281. (In Turkish).
39. Zheng, C.; Jiang, D.; Liu, F.; Dai, T.; Liu, W.; Jing, Q.; Cao, W. Exogenous Nitric Oxide improves seed germination in wheat against mitochondrial oxidative damage induced by high salinity. *Environ. Exp. Bot.* **2009**, *67*, 222–227. [CrossRef]
40. Hendawey, M.H. Effect of salinity on proteins in some wheat cultivars. *Aust. J. Basic Appl. Sci.* **2009**, *3*, 80–88. Available online: http://ajbasweb.com/old/ajbas/2009/80-88.pdf (accessed on 15 October 2020).
41. Yazar, S.; Salantur, A.; Özdemir, B.; Alyamaç, M.E.; Evlice, A.K.; Pehlivan, A.; Akan, K.; Aydoğan, S. Investigation of some agricultural characters in bread wheat improvement studies in the Central Anatolia Region. *J. Field Crop. Cent. Res. Inst.* **2013**, *22*, 32–40. Available online: https://dergipark.org.tr/tr/pub/tarbitderg/issue/11498/136980 (accessed on 15 May 2020).
42. Egesel, C.Ö.; Kahrıman, F.; Tayyar, Ş.; Baytekin, H. Mutual interaction of flour quality characteristics and grain yield in bread wheat and selection of appropriate variety. *Anadolu J. Agric. Sci.* **2009**, *24*, 76–83. Available online: https://dergipark.org.tr/tr/pub/omuanajas/issue/20221/214397 (accessed on 12 July 2020).
43. Sharma, B.R.; Minhas, P.S. Strategies for managing saline/alkali waters for sustainable agricultural production in South Asia. *Agric. Water Manag.* **2005**, *78*, 136–151. [CrossRef]
44. Torbica, A.; Antov, M.; Mastilovic, J.; Knezevic, D. The influence of changes in gluten complex structure on technological quality of wheat (*Triticum aestivum* L.). *Food Res. Int.* **2007**, *40*, 1038–1045. [CrossRef]
45. Singh, R.B.; Chauhan CP, S.; Minhas, P.S. Water production functions of wheat (*Triticum aestivum L.)* irrigated with saline and alkali waters using double-line source sprinkler system. *Agric. Water Manag.* **2009**, *96*, 736–744. [CrossRef]
46. Öztürk, A.; Aydın, F. Effect of Water stress at various growth stages on some quality characteristics of winter wheat. *J. Agron. Crop Sci.* **2004**, *190*, 93–99. [CrossRef]
47. Tilic, S.; Dodig, D.; Milašinovic Šeremešic, M.; Kandic, V.; Kostadinovic, M.; Prodanovic, S.; Saviç, D. Small grain cereals compared for dietary fibre and protein contents. *Genetika* **2012**, *43*, 381–395. [CrossRef]

 *agriculture*

Article

Hydropriming with Moringa Leaf Extract Mitigates Salt Stress in Wheat Seedlings

Talaat Ahmed *, Ahmed Abou Elezz and Muhammad Fasih Khalid

Environmental Science Center, Qatar University, Doha 2713, Qatar; a.hassan@qu.edu.qa (A.A.E.); mfasih.khalid@qu.edu.qa (M.F.K.)

* Correspondence: author: t.alfattah@qu.edu.qa

Abstract: Salinity is the major constraint that decreases the yield and production of crops. Wheat has a significant value in agricultural food commodities. The germination and growth of wheat seedlings are a big challenge in salt-affected soils. The seed priming technique is used to mitigate salt stress and enhance the germination and growth of the crops. Therefore, the current study was conducted to evaluate the hydropriming of natural plant extract (moringa leaf extract) and water on wheat seeds and grown under different saline (0, 0.05, 0.1, 0.15, and 0.2 M NaCl) environments. The germination attributes (germination percentage, germination index, mean germination day, coefficient of variance, vigor index) and seedling growth (fresh weight, dry weight, root length, shoot length) were enhanced in the plants primed by moringa leaf extract. The germination percentage was observed 10% more at 0.2 M NaCl stress in seeds treated with moringa leaf extract than seeds treated with water. The nutrient (K, Ca, Mg, P, S, Fe, B, Mn, Zn, Cu) uptake was also observed more in the shoots and roots of wheat seedlings soaked in moringa leaf extract as compared to soaked in water. Controlled plants showed higher concentrations of toxic ions (Na) and reactive oxygen species (H_2O_2) in shoots and roots of wheat seedlings. The use of moringa leaf extract for priming wheat seeds will enhance their germination and growth by maintaining efficient nutrient uptake and restricting the toxic ions and reactive oxygen species accumulation.

Keywords: abiotic stress; germination; plant growth; reactive oxygen species; toxic ions

Citation: Ahmed, T.; Abou Elezz, A.; Khalid, M.F. Hydropriming with Moringa Leaf Extract Mitigates Salt Stress in Wheat Seedlings. *Agriculture* **2021**, *11*, 1254. https://doi.org/10.3390/agriculture11121254

Academic Editors: Daniele Del Buono, Primo Proietti and Luca Regni

Received: 10 November 2021
Accepted: 7 December 2021
Published: 10 December 2021

1. Introduction

In arid and semiarid environments, due to climate change, abiotic stresses are a major threat to plants, specifically commercial crops. Among different abiotic stresses, salinity is one of the major factors that affect plant health and its production [1]. Wheat is one of the main food security crops in many countries around the world, including Qatar. However, Qatar depends mainly on imports to meet the domestic demand for wheat. Qatar's wheat production is deficient, and the productivity of the improved local bread wheat cultivar, Doha-88, is estimated at 2 tons/ha [2]. This low productivity is mainly due to the lack of well-adapted wheat varieties and suitable production technologies. Sustainable wheat production in Qatar can be obtained after developing new cultivars and advanced production technology. In Qatar, groundwater is the primary source of water irrigation that contains a significant amount of salt. Therefore, salt is accumulated in the surface layer of soil because of over-irrigation [3]. Moreover, the little rainfall and high rate of evaporation lead to salinity-related problems in these environments. When the seeds are exposed to abiotic stresses, crop germination capability and seed vigor are significantly affected, which further affects the crop growth, development, and yield [4].

To increase seed germination, different mechanisms have been analyzed by different scientists, and seed priming is most applicable among them [5–7]. Among different seed priming techniques, hydropriming is one of the most used techniques. Hydropriming allows the seeds to swiftly attain a sufficient amount of water and constant oxygen, which enhances germination by overcoming germination barriers specifically under unfavorable conditions, i.e., salinity, drought, etc. [7]. In hydropriming, different priming agents were used (i.e., salicylic acid, gibberellic acid, etc.), but the extract of leaves of many plants is found to be cheaper and more useful.

Moringa (*Moringa oleifera*) is well known as a miracle tree. It has been reported that moringa leaf extract has antioxidant properties [8]. In addition, the moringa leaf is also rich in many plant growth promoters such as cytokinin and zeatin [9,10]. Zeatin stimulates cell division and cell elongation. It boasts several enzymes' antioxidant properties and protects the plants' cells from the aging effects of the reactive oxygen species [11,12]. Therefore, the current study was aimed to investigate the potential effect of hydropriming on wheat seeds in moringa leaf extract and their responses on seed germination, growth, and nutrient uptake in leaves and roots of wheat seedlings under different salt (NaCl) stress conditions.

2. Material and Methods

2.1. Moringa Leaf Extract Preparation

Moringa leaves were collected from the Qatar University Agriculture Research Station located 60 km north of the State of Qatar in Rawdat Al-Faras. Fresh moringa leaves were collected, washed three times with distilled water, and dried in the air-dried oven at 45 °C for the next 48 h. The dried leaves were cut into small pieces before being ground into a fine powder. About 10 g of dried leaves powder was soaked into 50 mL polypropylene conical centrifuge tubes (Falcon 50 mL, Fisher Scientific, Pittsburgh, PA, USA) with distilled water and shacked for the next 48 h at lab conditions. The solution was centrifuged (SL 8 Benchtop Centrifuge, Thermo Fisher, Loughborough, UK) at 5000 rpm for 30 min. After that, the supernatant was filtered using a vacuum filter unit (1000 mL Buchner apparatus). The extract brownish and dark color solution was stored in the fridge at 4 °C until required.

2.2. Seed Germination and Seedling Growth

Sodium chloride (ACS reagent > 99%, Merck, Kenilworth, NJ, U.S.) was used to prepare saline solutions in distilled water. Five treatments, i.e., 0 M, 0.05 M, 0.1 M, 0.15 M and 0.2 M of NaCl were examined. Wheat seeds of *Triticum aestivum* (Cultivar Norin-*61*) were obtained from Tottori University-Japan. The healthy seeds with uniform size were sterilized with 20% (v/v) sodium hypochlorite for two min before washing three times with double distilled water. The wheat seeds were soaked in moringa leaf extract and water for 8 h and then placed in Petri dishes to germinate. A completely randomized design (CRD) with three replication was applied for this study.

The germination test was performed using Petri dishes (150 × 15 mm, Falcon) lined with Whatman filter paper No. 1 in a controlled growth chamber (day/night cycle of 16/8 h and R-humidity 50–60% at 25 ± 2 °C) for seven days. Twenty seeds were distributed on each dish with 30 mL of saline dilution treatments with three replicates. After maximum elongation, the germination parameters (germination percentage (*GP*), germination index (*GI*), mean germination rate (*MGR*), coefficient of variation (Cv_i), and vigor index (*VI*)) were calculated according to the work of [13].

2.3. Determination of Growth

Seedling growth under different salt and moringa extract treatments was assessed in terms of roots length and shoots in addition to fresh and dry weights. Seedlings were selected randomly from control and treated samples, divided into root and shoot before measuring the fresh weight and length. The length of seedlings was determined using a Digimizer software version 5.4.7 in mm. For estimation of dry weight, seedlings were freeze-dried (SP VirTis AdVantage Pro, Stone Ridge, NY, USA) for 48 h. The weight of samples was determined by analytical balance (AS 220.R2 PLUS, RADWAG, Radom, Poland).

2.4. Determination of Macro and Micronutrients

To prepare seedlings for nutrients and metal analysis, seedlings were washed three times with distilled water, and then samples were freeze-dried. Approximately 0.15 g of dried samples were weighed and digested with 5 mL HNO_3 and 2 mL H_2O_2 at 135 °C using a hot block for 2 h to determine the macronutrients; Ca, Mg, K, P, S, and micronutrients; B, Cu, Fe, Mn, Zn, and Na, in addition to Hg, a definite amount. The volume of digested sample was maintained up to 25 mL with deionized water. The elements were measured by inductively coupled plasma-optical emission spectroscopy (OPTIMA-5300DV, Perkin Elmar, Waltham, MA, USA), and contents of Hg were determined by cold vapor atomic absorption spectroscopy (AULA-254 Gold, Mercury Instruments, Karlsfeld, Germany).

2.5. Determination of Oxidative Stress Indices; Hydrogen Peroxide

According to Velikova et al. (2000), Hydrogen peroxide (H_2O_2) content was estimated in control and treated seedling samples [14]. Approximately 0.15 g of fresh tissue was homogenized in an ice bath with 2 mL 0.1% (*w/v*) trichloroacetic acid (TCA) before centrifugation at 12,000× *g* for 20 min at 4 °C. A total of 0.5 mL of the supernatant was added to 0.5 mL 10 mM potassium phosphate buffer (pH 7.0) and 1 mL 1 M KI. The absorbance of the reaction mixture was read at 390 nm using UV-VIS spectrometry. Hydrogen peroxide concentration was calculated by using a standard curve prepared with H_2O_2 [14,15].

2.6. Statistical Analysis

Data were examined for normality and homogeneity before starting the statistical analysis. One-way analysis of variance (ANOVA) followed by Tukey test to examine the significant differences between control and other treatments at 0.05 level. Linear regression was performed on excel. Principal component analysis (PCA) was constructed on SPSS by using a correlation matrix.

3. Results

3.1. Germination

Germination percentage (GP), germination index (GI), mean of the germination rate (MGR), coefficient of variation (CVt), and vigor index (VI) were decreased gradually through treatments in both soaking methods (MLE and water). Compared to the control, germination parameters (GP) (GI), (MGR), (CVt), and (VI) were dropped with a percentage difference 37%, 63%, 32%, 1%, and 93%, respectively, at 0.2 M NaCl under the effect of moringa soaking. While under the effect of water soaking, germination parameters were dropped more than the moringa soaking method. It was reduced up to 52%, 69%, 17%, 42%, and 97%, in (GP) (GI), (MGR), (CVt), and (VI) respectively, at 0.2 M NaCl (Table 1). Significant differences ($p < 0.05$) between treatments in GP, GI, VI, fresh weight, dry weight, root length, and shoot length under both soaking methods according to the Tukey test.

Table 1. Effect of salinity stress (NaCl) on the germination of wheat primed with moringa leaf extract (MLE) and water and grown under different saline conditions.

Treatments (NaCl)	Germination (Soaking in Water)					Germination (Soaking in MLE)				
	GP (%)	GI (Day)	MGR (Day^{-1})	CVt (%)	VI	GP (%)	GI (Day)	MGR (Day^{-1})	CVt (%)	VI
Control	92 ± 10	8.17 ± 0.66	0.42 ± 0.05	72.90 ± 10.39	2127.72 ± 463.95	87 ± 5	7.72 ± 0.51	0.44 ± 0.02	55.31 ± 4.11	2158.71 ± 260.38
0.05 M	84 ± 7	6.60 ± 0.53	0.42 ± 0.06	48.63 ± 10.91	1228.63 ± 177.21	77 ± 7	4.13 ± 0.70	0.30 ± 0.04	54.40 ± 9.09	1040.07 ± 60.39
0.15 M	49 ± 13	2.87 ± 0.65	0.37 ± 0.04	38.83 ± 7.61	465.05 ± 140.48	67 ± 5	3.98 ± 0.49	0.36 ± 0.03	35.91 ± 3.11	624.96 ± 112.75
0.1 M	58 ± 4	3.49 ± 0.55	0.36 ± 0.05	40.61 ± 12.17	306.18 ± 10.38	57 ± 12	3.20 ± 0.83	0.33 ± 0.03	49.88 ± 13.42	390.79 ± 104.77
0.2 M	44 ± 18	2.55 ± 1.06	0.35 ± 0.06	42.46 ± 20.77	59.83 ± 53.19	55 ± 10	2.86 ± 0.69	0.30 ± 0.05	54.61 ± 13.42	148.92 ± 81.19

GP = germination percentage; GI = germination index; MGR = mean germination day; CVt = coefficient of variance; VI = vigor index. ± standard deviation.

3.2. Growth

Increasing the salinity (NaCl) stress levels declined the growth parameters of the wheat seedlings that were soaked in water and leaf extract of *Moringa oliefera*. Even though seeds were soaked in moringa leaf extract represented even better growth than those soaked in water. Compared to the control, fresh (FW) and dry weight (DW) was dropped steadily with percentage difference up to 77% and 72%, respectively, and length of roots and shoots by 96% and 86%, at 0.2 M NaCl under the effect of moringa soaking. Instead, Fresh (FW) and dry weight (DW) were dropped steadily up to 93% and 50%, respectively, and length of roots and shoots by 97% and 89%, at 0.2 M NaCl under the effect of water soaking (Table 2).

Table 2. Effect of salinity stress (NaCl) on the growth of wheat primed with moringa leaf extract (MLE) and water and grown under different saline conditions.

Treatments (NaCl)	Growth (Soaking in Water)				Growth (Soaking in MLE)			
	FW (g)	DW (g)	RL (cm)	SL (cm)	FW (g)	DW (g)	RL (cm)	SL (cm)
Control	1.74 ± 0.15	0.22 ± 0.02	63.92 ± 19.3	166.3 ± 19.0	2.31 ± 0.58	0.29 ± 0.03	95.8 ± 18.9	152.58 ± 11.9
0.05 M	0.92 ± 0.34	0.11 ± 0.04	21.4 ± 3.1	125.5 ± 10.8	1.32 ± 0.39	0.25 ± 0.06	18.01 ± 3.38	118.4 ± 12.4
0.15 M	0.36 ± 0.16	0.06 ± 0.02	3.9 ± 1.5	52.2 ± 5.7	0.70 ± 0.49	0.22 ± 0.17	6.53 ± 1.70	62.74 ± 13.3
0.1 M	0.22 ± 0.14	0.04 ± 0.02	10.8 ± 3.9	85.2 ± 14.1	0.81 ± 0.40	0.24 ± 0.13	10.0 ± 2.97	83.84 ± 13.2
0.2 M	0.12 ± 0.05	0.11 ± 0.08	2.23 ± 1.3	18.29 ± 11.9	0.53 ± 0.36	0.08 ± 0.06	4.1 ± 0.93	21.97 ± 11.9

FW = fresh weight; DW = dry weight; RL = root length; SL = shoot length; ± standard deviation.

3.3. *Macronutrients*

Seedlings treated with moringa leaf extract showed a reduction in K, Ca, Mg, P, and S by 76%, 81%, 45%, 38%, and 29% in roots and K, Mg, and S by 52%, 27%, and 19% in shoots, respectively compared to the control (Table 3). In contrast, Ca and P were enhanced by 40%, 37% in shoots. On the other hand, seedlings soaked in water only showed a decreased content in K, Mg, and S by 45%, 5%, 8% in roots and by 77%, 33%, 16% in shoots, while Ca and P have a stimulatory effect by 81%, 20%, and 113%, 42% in roots and shoots, respectively compared to the control (Table 3).

3.4. *Micronutrients*

The micronutrient's concentrations of Fe, B, Mn, and Zn in seedlings treated with MLE declined by 46%, 65%, 97%, and 109%, respectively, while Cu showed enhancement by 38% in roots, compared to the control values (Table 4). In contrast, Fe, B, and Cu concentrations improved gradually within treatments by 15%, 166%, and 31%, respectively, while Zn and Mn were dropped by 22%, 64% in shoots compared to the control (Table 4). The concentrations of Fe, Mn, and Zn in seedlings that soaked in water dropped by 7%, 52%, and 23% in roots, respectively, while B and Cu indicated improvement by 46%, 57% in roots, compared to the control (Table 4). Conversely, Fe, B, Zn, and Cu content enhanced steadily within treatments by 41%, 132%, 45%, and 120%, respectively. In comparison, Mn dropped by 66% in shoots compared to the control, although the maximum concentrations of the micronutrients were observed in shoots at 0.1 M NaCl level (Table 4).

Table 3. Mean values of macronutrients (mg/Kg) in wheat primed with water and moringa leaf extract (MLE) under different saline conditions.

Treatments (NaCl)		Macronutrients (Soaking in Water)					Macronutrients (Soaking in MLE)				
		K	Ca	Mg	P	S	K	Ca	Mg	P	S
Roots	Control	5936.30 ± 134.90	1149.25 ± 25.15	2338.48 ± 47.20	5879.25 ± 34.88	3284.83 ± 18.73	6698.21 ± 99.74	1185.17 ± 17.41	2312.42 ± 33.16	6109.07 ± 67.53	2703.44 ± 19.64
	0.05 M	5387.50 ± 195.64	849.33 ± 34.91	3055.41 ± 116.35	7224.52 ± 81.25	3516.45 ± 93.39	5200.03 ± 22.23	765.14 ± 4.54	2281.59 ± 8.65	6412.63 ± 60.94	2730.08 ± 8.56
	0.1 M	6550.47 ± 162.17	1944.82 ± 50.04	2527.56 ± 59.78	8184.88 ± 175.87	3784.04 ± 75.38	4224.73 ± 15.65	574.94 ± 1.26	2206.44 ± 4.95	6049.89 ± 116.07	2675.15 ± 42.18
	0.15 M	5160.74 ± 223.33	955.82 ± 37.89	2530.00 ± 94.07	7530.22 ± 148.24	3365.00 ± 35.31	4795.13 ± 19.91	598.58 ± 1.55	2011.18 ± 4.74	5781.52 ± 120.10	2597.09 ± 14.67
	0.2 M	3742.33 ± 77.69	2729.63 ± 59.54	2230.23 ± 38.03	7156.66 ± 157.11	3030.40 ± 70.09	3014.70 ± 58.15	501.11 ± 11.34	1466.79 ± 18.23	4163.86 ± 48.69	2013.21 ± 33.62
Shoots	Control	8484.92 ± 117.16	1607.88 ± 17.12	3104.31 ± 26.98	7450.38 ± 123.50	3985.20 ± 15.25	8222.75 ± 122.19	1418.96 ± 27.51	1760.37 ± 30.11	4944.41 ± 23.62	3005.07 ± 15.30
	0.05 M	7881.03 ± 136.07	757.86 ± 9.26	1468.28 ± 17.14	6812.66 ± 145.95	4202.46 ± 36.56	7945.67 ± 60.19	1046.09 ± 13.38	1447.22 ± 14.33	5777.16 ± 102.46	3273.34 ± 72.39
	0.1 M	10,654.71 ± 213.16	4031.75 ± 108.80	2143.13 ± 14.38	11,174.58 ± 281.59	5050.39 ± 229.14	5372.57 ± 22.30	561.76 ± 4.53	1139.84 ± 4.75	4940.49 ± 30.86	2390.87 ± 33.75
	0.15 M	7594.40 ± 100.63	1537.39 ± 21.95	1835.31 ± 27.25	8259.17 ± 173.49	3795.11 ± 76.60	6844.69 ± 100.74	1018.20 ± 21.59	1159.64 ± 20.19	5720.02 ± 49.76	3023.21 ± 16.21
	0.2 M	3749.04 ± 158.77	5798.88 ± 127.11	2220.99 ± 7.95	11,395.67 ± 314.41	3409.75 ± 158.12	4806.81 ± 135.61	2122.32 ± 7.84	1343.90 ± 27.44	7186.65 ± 132.31	2494.86 ± 69.27

K = potassium; Ca = calcium; Mg = magnesium; P = phosphorous; S = sulfur.

Table 4. Mean values of micronutrients (mg/Kg) in wheat primed with water and moringa leaf extract (MLE) under different saline conditions.

Treatments (NaCl)		Micronutrients (Soaking in Water)					Micronutrients (Soaking in MLE)				
		Fe	B	Mn	Zn	Cu	Fe	B	Mn	Zn	Cu
Roots	Control	90.36 ± 1.66	17.06 ± 0.10	84.56 ± 2.06	158.94 ± 4.35	14.38 ± 0.28	76.19 ± 0.76	15.41 ± 0.62	77.71 ± 0.81	220.04 ± 1.93	12.12 ± 0.20
	0.05 M	76.05 ± 1.75	17.21 ± 0.32	88.57 ± 0.93	135.14 ± 1.43	13.63 ± 0.26	60.59 ± 1.00	6.94 ± 0.37	53.40 ± 0.64	104.73 ± 0.93	11.41 ± 0.22
	0.1 M	87.72 ± 2.00	54.78 ± 1.82	63.40 ± 1.20	148.55 ± 3.05	22.93 ± 0.16	60.08 ± 1.20	15.59 ± 0.18	47.83 ± 0.85	89.58 ± 0.88	11.47 ± 0.17
	0.15 M	98.13 ± 0.50	27.35 ± 1.42	56.98 ± 1.06	110.18 ± 1.74	16.65 ± 0.74	61.28 ± 0.60	23.80 ± 0.56	41.10 ± 0.69	85.47 ± 2.08	11.03 ± 0.05
	0.2 M	84.34 ± 2.81	27.18 ± 1.28	49.52 ± 0.90	125.47 ± 2.98	25.78 ± 1.38	47.61 ± 0.56	7.84 ± 0.24	27.04 ± 0.32	65.17 ± 0.74	8.79 ± 0.42
Shoots	Control	114.69 ± 1.47	14.56 ± 0.33	76.05 ± 0.94	109.42 ± 1.89	15.30 ± 0.17	99.44 ± 0.19	12.22 ± 0.33	42.55 ± 0.21	140.02 ± 1.05	13.00 ± 0.07
	0.05 M	93.68 ± 0.62	18.68 ± 0.22	34.61 ± 0.22	94.00 ± 0.71	17.91 ± 0.11	96.77 ± 2.82	21.35 ± 0.48	27.11 ± 0.37	125.06 ± 2.35	17.86 ± 0.10
	0.1 M	106.08 ± 1.74	51.3 ± 1.48	41.91 ± 1.13	189.63 ± 4.87	44.62 ± 1.79	75.07 ± 1.27	28.03 ± 0.50	20.92 ± 0.07	70.02 ± 0.57	12.45 ± 0.34
	0.15 M	118.03 ± 2.15	27.35 ± 1.42	29.00 ± 0.60	98.51 ± 1.61	15.63 ± 0.73	78.59 ± 0.52	81.13 ± 0.93	19.06 ± 0.25	73.41 ± 0.90	18.52 ± 0.46
	0.2 M	174.74 ± 4.21	71.40 ± 7.07	38.52 ± 0.79	173.48 ± 5.20	61.08 ± 5.01	115.32 ± 2.64	133.28 ± 3.66	21.82 ± 0.39	112.67 ± 2.01	15.51 ± 0.36

Fe = iron; B = boron; Mn = manganese; Zn = zinc; Cu = copper. ± standard deviation.

3.5. H_2O_2 and Na^+ Contents

The H_2O_2 content in shoots and roots of wheat seedlings was increased by increasing salt concentration (Figure 1). The H_2O_2 was observed more in the shoots and roots of wheat seedlings when treated with water as compared to when treated with moringa leaf extract (Figure 1).

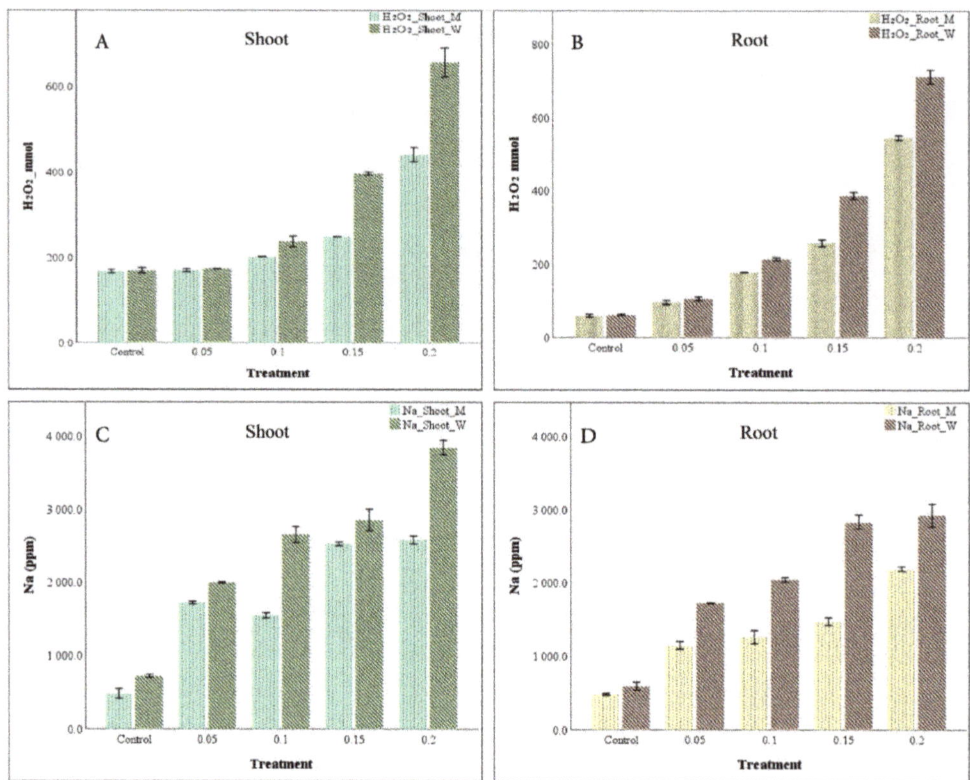

Figure 1. The effect of salt stress on wheat seedlings treated with moringa leaf extract and water on hydrogen peroxide and sodium content in leaves and roots. (**A**) Hydrogen peroxide in shoots; (**B**) hydrogen peroxide in roots; (**C**) Na in shoots; (**D**) Na in roots. Light color bar = moringa leaf extract; dark color bar = water.

Similarly, Na concentration was also increased under different saline conditions. The shoots and roots of wheat seedlings soaked in water exhibited more Na as compared to soaked in moringa leaf extract.

3.6. Linear Regression and PCA

The linear regression of Na and H_2O_2 was constructed and shown in Figure 2. The linear regression of Na and H_2O_2 indicated that by increasing the Na concentration, the production of H_2O_2 also increased, and the steep of the regression was higher in the shoots and roots of seeds treated with water as compared to moringa leaf extract (Figure 2).

PCA was applied on the 11 nutrient concentrations, in addition to H_2O_2, GP, and VI in to investigate the effect of factors on the wheat roots and shoots soaked in water and *Moringa oleifera* leaf extract individually, under the stress of saline (Figure 3, Table 5). The first two principal components (PCs) explained 46.95% (PC1), 31.88% (PC2) in roots, and 43.92% (PC1), 29.88% (PC2) in shoots (Figure 3a,b). PC1 in root treatments is robustly weighted in P, Cu, B, S, Fe, Ca, Na, and Mg, contributing 46.95% to the total variance, VI, Mn,

Zn, GP, K, and H_2O_2 dominated PC2, contributing 31.88% to the total variance (Table 5). The data set showed, two independent components were extracted with a cumulative percent of 78.82% and 73.81% of the total variance in seedlings root and shoot, respectively, under different treatments. The first component in Table 5 demonstrated significantly strong positive loadings of P (0.92), Cu (0.91), B (0.87), S (0.84), Fe (0.77), and Ca (0.75), and moderate positive loadings of Na (0.67), and Mg (0.58). In contrast, the second component of the roots data set showed significantly strong positive loadings of VI (0.95), Mn (0.91), Zn (0.85), GP (0.82), K (0.77), and H_2O_2 (0.76). In the same manner, the shoots data set has two independent components (Table 5). The first component demonstrated significantly strong positive loadings of P (0.95), Fe (0.94), Ca (0.87), Zn (0.87), Cu (0.85), B (0.75), S (0.71), and moderate positive loading of Mg (0.58). In contrast, PC2 explained significantly strong positive loadings of VI (0.93), GP (0.82), Mn (0.79), K (0.72), although moderate negative loadings were characterized in Na (−0.76), H_2O_2 (−0.50).

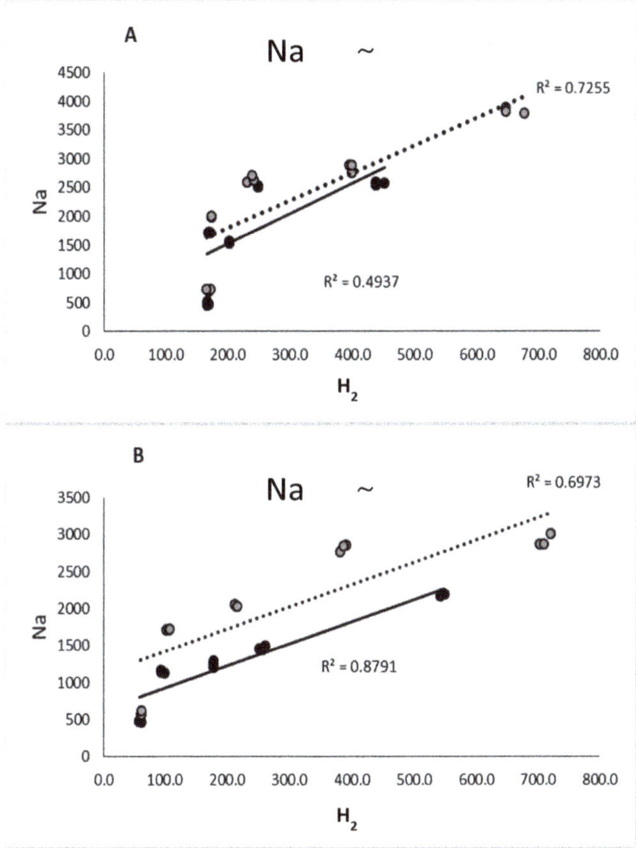

Figure 2. The linear regression shows the interaction between sodium (Na) and hydrogen peroxide (H_2O_2) in shoots (**A**) and roots (**B**) of wheat seedlings treated with moringa leaf extract (black filled, solid line) and water (gray filled, dotted line) under different saline conditions.

Table 5. PCA for germination, nutrients, toxic ions and reactive oxygen species in roots (a) and shoots (b) of wheat soaked in moringa leaf extract and water under NaCl stress.

Component	Total Variance Explained in Roots						Total Variance Explained in Shoots					
	Initial Eigenvalues			Rotated Component Matrix			Initial Eigenvalues			Rotated Component Matrix		
	Total	Variance %	Cumulative %	Variable	Loading PC1	Loading PC2	Total	Variance %	Cumulative %	Variable	Loading PC1	Loading PC2
1	6.572	46.946	46.946	P	0.918		6.150	43.929	43.929	P	0.945	
2	4.462	31.875	78.821	Cu	0.907		4.184	29.883	73.812	Fe	0.943	
3	1.500	10.717	89.538	B	0.876		1.917	13.693	87.505	Ca	0.874	
4	0.619	4.418	93.955	S	0.842		0.725	5.175	92.680	Zn	0.873	
5	0.340	2.430	96.385	Fe	0.768		0.527	3.766	96.446	Cu	0.850	
6	0.181	1.295	97.681	Ca	0.752		0.235	1.679	98.125	B	0.750	
7	0.165	1.175	98.856	Na	0.668		0.128	0.911	99.036	S	0.708	
8	0.105	0.752	99.608	Mg	0.586		0.083	0.594	99.630	Mg	0.588	
9	0.022	0.157	99.765	VI		0.956	0.021	0.151	99.780	VI		0.931
10	0.014	0.096	99.862	Mn		0.906	0.019	0.133	99.913	GP		0.821
11	0.011	0.076	99.938	Zn		0.846	0.007	0.051	99.964	Mn		0.792
12	0.005	0.037	99.974	GP		0.821	0.003	0.018	99.982	Na		−0.76
13	0.003	0.018	99.993	K		0.776	0.002	0.015	99.997	K		0.722
14	0.001	0.007	100.00	H_2O_2		0.757	0.00	0.003	100.00	H_2O_2		−0.50

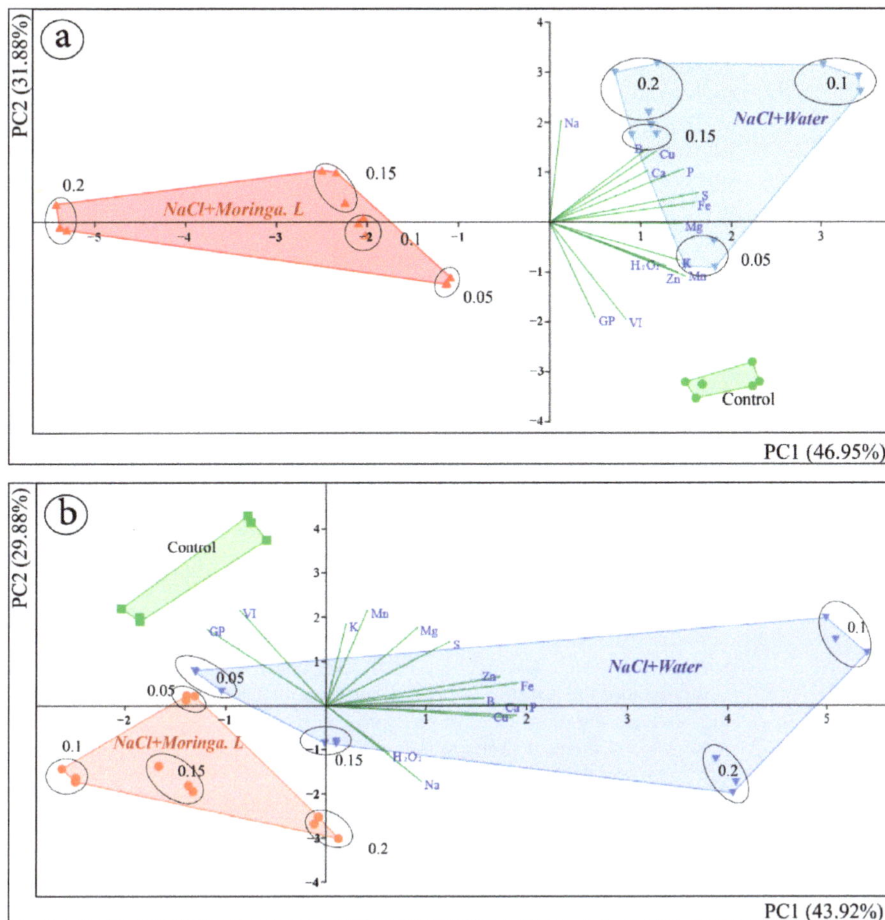

Figure 3. PCA of germination, nutrients, toxic ions, and reactive oxygen species in roots (**a**) and shoots (**b**) of wheat soaked in moringa leaf extract and water under NaCl stress.

4. Discussion

Salinity is one of the major factors that affect plant growth and development [16]. In our study, germination attributes of wheat were strictly affected by increasing the salt concentration. However, the seeds hydroprimed with moringa leaf extract showed more germination percentage, germination index, and vigor index as compared to seeds hydroprimed by water under different saline conditions (Table 1). Cytokinins have a great impact on plant germination and growth under stress conditions [17]. Moringa leaf extract has the cheapest and valuable source of cytokinins in the form of zeatin [10]. Hydropriming mainly changes physiological attributes, and accumulated ions with an ample amount of water help the plant to germinate under unfavorable conditions [18]. Hydropriming mainly increases the adenosine triphosphate (ATP) production and induced RNA activity, which helps the plant to germinate [19]. Demir and Ermis (2003) and Yagmur and Kaydan (2008) also observed similar findings on sunflower, potato seeds, and pyrethrum [20,21]. It was observed that after germination, the plant growth and development also enhanced in primed seeds as compared to non-primed seeds (Khalid et al., 2019b). The hydropriming enhances the cell division process in the apical meristem of seedlings which ultimately increases the plant growth and development [22]. In our study, the seedling growth and development (fresh weight, dry weight, root length, and shoot length) of wheat was decreased by increasing salt concentration (Table 2), but the decrease was more in plants primed with water as compared to plants primed with moringa leaf extract. Our findings are in line with Meriem et al. (2014) and Mirabi and Hasanabadi (2012), which found that primed seeds showed more growth in coriander and tomato seedlings under salinity as compared to non-primed [23,24]. Yasmeen et al. (2013) reported that by application of moringa leaf extract, wheat plants showed more growth by modulating their antioxidant defense mechanism under saline conditions [12].

Salinity affects the nutrient uptake in plants [25]. The decrease in nutrient uptake is associated with the accumulation of toxic ions. The nutrient uptake was affected in shoots and roots of wheat seedlings under different saline conditions (Table 3, 4). Seeds primed with moringa leaf extract cope with salt stress more efficiently and maintain their nutrients somehow as compared to seeds with water. The nutrients play an important role in plant different mechanisms.

Hydropriming significantly enhances the antioxidant defense mechanism under salt stress conditions [25]. The accumulation of toxic ions Na significantly decreases the uptake of K and Ca ions, which ultimately affects the physiological attributes, i.e., photosynthesis, stomatal conductance, etc. [16]. However, the different mechanisms, including maintaining the uptake of toxic ions and compartmentalization of toxic ions, are the indication of tolerance mechanisms in plants. So, in our study, seeds primed with moringa leaf extract showed less uptake of Na ions in shoots and roots of wheat seedlings. The results indicated that the tolerance mechanism of wheat seedlings primed with moringa leaf extract compartmentalizes the Na ions and restricts their uptake more efficiently as compared to the seeds primed with water. Khalid et al. (2020) also reported that restricted accumulation of toxic ions enhances the accumulation of other minerals, which maintains the plant metabolism and anchors the plant to fight more efficiently against salt stress [7]. The accumulation of toxic ions causes an imbalance in the plant nutrients, which affects its physiological attributes and production of different reactive oxygen species (ROS) [16]. H_2O_2 is one of the major ROS. Its accumulation in the cell results in disturbed cellular metabolism in mitochondria and chloroplast [26]. The degree of tolerance is identified by the H_2O_2 content in leaves and roots [12,27].

Wheat seeds treated with moringa leaf extract cope with the saline environment more efficiently by reducing the production of H_2O_2 as compared to those treated with water under different saline conditions. Similar findings were observed by the authors of [28]. Similar findings were observed that moringa leaf extract enhanced the seed germination of maize [29] and common bean [30] under salt stress condition

PCA illustrated that increasing the NaCl level (from 0.05 to 0.2 M) is often associated with decreasing germination parameters of wheat seedlings (Figure 3; Table 5). A positive relationship between nutrients and roots elongation of seeds soaked in water. However, the higher levels of NaCl (0.2 M) are associated with root inhibition due to the accumulation of Na, while roots of seeds are soaked in moringa leaf extract represents lower inhibition, especially at 0.05 and 0.1 M treatments. The contribution of shoots elongation of seeds soaked in moringa leaf extract and water. Germination parameters (GP and VI) were correlated with low levels (0.05, 0.1 M) in both treatments while increasing the level of Na associated with shoots inhibition. However, seeds treated with water showed the highest level of inhibition due to the strong accumulation of Na in shoot cells during germination of wheat seedlings.

5. Conclusions

It is concluded from the results that seed priming of wheat is very much useful to cope with the adverse saline condition (Figure 4). The right choice of primed solution will also play an important role under unfavorable conditions. Wheat seeds treated with moringa leaf extract showed more strong response by maintaining their germination and growth attributes under different saline conditions as compared to the seeds primed with water. Moringa leaf extract can restrict the accumulation of toxic ions more efficiently and decrease the production of ROS, which ultimately makes the plant tolerate adverse saline conditions. Therefore, the hydropriming of wheat seeds with moringa leaf extract will enhance their growth and development under saline conditions. Further studies on wheat seeds will be performed by hydropriming with moringa leaf extract in the open field to assess the response of mature wheat plants and the yield and quality of the grains.

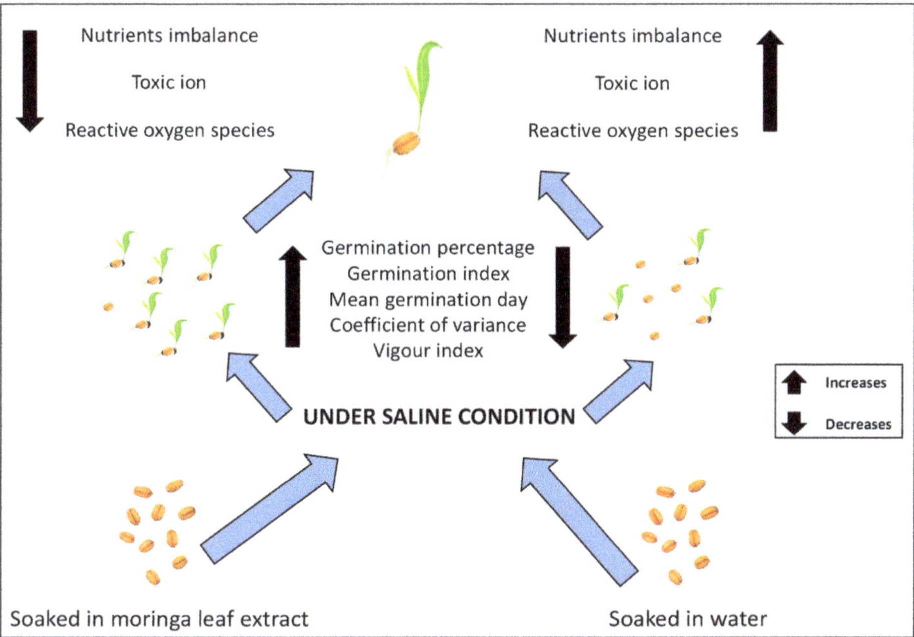

Figure 4. Schematic diagram showed the hydropriming effect of moringa leaf extract and water on wheat seeds under saline condition.

Author Contributions: Conceptualization, T.A.; methodology A.A.E.; software and formal analysis, A.A.E. and M.F.K.; writing—original draft preparation, A.A.E. and T.A.; writing—review and editing, T.A., M.F.K. and A.A.E. All authors have read and agreed to the published version of the manuscript.

Funding: This research was made possible by M-QJRC-2020-011. The statements made herein are solely the responsibility of the authors.

Institutional Review Board Statement: Not applicable.

Informed Consent Statement: Not applicable.

Data Availability Statement: Data is contained within the article.

Conflicts of Interest: The authors have no conflict of interest to declare.

References

1. Khalid, M.F.; Hussain, S.; Ahmad, S.; Ejaz, S.; Zakir, I.; Ali, M.A.; Ahmed, N.; Anjum, M.A. Impacts of Abiotic Stresses on Growth and Development of Plants. In *Plant Tolerance to Environmental Stress*; CRC Press: Boca Raton, FL, USA, 2019; pp. 1–8.
2. ALWawi, H. Selection Some Promise Genotypes of Wheat (*Triticum aestivum* L.) Under Local Conditions of Qatar. *J. Agric. Sci.* **2011**, *4*, 189–192. [CrossRef]

3. Gama, P.B.S.; Inanaga, S.; Tanaka, K.; Nakazawa, R. Physiological response of common bean (*Phaseolus vulgaris* L.) seedlings to salinity stress. *African J. Biotechnol.* **2007**, *6*, 79–88. [CrossRef]
4. Fazlali, R.; Eradatmand Asli, D.; Moradi, P. The effect of seed priming by ascorbic acid on bioactive compounds of naked Seed pumpkin (*Cucurbita pepovar.* styriaca) under salinity stress. *Int. J. Farming Allied Sci.* **2013**, *2*, 587–590.
5. Rouhi, H.R.; Aboutalebian, M. Effects of hydro and osmopriming on drought stress tolerance during germination in four grass species. *Int. J. AgriScience* **2011**, *1*, 107–114.
6. Omid, L.; Ali, A.G.; Ahmad, S.; Rad Amirhosein, S.; Samaneh, M.; Leila, M. The Effect of Low Temperature during Germination Stage on Seeds Germination Characteristics of 12 Cultivars of Spring Rapeseed (*Brasicca napus* L.). Available online: https://www.gcirc.org/fileadmin/documents/Proceedings/IRCWuhan2007vol3/316-318.pdf (accessed on 1 November 2021).
7. Khalid, M.F.; Hussain, S.; Akbar Anjum, M.; Arif Ali, M.; Ahmad, S.; Ejaz, S.; Ali, S.; Usman, M.; Ul Haque, E.; Morillon, R. Efficient compartmentalization and translocation of toxic minerals lead tolerance in volkamer lemon tetraploids more than diploids under moderate and high salt stress. *Fruits* **2020**, *75*, 204–215. [CrossRef]
8. Siddhuraju, P.; Becker, K. Antioxidant properties of various solvent extracts of total phenolic constituents from three different agroclimatic origins of drumstick tree (*Moringa oleifera* Lam.) leaves. *J. Agric. Food Chem.* **2003**, *51*, 2144–2155. [CrossRef] [PubMed]
9. Foidl, N.; Makkar, H.P.S.; Becker, K.; Foild, N.; Km, S. The Potential of *Moringa oleifera* for Agricultural and Industrial Uses. *What Dev. Potential Moringa Prod.* **2001**, *20*, 1–20.
10. Yang, R.Y.; Tsou, S.C.S.; Lee, T.C.; Chang, L.C.; Kuo, G.; Lai, P.Y. Moringa, a novel plant rich in antioxidants, bioavailable iron, and nutrients. In *Proceedings of the ACS Symposium Series*; ACS Publication: Washington, DC, USA, 2006; Volume 925, pp. 224–239.
11. Taiz, L.; Zeiger, E. Responses and Adaptations to Abiotic Stress. In *Plant Physiology*; Sinauer Associates, Inc.: Sunderland, MA, USA, 2010; pp. 755–788.
12. Yasmeen, A.; Basra, S.M.A.; Farooq, M.; Rehman, H.U.; Hussain, N.; Athar, H.U.R. Exogenous application of moringa leaf extract modulates the antioxidant enzyme system to improve wheat performance under saline conditions. *Plant Growth Regul.* **2013**, *69*, 225–233. [CrossRef]
13. Elezz, A.A.; Ahmed, T.A. The efficacy data of two household cleaning and disinfecting agents on Lens culinaris Medik and Vicia faba seed germination. *Data Br.* **2021**, *35*, 106811. [CrossRef] [PubMed]
14. Velikova, V.; Yordanov, I.; Edreva, A. Oxidative stress and some antioxidant systems in acid rain-treated bean plants. *Plant Sci.* **2000**, *151*, 59–66. [CrossRef]
15. Tripathi, D.K.; Singh, V.P.; Prasad, S.M.; Dubey, N.K.; Chauhan, D.K.; Rai, A.K. LIB spectroscopic and biochemical analysis to characterize lead toxicity alleviative nature of silicon in wheat (*Triticum aestivum* L.) seedlings. *J. Photochem. Photobiol. B Biol.* **2016**, *154*, 89–98. [CrossRef]
16. Khalid, M.F.; Hussain, S.; Anjum, M.A.; Ahmad, S.; Ali, M.A.; Ejaz, S.; Morillon, R. Better salinity tolerance in tetraploid vs. diploid volkamer lemon seedlings is associated with robust antioxidant and osmotic adjustment mechanisms. *J. Plant Physiol.* **2020**, *244*, 153071. [CrossRef] [PubMed]
17. Makkar, H.P.S.; Francis, G.; Becker, K. Bioactivity of phytochemicals in some lesser-known plants and their effects and potential applications in livestock and aquaculture production systems. *Animal* **2007**, *1*, 1371–1391. [CrossRef]
18. Alvarado, A.D.; Bradford, K.J.; Hewitt, J.D. Osmotic priming of tomato seeds: Effects on germination, field emergence, seedling growth, and fruit yield. *J. Am. Soc. Hortic. Sci.* **1987**, *112*, 427–432.
19. DELL'AQUILA, A.; BEWLEY, J.D. Protein Synthesis in the Axes of Polyethylene Glycol-Treated Pea Seed and during Subsequent Germination. *J. Exp. Bot.* **1989**, *40*, 1001–1007. [CrossRef]
20. Demir, I.; Ermis, S. Effect of controlled hydration treatment on germination and seedling growth under salt stress during development in tomato seeds. *Eur. J. Hortic. Sci.* **2003**, *68*, 53–58.
21. Yağmur, M.; Kaydan, D. Alleviation of osmotic stress of water and salt in germination and seedling growth of triticale with seed priming treatments. *African J. Biotechnol.* **2008**, *7*, 2156–2162. [CrossRef]
22. Sakhabutdinova, A.R.; Fatkhutdinova, D.R.; Bezrukova, M.V.; Shakirova, F.M. Salicylic Acid Prevents the Damaging Action of Stress Factors on Wheat Plants. *Bulg. J. Plant Physiol.* **2003**, 314–319.
23. Meriem, B.F.; Kaouther, Z.; Chérif, H.; Tijani, M.; André, B.; Meriem, B.F.; Kaouther, Z.; Chérif, H.; Tijani, M.; André, B. Effect of priming on growth, biochemical parameters and mineral composition of different cultivars of coriander (*Coriandrum sativum* L.) under salt stress. *J. Stress Physiol. Biochem.* **2014**, *10*, 84–109.
24. Mirabi, E.; Hasanabadi, M. Effect of seed priming on some characteristic of seedling and seed vigor of tomato (*Lycopersicon esculentum*). *J. Adv. Lab. Res. Biol.* **2012**, *3*, 237–240.
25. Hussain, S.; Khalid, M.F.; Hussain, M.; Ali, M.A.; Nawaz, A.; Zakir, I.; Fatima, Z.; Ahmad, S. Role of Micronutrients in Salt Stress Tolerance to Plants. In *Plant Nutrients and Abiotic Stress Tolerance*; Springer: Singapore, 2018; pp. 363–376, ISBN 9789811090448.
26. Gill, S.S.; Tuteja, N. Reactive oxygen species and antioxidant machinery in abiotic stress tolerance in crop plants. *Plant Physiol. Biochem.* **2010**, *48*, 909–930. [CrossRef] [PubMed]
27. Khalid, M.F.; Morillon, R.; Anjum, M.A.; Ejaz, S.; Rao, M.J.; Ahmad, S.; Hussain, S. Volkamer Lemon Tetraploid Rootstock Transmits the Salt Tolerance When Grafted with Diploid Kinnow Mandarin by Strong Antioxidant Defense Mechanism and Efficient Osmotic Adjustment. *J. Plant Growth Regul.* **2021**, 1–13. [CrossRef]

28. Umair, A.; Ali, S.; Tareen, M.J.; Ali, I.; Tareen, M.N. Effects of Seed Priming on the Antioxidant Enzymes Activity of Mungbean (Vigna radiata) Seedlings. *Pakistan J. Nutr.* **2012**, *11*, 140–144. [CrossRef]
29. Ali, A.; Abbas, M.N.; Maqbool, M.M.; Haq, T.U.; Mahpara, S.; Mehmood, R.; Arshad, M.I.; Lee, D.J. Optimizing the various doses of moringa (*Moringa oleifera*) leaf extract for salt tolerance in maize at seedling stage. *Life Sci. J.* **2021**, *7*, 18. [CrossRef]
30. Rady, M.M.; Varma, B.; Howladar, S.M. Common bean (*Phaseolus vulgaris* L.) seedlings overcome NaCl stress as a result of presoaking in Moringa oleifera leaf extract. *Sci. Hortic.* **2013**, *162*, 63–70. [CrossRef]

 *agriculture* 

Article

Biostimulant Effects of an Aqueous Extract of Duckweed (*Lemna minor* L.) on Physiological and Biochemical Traits in the Olive Tree

Luca Regni [1], Daniele Del Buono [1,*], Begoña Miras-Moreno [2], Biancamaria Senizza [2], Luigi Lucini [2], Marco Trevisan [2], Diletta Morelli Venturi [3], Ferdinando Costantino [3] and Primo Proietti [1]

1 Dipartimento di Scienze Agrarie, Alimentari e Ambientali, University of Perugia, Borgo XX Giugno, 06121 Perugia, Italy; luca.regni@unipg.it (L.R.); primo.proietti@unipg.it (P.P.)
2 Department for Sustainable Food Process, Università Cattolica del Sacro Cuore, Via Emilia Parmense 84, 29122 Piacenza, Italy; mariabegona.mirasmoreno@unicatt.it (B.M.-M.); biancamaria.senizza@unicatt.it (B.S.); luigi.lucini@unicatt.it (L.L.); marco.trevisan@unicatt.it (M.T.)
3 Dipartimento di Chimica, Biologia e Biotecnologia, University of Perugia, Via Elce di Sotto 8, 06123 Perugia, Italy; diletta.morelliventuri@studenti.unipg.it (D.M.V.); ferdinando.costantino@unipg.it (F.C.)
* Correspondence: daniele.delbuono@unipg.it

Abstract: Biostimulants are becoming increasingly popular in agriculture for their ability to induce beneficial effects in crops, paving the way towards the identification of new materials with biostimulant potential. This study evaluated the potential of different concentrations of an aqueous extract (0.25%, 0.50%, and 1.00%, dry weight/water volume, respectively) obtained from duckweed (*Lemna minor* L.) to stimulate olive plants. Leaf net photosynthesis (Pn), leaf transpiration rate (E), stomatal conductance (gs), sub-stomatal CO_2 concentration (Ci), chlorophyll content and other plant growth parameters were investigated. As a result, the extract improved Pn, gs, Ci, chlorophyll content and plant biomass production (leaf fresh and dry weight). Furthermore, the duckweed extract generally increased the uptake of nitrogen (N), potassium (K), calcium (Ca), magnesium (Mg), iron (Fe) and zinc (Zn), while it did not influence the content of sodium (Na), manganese (Mn) and copper (Cu). The untargeted metabolomic profiling of the extract revealed the presence of signalling compounds (including phytohormones), phenolics and glutathione. Such broad diversity of bioactives may support the stimulatory potential observed in olive. In summary, this study revealed for the first time that duckweed could be seen as a promising species to obtain extracts with biostimulant properties in olive trees.

Keywords: biostimulant; aquatic species; photosynthesis; plant growth; plant nutrition; bioactive metabolites

Citation: Regni, L.; Del Buono, D.; Miras-Moreno, B.; Senizza, B.; Lucini, L.; Trevisan, M.; Morelli Venturi, D.; Costantino, F.; Proietti, P. Biostimulant Effects of an Aqueous Extract of Duckweed (*Lemna minor* L.) on Physiological and Biochemical Traits in the Olive Tree. *Agriculture* **2021**, *11*, 1299. https://doi.org/10.3390/agriculture11121299

Academic Editor: Cinzia Margherita Bertea

Received: 16 November 2021
Accepted: 16 December 2021
Published: 20 December 2021

1. Introduction

The use of different formulations of certain organic materials and microorganisms, defined with the term biostimulants, aims to stimulate nutrition and crops growth, increase their tolerance to environmental stress, and improve the efficiency in the use of natural resources of agroecosystems [1,2]. In addition, biostimulants are gaining attention for the possibility of reducing chemical inputs as fertilizers [3]. The first definition of biostimulants can be found in a web journal of 1997 called "Ground Maintenance", in which Zhar and Schmidt, from the Department of Crop Science and Soil Environment at Virginia Polytechnic Institute, defined biostimulants as "materials that, in minute amounts, promote plant growth" [1]. In the following years, the use of the term "biostimulant" has become increasingly popular in the literature, expanding the range of substances and the inherent modes of action.

Biostimulants are successfully utilized in both cereal and horticultural crops, as they are materials capable of promoting plant growth without being fertilizers, soil conditioners or pesticides [4,5]. Currently, biostimulants are seen as an interesting and innovative technology to increase the ability of crops to cope with some adverse environmental conditions [4]. These materials can exert beneficial effects on crops both when applied to plants grown under optimal environmental conditions and when administered to species exposed to abiotic and biotic stresses [4,5]. In this context, the effectiveness of biostimulants does not only derive from increased crop yields under stressful environmental conditions, but also from their ability to maintain high product quality traits [6]. In this sense, biostimulants are the object of intense research to ascertain their capacity to improve fruit quality and nutraceutical value [7]. For instance, in a recent study, it was found that a commercial biostimulant increased tomato fruit yield and size, nutritional composition and antioxidant properties [8]. Additionally, the effect of an extract of *Moringa oleifera* L on two genotypes of *Brassica* was investigated [7]. The authors of this study found that the biostimulant improved some nutraceutical aspects, depending on the species treated.

Given the high number of materials capable of stimulating crops, biostimulants are usually grouped into different families depending on the raw materials from which they are derived: humic substances, complex organic materials, beneficial chemical elements (e.g., silicon), inorganic salts, algae and plant extracts, protein hydrolysates, chitin and chitosan derivatives, antiperspirants (e.g., kaolin), amino acids and other compounds [1,9,10].

When applied to plants or soil, biostimulants can regulate and/or improve crops' physiological and biochemical processes, increasing their productivity and quality [3,9,11]. Biostimulants can also promote plant growth by modifying root development and architecture, thus predisposing the treated crop to absorb and translocate nutrients more efficiently [12]. In addition, these materials can increase photosynthetic efficiency, promote the accumulation of sugars in fruits, fruit set and storability. Particularly interesting is also the ability of biostimulants to make crops less sensitive to abiotic stresses, such as extreme temperatures, drought, salinity, excessive moisture in the rhizosphere, or over-or under-exposure to light [13].

As for the olive tree (*Olea europaea* L.), the application of biostimulants results in some beneficial effects, as these substances can enhance the leaf area and the total chlorophyll content [14], the nutritional status, the olive production and some olive oil quality parameters [15]. Differently, contrasting results can be found in the literature in young olive. Molina Soria [16] reported no significant effects of biostimulants on the growth of young trees, whereas Saour [17] found that the use of a combination of biostimulant/kaolin particle film enhanced growth, resulting in the production of higher quality olive seedlings. Positive effects on photosynthetic activity and growth of young olive trees were also observed by Almadi et al. [18], who applied a biostimulant consisting of a complex of amino acids (glycine, proline, hydroxyproline, etc.). Furthermore, during their production cycle, olive trees can often be subjected to environmental stress conditions, whose frequency and intensity are increasing due to climate change, which could lead to a lower yield and, in some cases, provoke plant death. For this, it is necessary to implement agronomic techniques, including cultivation operations, to increase and encourage those physiological mechanisms of tolerance to stress triggered by processes activated at the molecular level. Among the techniques that can be used to enhance the tolerance of the olive trees to abiotic stress, there is also the biostimulant application.

In this context, the research is also interested in finding new substances with biostimulant activity from a wide range of starting materials; to this end, particular interest is shown in plant extracts showing biostimulatory potential for their bio- and eco-compatibility [19,20]. Furthermore, some plant extracts obtained from terrestrial species can improve crop growth and productivity, dry matter, nutrient concentration and antioxidant activities, thus representing a sustainable and effective tool for crop systems [20–22]. As for aquatic species, the ability of extracts obtained from seaweeds on crop growth and stress resistance has been demonstrated in many studies [23–25].

In general, no studies have investigated the effects of plant extracts obtained from freshwater aquatic species except for a recent one on *Lemna minor* L. (duckweed), which showed the biostimulant potential of this plant in maize [26]. In particular, the beneficial effects of the duckweed extract, found in this study, were attributed to the high abundance of phytochemicals with bioactive properties [26].

Duckweed is a free-floating plant of *Lemnaceae*, widely distributed in lagoons, wetlands, and ponds, which shows rapid growth and adaptability to adverse environmental conditions [27]. Duckweed is also excellent in removing toxic substances from polluted water, and its ability to tolerate toxicants has been attributed to its antioxidant activities, which can be easily induced by some compounds [28]. *Lemnaceae* are plants rich in metabolites that exhibit antioxidant and antibacterial properties [29]. In addition, duckweed has been recently demonstrated to possess a high content of phenolic acids, phenols and flavonoids [26,30]. It is well known that certain compounds could benefit plants when exogenously applied [31]. In light of the above, duckweed can be seen as a biological stock of metabolites with potential bioactive properties.

Based on these premises, this research aimed at ascertaining whether an aqueous extract obtained from duckweed could exert biostimulatory activity on olive plants. To this scope, a duckweed aqueous extract was administrated at different concentrations to olive plants (cv. Arbequina) grown in hydroponic, and some physiological and nutritional aspects were investigated in treated samples compared to untreated controls. The cv. Arbequina was chosen for the experiment since its use in the world is rapidly increasing due to its adaptability to new high-density olive planting [32–35].

Finally, this research aimed to identify the metabolites with biostimulant potential in the duckweed extract by an untargeted metabolomic approach.

2. Materials and Methods

2.1. Olive Material and Growing Conditions

Rooted olive cuttings cv. Arbequina (about 18 cm height) were removed from the perlite substrate of the mist propagation system. After root washing with distilled water, the plants were placed in 800 mL pots containing expanded clay (10 g per pot) and put in a hydroponic system for an adaptation period of 60 days. Clay is an inert and commonly used substrate for hydroponic [36].

The hydroponic system was maintained in a growing chamber (Figure 1), and plants were exposed to light with an active photosynthetic radiance by a system equipped with lamps (PHILIPS SON-T AGRO 400 W) producing 200 μmol m^{-2} s^{-1} photon flux density, under a photoperiod of 16 h d^{-1}. The temperature was constantly set at 23 °C (±1 °C) and relative humidity at about 60%. The hydroponic system consists of PVC containers comprising five plastic hydroponic pots and five plants each. Each container is connected to a tank (volume 3.5 L) containing the nutrient solution (half strength Hoagland solution, pH 7.5). An automated system due to pressurized air ensured the flux of the nutrient solution from the tank to the PVC containers with the plants three times per day. The nutrient solution was replaced every 30 days, while the evapotranspirated water was reintegrated every 2 days.

2.2. Lemna Minor Growth Conditions and Preparation of the Extract

Duckweed was harvested from a freshwater basin near the city of Perugia (Italy). First, the plants were sterilized with a 0.5% sodium hypochlorite solution for 2 min. After that, the plants were copiously rinsed twice with distilled water. Duckweed plants were then transferred to polyethylene trays (35 × 28 × 14 cm) and grown according to a published protocol [28]. The culture media were renewed every two weeks.

Ten grams of duckweed were collected, rinsed with water, and dried at 40 °C until constant weight. After that, 1 g of dried plant material was extracted with a mortar and pestle and 100 mL water (pH 7.0). The resulting suspension was maintained in an orbital shaker (100 rpm) for 24 h. After this time, the extract was vacuum filtered on a Buchner filter, and the liquid phase was brought to 100 mL. This solution was the most concentrated (1.00% duckweed extract).

Figure 1. Hydroponic system used for the experiments.

2.3. *Olive Treatments with Duckweed Extract*

At the end of the adaptation period to hydroponic conditions, plants were treated with different biostimulant concentrations (1.00, 0.50 and 0.25% *w*/*v*, named BIO 1, BIO 0.5 and BIO 0.25, respectively) through the foliar application (time 0 days after the treatment—0 DAT). In particular, for each biostimulant concentration, 15 plants were treated, using 15 mL of solution for each plant. Another 15 plants were left as control and not treated with the biostimulant. After 10 days (10 DAT), the biostimulant treatment was repeated.

2.4. *Olive Leaf Gas Exchange, Chlorophyll Content and Growth*

Leaf net photosynthesis (Pn), leaf transpiration rate (E), stomatal conductance (gs) and sub-stomatal CO_2 concentration (Ci) were determined for each treatment at 7, 14 and 21 DAT. Leaf gas exchange rates were measured using a portable IRGA (ADC-LCA-3, Analytical Development, Hoddesdon, UK) and a Parkinson-type assimilation chamber. Leaves were enclosed in the chamber and exposed to the same light as in the hydroponic system. The flow rate of air passing through the chamber was kept at 5 $cm^3\ s^{-1}$. During gas-exchange measurements, the external CO_2 concentration was about 375 $cm^3\ m^{-3}$, and the air temperature inside the leaf chamber was about 1 °C higher than the hydroponic room temperature. The chlorophyll content was measured on the middle part of 15 leaves for each treatment, using a SPAD-502 Chlorophyll Meter (Minolta Camera Co. Ltd., Osaka, Japan) at 7, 14 and 21 DAT.

At the end of the experiment, 30 DAT, five plants from each treatment were selected, and roots, shoots, stems and leaves of each plant were weighed fresh (FW) and then oven-dried at 95 °C until constant weight to determine dry weight (DW). Moreover, the number of leaves was also determined.

2.5. *Nutrient Determination in Olive Leaves*

The nutrient determination in olive leaves was run in triplicate in samples dried in an oven at 60 °C until a constant weight had been reached. Nitrogen quantification was carried out on leaf samples (1.0 g) digested with 12.5 mL H_2SO_4 96% (*v*/*v*), 7.0 mL H_2O_2 30% (*v*/*v*) and a Kjeldahl tablet. Then, digested tissues were left to cool and added with 80.0 mL of NaOH 32.5% (*w*/*v*). Total nitrogen was determined by titration with H_2SO_4 0.1 N [37].

Furthermore, 0.25 g of dried leaves were added with 7.0 mL of HNO_3 65% (*v*/*v*) and 3.0 mL of H_2O_2 30% (*v*/*v*) and left at 90 °C for 90 min. The acid digested samples were filtered, and K, Ca, Mg, Na, Fe, Mn, Zn and Cu were quantified by Inductive Coupling Plasma spectrometry (ICP) [38].

2.6. Duckweed Extract Profiling

The phytochemical profile of the duckweed extract was characterized through an untargeted metabolomics approach, based on ultrahigh-pressure liquid chromatography quadrupole-time-of-fight mass spectrometry (UHPLC-ESI/QTOF-MS) as reported by Del Buono et al. [26]. Briefly, the chromatographic separation was achieved using an Agilent Zorbax eclipse plus column (50 × 2.1 mm, 1.8 μm) and a binary mixture of water and acetonitrile (4–96% in 33 min linear gradient). QTOF-MS acquisition used positive polarity and full scan mode (100–1200 m/z, 1 Hz scan rate, absolute peak height threshold 3000 counts), and the injection volume was 4 μL. Triplicate samples were analysed.

The annotation of raw mass features was performed as previously reported using the software Profinder B.07 (from Agilent Technologies, CA, USA), according to monoisotopic accurate mass and the whole isotopic pattern [26]. The subsequent annotation was carried out using the plant metabolome database PlantCyc (https://plantcyc.org/ access date: 14 October 2021), and only the compounds annotated within 100% of replications were retained and annotated.

2.7. Statistical Analysis

The trials were organized according to a randomized block design, with 4 treatments and 15 plants for each treatment. The experiments were carried out in triplicate. Statistical analysis was performed by analysis of variance (one-way ANOVA). Significant differences between the values were determined at $p \leq 0.05$, according to the Tukey test. The statistical environment R was used to perform the analysis [39].

3. Results

3.1. Leaf Net Photosynthesis (Pn), Leaf Transpiration Rate (E), Stomatal Conductance (gs) and Sub-Stomatal CO_2 Concentration (Ci)

The photosynthetic activity was recorded at 7, 14 and 21 DAT in plants treated with the duckweed extract at three concentrations (BIO 0.25, BIO 0.5 and BIO 1). Significant Pn increases were observed at 21 DAT for all the duckweed extract concentrations used in biostimulated olive leaves compared to untreated controls (Figure 2). Furthermore, an increase in gs for samples treated with the two highest dosages, BIO 0.5 and BIO 1 was observed (Figure 2). All the duckweed extract concentrations used significantly increased Ci with respect to the control. In contrast, the leaf transpiration rate (E) resulted unaffected by the treatment with the duckweed extract. Therefore, the biostimulated plants at 21 DAT exhibited greater Pn, gs, and Ci values than untreated plants (control).

The olive treatment with the duckweed extract enhanced the leaves chlorophyll significantly compared to untreated samples throughout the experimental period (Figure 3). In particular, inductive effects on chlorophyll were found for all the dosages applied BIO 0.25, BIO 0.5 and BIO 1 at 14 and 21 DAT. The biostimulated plant leaves, at all the duckweed extract concentrations applied, showed higher SPAD values than those of control plants.

3.2. Plant Growth and Biomass Development

Regardless of the concentration applied, the treatment with the duckweed plant extract prompted significant increases in the leaves number, fresh and dry weight and shoot fresh and dry weight compared to the untreated controls (Table 1). The BIO 0.25, BIO 0.5, and BIO 1 treatments showed no significant difference between them regarding the effect on the above parameters. Finally, the duckweed extract did not affect the growth and development of the other plant organs such as roots and stem (Table 1).

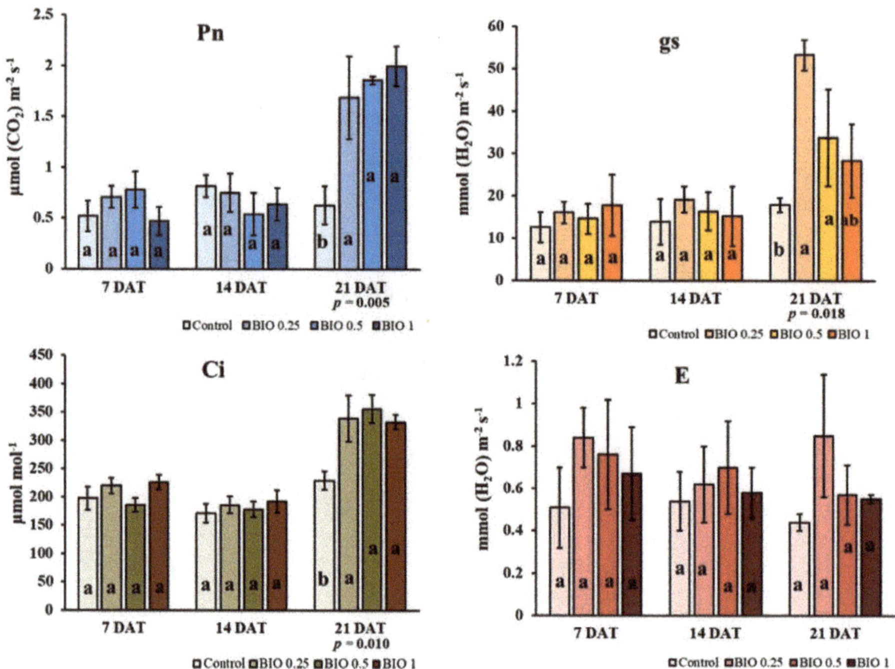

Figure 2. Leaf net photosynthesis (**Pn**) (µmol (CO_2) m^{-2} s^{-1}), stomatal conductance (**gs**) (mmol (H_2O) m^{-2} s^{-1}), sub-stomatal CO_2 concentration (**Ci**) (µmol mol^{-1}) and leaf transpiration rate (**E**) (mmol (H_2O) m^{-2} s^{-1}) measured at 7, 14 and 21 days after duckweed extract treatment (DAT). For each DAT and for each parameter, means with different letters are significantly different ($p < 0.05$) as indicated by one-way ANOVA followed by Tuckey test. The bars reported SD (standard deviation).

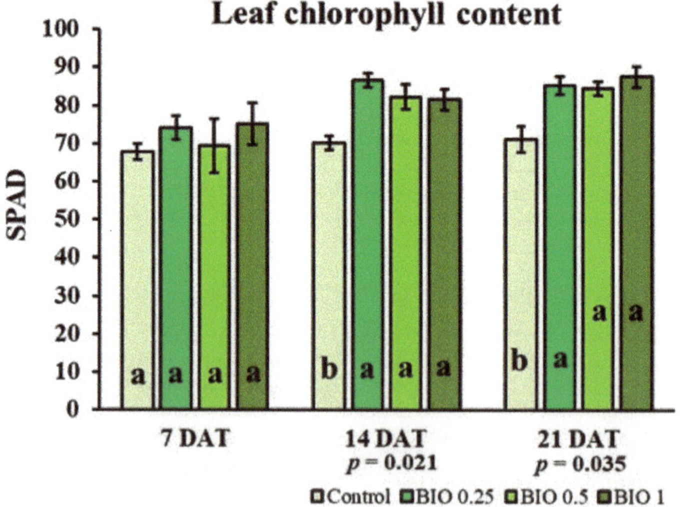

Figure 3. Leaf chlorophyll content was measured by SPAD at 7, 14 and 21 days after duckweed extract treatment (DAT). For each DAT, means with different letters are significantly different ($p < 0.05$) as indicated by one-way ANOVA followed by Tuckey test. The bars reported SD (standard deviation).

Table 1. Fresh weight (FW) and dry weight (DW) of leaves, roots, stem and lateral shoots and total number of leaves at 30 days after duckweed extract treatments (DAT).

	Leaves FW	Leaves DW	Roots FW	Roots DW	Stem FW	Stem DW	Lateral Shoots FW	Lateral Shoots DW	Number of Leaves
	(g)	(g)	(g)	(g)	(g)	(g)	(g)	(g)	(n)
Control	1.19 (0.27) b	0.59 (0.08) b	20.61 (3.23) a	3.19 (0.34) a	2.46 (0.25) a	1.37 (0.10) a	0.28 (0.04) b	0.09 (0.03) b	18.4 (3.11) b
BIO 0.25	3.68 (0.30) a	1.48 (0.10) a	19.23 (4.43) a	3.24 (0.62) a	2.95 (0.41) a	1.48 (0.21) a	0.72 (0.19) a	0.24 (0.02) a	46.0 (7.48) a
BIO 0.5	1.76 (0.27) a	0.81 (0.10) a	22.18 (4.93) a	3.80 (0.85) a	3.02 (0.30) a	1.62 (0.13) a	0.47 (0.04) a	0.13 (0.03) a	24.8 (2.52) a
BIO 1	2.62 (0.58) a	1.17 (0.21) a	17.86 (4.00) a	3.06 (0.54) a	2.55 (0.10) a	1.36 (0.08) a	0.45 (0.09) a	0.21 (0.04) a	33.4 (7.32) a
	$p = 0.0015$	$p = 0.0013$					$p = 0.0034$	$p = 0.0042$	$p = 0.0027$

In each column and for each parameter, mean values followed by different letters are significantly different ($p < 0.05$) as indicated by one-way ANOVA followed by Tuckey test. In parenthesis, SD (standard deviation) is reported.

3.3. *Effect of the Duckweed Extract, Applied at the Three Different Concentrations, on Olive Nutrient Content*

The content of some macro- and micronutrients was investigated in olive leaves treated with the duckweed extract applied at the three different concentrations and compared with untreated control samples (Table 2). Regarding the N content, it was found that the samples treated with the duckweed extract, regardless of the concentration applied, showed significant increases in the content of this element with respect to the control samples. However, the different concentrations of the duckweed extract showed no significant difference in the N content between them. With regard to K, BIO 0.25, BIO 0.5 and BIO 1 significantly increased the content of this element compared to the control samples. However, as with N, no significant differences were found between the different concentrations of duckweed applied to olive samples. Concerning Ca, the BIO 1 was the only treatment effective in raising the content of this element in the biostimulated olive compared to the control samples. Differently, all the treatments with the duckweed extract significantly stimulated the Mg content, but no significant differences were found between BIO 0.25, BIO 0.5 and BIO 1. Finally, the last macronutrient investigated, Na, was unchanged in olive leaves following the treatments with the duckweed extracts.

Table 2. Content of mineral elements (N, K, Ca, Mg, Na, Fe, Mn, Zn, Cu) determined in olive leaves at 30 days after duckweed treatment (DAT).

	N	K	Ca	Mg	Na	Fe	Mn	Zn	Cu
			(mg g^{-1} DW)				(μg g^{-1} DW)		
Control	1.84 (0.60) b	59.8 (7.1) b	3.68 (0.15) b	0.52 (0.07) b	0.50 (0.10) a	23.8 (2.4) b	13.7 (3.5) a	9.1 (0.10) b	16.8 (4.8) a
BIO 0.25	3.10 (0.15) a	78.9 (4.5) a	3.70 (0.07) b	0.70 (0.01) a	0.51 (0.12) a	25.0 (2.0) b	17.8 (4.4) a	8.8 (0.6) b	15.6 (3.0) a
BIO 0.5	2.80 (0.10) a	81.6 (5.7) a	4.56 (0.86) ab	0.76 (0.03) a	0.65 (0.11) a	43.2 (7.7) a	12.6 (5.6) a	9.3 (0.3) b	18.2 (2.0) a
BIO 1	3.00 (0.01) a	95.7 (15.4) a	5.30 (0.3) a	0.71 (0.01) a	0.64 (0.07) a	51.8 (7.8) a	14.0 (5.4) a	11.9 (0.3) a	17.5 (1.3) a
	$p = 0.0051$	$p = 0.0093$	$p = 0.0079$	$p = 0.0004$		$p = 0.0006$		$p = 0.00004$	

In each column and for each parameter, mean values followed by different letters are significantly different ($p < 0.05$) as indicated by one-way ANOVA followed by Tukey test. In parentheses, SD (standard deviation) is reported.

Regarding micronutrients, significant effects were found for Fe and Zn, while the treatments did not influence the Cu and Mn content. (Table 2). In particular, BIO 1 and BIO 0.5 significantly increased the Fe content in olive leaves compared to the control sample; in contrast, the lowest duckweed dosage was ineffective in eliciting such an effect. The Zn content was stimulated only in samples subjected to the highest duckweed concentration (BIO 1), while the other two dosages, BIO 0.5 and BIO 0.25, were ineffective in stimulating the content of this element in olive leaves.

3.4. *Duckweed Extract Phytochemical Profile*

In a previous study [26], duckweed extract was quantitatively analysed to determine phenolics and glucosinolates content and identify other bioactive compounds. In the present study, a broader screening of metabolites of the aqueous duckweed extract was performed using a plant metabolome database to comprehensively highlight the molecules that may help explain the extract biostimulant capacity (Supplementary Table S1). This untargeted approach revealed the presence of several compounds belonging to phytohormones (auxins, cytokinins, brassinosteroids), amino acids, phenylpropanoids and their glycosides (mainly flavonoids such as hesperidin, kaempferol and quercetin, and phenolic acids such as caffeic acid), and glucosinolates, as previously reported [26]. Besides glucosinolates, other nitrogen-containing secondary metabolites (namely alkaloids) were found in the extract. Moreover, isoprenoids were well represented, including triterpenoids, sesquiterpenes, and terpene hormones (gibberellins and their precursors, abscisic acid derivatives and brassinosteroids). Tetraterpenes (carotenes and xanthophylls) could also be detected in the extract, with pigments such as chlorophylls and related compounds. The duckweed extract showed an accumulation of several molecules involved in plant signalling and communication. For instance, the results indicated the presence of choline and phosphatidylcholine related compounds, jasmonates, dopamine and L-dopa, methylsalicylate and proline. Finally, compounds related to plant stress and detoxification (ascorbates or glutathione) were identified in the plant extract.

4. Discussion

Currently, plant biostimulants are gaining increasing attention, as this category of materials is considered an innovative agronomic tool for improving crop productivity [40]. In particular, it has been reported that biostimulants can act in plants at different levels, showing the main effects in increasing plant metabolic and photosynthetic activities, nutrient absorption, growth, biomass production and yield [3,41–43].

This study suggests a significant potential of the duckweed extract in promoting beneficial effects in olive in terms of nutritional status, leaf photosynthetic activity and chlorophyll content and, consequently, on the plant growth. On this account, it has been well documented that biostimulant treatments often increase leaf chlorophyll content [44]. In particular, different biostimulants such as a *Moringa oleifera* extract, Actiwave®, the commercial product ONE® and borage extracts enhanced chlorophyll and carotenoid contents in some horticultural crops such as rocket, lettuce and endive [44–46]. Photosynthesis is an integrated and symptomatic result of the general status of the plant [47]. In particular, this activity can give important information on the productive potential of plants and their capacity to react to environmental factors [47]. The increase in photosynthetic activity found in our experiments was associated with increased stomatal conductance (gs) and intercellular CO_2 concentration (Ci), suggesting that the duckweed extract enhanced photosynthesis also by positively affecting the stomatal aperture (Figure 2). Our results agree with Kuluzewicz et al. [48] and Almadi et al. [18], who found that in broccoli and in olive tree, the use of biostimulants can increase stomatal conductance and photosynthetic activity. In vine, humic acids improved physiological parameters related to the whole plant photosynthesis, such as the increased leaf net CO_2 and chlorophyll concentration and total leaf area [49]. The greenhouse jute treated with a commercial vegetal-derived biostimulant from a tropical plant extract (PE; Auxym ®, Italpollina, Rivoli Veronese, Italy) enhanced photosynthetic activity, SPAD index, and especially the nutritional status [50]. The use of borage extracts increased the net photosynthesis in lettuce, while Actiwave® increased the photosynthetic activity by 27% in strawberries [51]. An increase in net photosynthesis was also observed by treating hibiscus and *Euphorbia* × *lomi* plants with a biowaste [52,53].

Furthermore, the stimulatory effects exerted by the duckweed extract on the photosynthetic activity can explain the increased leaf fresh and dry weights (Table 1). These results agree with other studies [41,46,54] that report as biostimulant treatments can enhance plant growth, determining higher dry matter accumulation in vegetable and ornamental crops. In particular, an in vitro experiment with an extract of brown marine algae evidenced significant stimulatory effects on the growth of spinach [55]. In the same way, the use of Bio-algeen S-90 determined an increase of about 30% on the aboveground biomass of lettuce 'Four Seasons' compared to control plants [56]. In addition, a biomass increase in lettuce was reported when the crop was treated with a mixture of extracts from different plant species associated with *Lactobacillus* and yeast [44]. Finally, in olive trees (cv. Arbequina) subjected to severe salt stress, the treatment with a commercial biostimulant Megafol improved plant dry weight and leaf area due to greater photosynthetic activity. Moreover, Megafol caused a reduction in leaf fall and an improvement in the chlorophyll content and antioxidant activities in the salt-stressed olive trees [13]. The improved vegetative activity, due to a higher photosynthetic activity promoted by the biostimulant treatment, deserves attention also for the opportunity of increasing the plant potential to sequester carbon in olive trees [57].

Biostimulants can strongly influence crops ability to acquire nutrients, making their uptake and use more efficient; such an effect, consequently, increases crop productivity and quality [3,10,58]. As already mentioned in the introduction, the potential to act as a biostimulant for a given material is also assessed on its ability to promote plant nutrition without providing nutrients per se [10,58,59]. It has been postulated that biostimulants improve nutrient acquisition by prompting the release from roots of specific substances capable of increasing the mobility and solubility of nutrients [57,58]. In addition, biostimulants can also affect root biomass or modify root architecture and organization [2].

This study showed that the duckweed extract generally increased at the three different dosages the N, K, Mg, Ca, Fe and Zn contents in treated olive (Table 2). Differently, Na, Mn and Cu contents were unaffected by the plant extract.

All the treatments significantly elevated N; this effect could be related to the higher photosynthetic activity, chlorophyll content, and biomass shown by olive samples treated with the extract (Figures 2 and 3, Table 1). The N supply is a key factor that can condition the activities mentioned above [2,60,61]. Generally, the impact of biostimulants on N content is attributed to their ability to stimulate the enzymes of the nitrogen metabolism and upregulate the root nitrate transporters, as shown in recent studies carried out in maize and soybean [62–64].

The duckweed extract also exerted a strong effect on the K acquisition; this can, in turn, stimulate the photosynthetic activity due to the K capacity of inducing the enzyme ribulose 1,5-bisphosphate carboxylase/oxygenase, maintaining high stomatal and gas exchange activities [65,66]. In addition, all the treatments increased the Mg content, making it possible to postulate that the duckweed extract, exerting a beneficial effect on this nutrient acquisition, activated the enzyme ribulose 1,5-bisphosphate carboxylase/oxygenase and stimulated the chlorophyll content [67]. Chloroplasts contain 35% of Mg, and of this, about 25% is bound to the pigment [68]. The effect on Ca was more modest than those found for the other elements mentioned; only the higher duckweed concentration increased its content (Table 2). However, Ca exerts protective and structural functions and affects stomatal conductance and photosynthetic activity [65].

Regarding the micronutrients, the highest dosages of the duckweed extract (BIO 0.5 and BIO 1) affected Fe content. Plant productivity depends on this nutrient for its involvement in photosynthesis, being part of the two photosystems and the Cyt-*b6f* complex [69]. Finally, Zn was slightly increased by the highest dosage, BIO 1. Increases in the content of this element could be of relevance as Zn is involved in chlorophyll biosynthesis and chloroplast development [30,70].

In general, the stimulatory effect of biostimulants on biomass development and plant growth is considered the mechanism which regulates the increased demand for nutrients [71]. On this account, Jannin et al. [72] showed that rapeseed elevates the expression of genes responsible for nutrient acquisition after applying an algal biostimulant. Therefore, the increases in K, Mg, Ca, Fe, and Zn contents in biostimulated olive samples can be seen as a crop response prompted by the biostimulant to support the increased demand for biomass production.

The potential of the duckweed extract in promoting the beneficial effects we observed could be linked to the presence of several bioactive compounds, as suggested by Del Buono et al. [26], but also to the presence of plant regulators and signal molecules that can trigger changes metabolic processes in plants. For instance, the untargeted profiling highlighted the presence of auxins and auxin-related compounds, which might partially explain the increase in photosynthetic performance and plant growth. Several studies indicate the benefits of applying exogenous auxins to plants and, in particular, the indoleacetic acid (IAA). Li et al. [73] revealed that the addition of exogenous IAA increased photosynthetic capacity in *Zizania latifolia*. These authors reported that exogenous IAA led to significant increases in biomass accumulation in *Z. latifolia* and contributed to higher stomatal conductance and transpiration rate [73]. Moreover, auxins are considered key regulators in plant root development, essential for water and nutrient acquisition [74]. Plant extracts having a biostimulant activity have been reported to contain cytokinins, auxins or hormone-like substances [44]. However, the extracts seemed to be more than just a plant regulator due to the presence of molecules such as phenolic compounds. The addition of exogenous phenolics has been previously reported to enhance plant performance [75]. In particular, Zhang et al. [75] showed that the addition of chlorogenic acid and hesperidin alleviates the impact of salt stress by improving photosynthetic performance. Moreover, Zhang et al. [76] pointed out that the addition of phenolics, including hesperidin, can modulate functional traits in lettuces, also modifying the endogenous phenolic content. In this sense, hesperidin has been found to be the most abundant flavone in duckweed extracts [26]. Exogenous phenolics have been reported to trigger the accumulation of electron carriers, increase stomatal conductance and elicit secondary metabolism in lettuce, both under normal and abiotic stress conditions [75].

On the other hand, the content of amino acids could also explain the enhancement of plant performance. Other authors observed that the effect of biostimulants on plant growth might be linked to the direct incorporation of amino acids used for protein biosynthesis [44]. Moreover, some amino acids found in the duckweed extract (proline) are also related to plant signalling. It has been reported that proline supplementation may ameliorate olive tolerance to salinity by increasing the activity of some antioxidant enzymes, photosynthetic activity, plant growth and plant water status [77]. Likewise, it has been proposed that the action of biostimulants could be linked to the presence of signal molecules, as in the case of protein hydrolysates. In this case, it has been proposed that the stimulatory effect is due to amino acids and small signalling peptides [78].

Besides, other signalling molecules such as L-dopa, dopamine, serotonin or phosphatidylcholine-related compounds could be detected in the duckweed extract. These compounds deserve future investigations in terms of their biostimulant potential. Particular attention should be paid to the presence of glutathione (GSH) in the extracts. GSH has numerous roles in plant cells in both primary and secondary metabolism [79]. Several authors showed that exogenous GSH could enhance abiotic stresses tolerance by restricting the entry of toxic ions, enhancing antioxidant defences, and modifying the photosynthetic parameters and photosystem II efficiency [80].

Further studies are needed since the duckweed extract contains many potential signal compounds. However, although it was not possible to identify a specific bioactive molecule, the biostimulant effects were evident and significant. Noteworthy, given the broad spectrum of bioactive compounds in the duckweed extract, a synergic action of different components can be postulated. This assumption would be in line with what is often observed for plant extracts.

5. Conclusions

In conclusion, this study showed for the first time the potential of an extract obtained from an aquatic species, duckweed (*Lemna minor* L.), to act as a biostimulant in olive for its capacity to improve leaves photosynthetic activity and chlorophyll content, plant growth and nutritional status at all the concentration used.

The metabolomic characterization of the extract evidenced a significant presence of several metabolites, which can support the beneficial effects found. In particular, plant regulators (including auxins) and signalling molecules, among others, were annotated in the extract, as discussed in the precedent section. Similarly, the presence of glutathione and the broad phenolic profile support the effects observed in olive. However, further investigations are needed to fully understand the stimulatory potential of duckweed.

Furthermore, the results of this research suggest that further studies should be carried out to ascertain the effect of duckweed extracts in mitigating the negative effects that biotic and abiotic stresses can have on plants, especially those related to climate change. Finally, this research highlighted that biostimulants could be found from resources readily available in nature. This aspect is relevant for finding new sustainable solutions to reduce the environmental impact of agriculture.

Supplementary Materials: The following supporting information can be downloaded at: https://www.mdpi.com/article/10.3390/agriculture11121299/s1, Table S1: untargeted metabolomics of the aqueous duckweed extract.

Author Contributions: L.R.: Conceptualization, Methodology, Formal analysis, Investigation, Data curation, Writing—Original Draft, Writing—Review and Editing. D.D.B.: Conceptualization, Methodology, Formal analysis, Investigation, Data curation, Writing—Original Draft, Writing—Review and Editing, Supervision. B.M.-M.: Methodology, Formal analysis, Investigation, Writing—Original Draft. B.S.: Methodology, Formal analysis, Investigation. L.L.: Conceptualization, Methodology, Formal analysis, Investigation, Data curation, Writing—Review and Editing. M.T.: Conceptualization, Methodology, Formal analysis, Investigation, Data curation, Writing—Review and Editing. F.C.: Methodology, Formal analysis, Investigation. D.M.V.: Methodology, Formal analysis, Investigation. P.P.: Conceptualization, Methodology, Formal analysis, Investigation, Data curation, Writing—Original Draft, Writing—Review and Editing, Supervision. All authors have read and agreed to the published version of the manuscript.

Funding: This study was funded by the project "Ricerca di Base 2020" of the Department of Agricultural, Food and Environmental Sciences of the University of Perugia (Coordinator: Primo Proietti).

Institutional Review Board Statement: Not applicable.

Informed Consent Statement: Not applicable.

Data Availability Statement: Data will be available on request to the corresponding author.

Acknowledgments: We are grateful to Giorgio Sisani for his technical support.

Conflicts of Interest: The authors declare no conflict of interest.

References

1. Du Jardin, P. Plant Biostimulants: Definition, Concept, Main Categories and Regulation. *Sci. Hortic.* **2015**, *196*, 3–14. [CrossRef]
2. Colla, G.; Nardi, S.; Cardarelli, M.; Ertani, A.; Lucini, L.; Canaguier, R.; Rouphael, Y. Protein Hydrolysates as Biostimulants in Horticulture. *Sci. Hortic.* **2015**, *196*, 28–38. [CrossRef]
3. Del Buono, D. Can Biostimulants Be Used to Mitigate the Effect of Anthropogenic Climate Change on Agriculture? It Is Time to Respond. *Sci. Total Environ.* **2021**, *751*, 141763. [CrossRef] [PubMed]
4. Agliassa, C.; Mannino, G.; Molino, D.; Cavalletto, S.; Contartese, V.; Bertea, C.M.; Secchi, F. A New Protein Hydrolysate-Based Biostimulant Applied by Fertigation Promotes Relief from Drought Stress in Capsicum Annuum L. *Plant Physiol. Biochem.* **2021**, *166*, 1076–1086. [CrossRef]
5. Malik, A.; Mor, V.S.; Tokas, J.; Punia, H.; Malik, S.; Malik, K.; Sangwan, S.; Tomar, S.; Singh, P.; Singh, N.; et al. Biostimulant-Treated Seedlings under Sustainable Agriculture: A Global Perspective Facing Climate Change. *Agronomy* **2021**, *11*, 14. [CrossRef]
6. Kocira, S. Effect of Amino Acid Biostimulant on the Yield and Nutraceutical Potential of Soybean. *Chil. J. Agric. Res.* **2019**, *79*, 17–25. [CrossRef]
7. Toscano, S.; Ferrante, A.; Branca, F.; Romano, D. Enhancing the Quality of Two Species of Baby Leaves Sprayed with Moringa Leaf Extract as Biostimulant. *Agronomy* **2021**, *11*, 1399. [CrossRef]
8. Mannino, G.; Campobenedetto, C.; Vigliante, I.; Contartese, V.; Gentile, C.; Bertea, C.M. The Application of a Plant Biostimulant Based on Seaweed and Yeast Extract Improved Tomato Fruit Development and Quality. *Biomolecules* **2020**, *10*, 1662. [CrossRef] [PubMed]
9. Calvo, P.; Nelson, L.; Kloepper, J.W. Agricultural Uses of Plant Biostimulants. *Plant Soil* **2014**, *383*, 3–41. [CrossRef]
10. Rouphael, Y.; Colla, G. Editorial: Biostimulants in Agriculture. *Front. Plant Sci.* **2020**, *11*, 40. [CrossRef]
11. Caradonia, F. Plant Biostimulant Regulatory Framework: Prospects in Europe and Current Situation at International Level. *J. Plant Growth Regul.* **2019**, *38*, 438–448. [CrossRef]

12. Nardi, S.; Pizzeghello, D.; Reniero, F.; Rascio, N. Chemical and Biochemical Properties of Humic Substances Isolated from Forest Soils and Plant Growth. *Soil Sci. Soc. Am. J.* **2000**, *64*, 639–645. [CrossRef]
13. Buono, D.D.; Regni, L.; Pino, A.M.D.; Bartucca, M.L.; Palmerini, C.A.; Proietti, P. Effects of Megafol on the Olive Cultivar 'Arbequina' Grown Under Severe Saline Stress in Terms of Physiological Traits, Oxid. *Front. Plant Sci.* **2021**, *11*, 603576.
14. Ali, A.H.; Aboohanah, M.A.; Abdulhussein, M.A. Impact of Foliar Application with Dry Yeast Suspension and Amino Acid on Vegetative Growth, Yield and Quality Characteristics of Olive (*Olea Europaea* L.) Trees. *Kufa J. Agric. Sci.* **2019**, *11*, 10.
15. Chouliaras, V.; Tasioula, M.; Chatzissavvidis, C.; Therios, I.; Tsabolatidou, E. The Effects of a Seaweed Extract in Addition to Nitrogen and Boron Fertilization on Productivity, Fruit Maturation, Leaf Nutritional Status and Oil Quality of the Olive (Olea Europaea L.) Cultivar Koroneiki. *J. Sci. Food Agric.* **2009**, *89*, 984–988. [CrossRef]
16. Soria, C.M. *Olive Tree (Olea europaea L.) Response to the Application of Biostimulants*; CIHEAM-IAMZ: Zaragoza, Spain, 2006.
17. Saour, G. Morphological Assessment of Olive Seedlings Treated with Kaolin-Based Particle Film and Biostimulant. *Adv. Hortic. Sci.* **2005**, *19*, 193–197.
18. Almadi, L.; Paoletti, A.; Cinosi, N.; Daher, E.; Rosati, A.; Di Vaio, C.; Famiani, F. A Biostimulant Based on Protein Hydrolysates Promotes the Growth of Young Olive Trees. *Agric. Switz.* **2020**, *10*, 618. [CrossRef]
19. El-Mageed, T.A.A.; Semida, W.M.; Rady, M.M. Moringa Leaf Extract as Biostimulant Improves Water Use Efficiency, Physio-Biochemical Attributes of Squash Plants under Deficit Irrigation. *Agric. Water Manag.* **2017**, *193*, 46–54. [CrossRef]
20. Zulfiqar, F.; Casadesús, A.; Brockman, H.; Munné-Bosch, S. An Overview of Plant-Based Natural Biostimulants for Sustainable Horticulture with a Particular Focus on Moringa Leaf Extracts. *Plant Sci.* **2020**, *295*, 110194. [CrossRef]
21. Caruso, G.; De Pascale, S.; Cozzolino, E.; Giordano, M.; El-Nakhel, C.; Cuciniello, A.; Cenvinzo, V.; Colla, G.; Rouphael, Y. Protein Hydrolysate or Plant Extract-Based Biostimulants Enhanced Yield and Quality Performances of Greenhouse Perennial Wall Rocket Grown in Different Seasons. *Plants* **2019**, *8*, 208. [CrossRef]
22. Saavedra, T.; Gama, F.; Correia, P.J.; Da Silva, J.P.; Miguel, M.G.; de Varennes, A.; Pestana, M. A Novel Plant Extract as a Biostimulant to Recover Strawberry Plants from Iron Chlorosis. *J. Plant Nutr.* **2020**, *43*, 2054–2066. [CrossRef]
23. Gebreluel, T.; He, M.; Zheng, S.; Zou, S.; Woldemicael, A.; Wang, C. Optimization of Enzymatic Degradation of Dealginated Kelp Waste through Response Surface Methodology. *J. Appl. Phycol.* **2020**, *32*, 529–537. [CrossRef]
24. Nanda, S.; Kumar, G.; Hussain, S. Utilization of Seaweed-Based Biostimulants in Improving Plant and Soil Health: Current Updates and Future Prospective. *Int. J. Environ. Sci. Technol.* **2021**, 1–14. [CrossRef]
25. Mzibra, A.; Aasfar, A.; Khouloud, M.; Farrie, Y.; Boulif, R.; Kadmiri, I.M.; Bamouh, A.; Douira, A. Improving Growth, Yield, and Quality of Tomato Plants (*Solanum Lycopersicum* L) by the Application of Moroccan Seaweed-Based Biostimulants under Greenhouse Conditions. *Agronomy* **2021**, *11*, 1373. [CrossRef]
26. Del Buono, D.; Bartucca, M.L.; Ballerini, E.; Senizza, B.; Lucini, L.; Trevisan, M. Physiological and Biochemical Effects of an Aqueous Extract of Lemna Minor L. as a Potential Biostimulant for Maize. *J. Plant Growth Regul.* **2021**. [CrossRef]
27. Lasfar, S.; Monette, F.; Millette, L.; Azzouz, A. Intrinsic Growth Rate: A New Approach to Evaluate the Effects of Temperature, Photoperiod and Phosphorus-Nitrogen Concentrations on Duckweed Growth under Controlled Eutrophication. *Water Res.* **2007**, *41*, 2333–2340. [CrossRef] [PubMed]
28. Panfili, I.; Bartucca, M.L.; Del Buono, D. The Treatment of Duckweed with a Plant Biostimulant or a Safener Improves the Plant Capacity to Clean Water Polluted by Terbuthylazine. *Sci. Total Environ.* **2019**, *646*, 832–840. [CrossRef] [PubMed]
29. Gülçin, I.; Kireçci, E.; Akkemik, E.; Fevzi, T.; Hisar, O. Antioxidant, Antibacterial, and Anticandidal Activities of an Aquatic Plant: Duckweed (Lemna Minor L. Lemnaceae). *Turk. J. Biol.* **2010**, *34*, 175–188. [CrossRef]
30. Del Buono, D.; Di Michele, A.; Costantino, F.; Trevisan, M.; Lucini, L. Biogenic ZnO Nanoparticles Synthesized Using a Novel Plant Extract: Application to Enhance Physiological and Biochemical Traits in Maize. *Nanomaterials* **2021**, *11*, 1270. [CrossRef]
31. Gülçin, I. The Antioxidant and Radical Scavenging Activities of Black Pepper (*Piper Nigrum*) Seeds. *Int. J. Food Sci. Nutr.* **2005**, *56*, 491–499. [CrossRef]
32. Caruso, T.; Campisi, G.; Marra, F.P.; Camposeo, S.; Vivaldi, G.A.; Proietti, P.; Nasini, L. Growth and Yields of "Arbequina" High-Density Planting Systems in Three Different Olive Growing Areas in Italy. *Acta Horticulturae* **2014**, *1057*, 341–348. [CrossRef]
33. Rosati, A.; Paoletti, A.; Al Harir, R.; Famiani, F. Fruit Production and Branching Density Affect Shoot and Whole-Tree Wood to Leaf Biomass Ratio in Olive. *Tree Physiol.* **2018**, *38*, 1278–1285. [CrossRef] [PubMed]
34. Rosati, A.; Paoletti, A.; Al Hariri, R.; Morelli, A.; Famiani, F. Resource Investments in Reproductive Growth Proportionately Limit Investments in Whole-Tree Vegetative Growth in Young Olive Trees with Varying Crop Loads. *Tree Physiol.* **2018**, *38*, 1267–1277. [CrossRef]
35. Famiani, F.; Farinelli, D.; Gardi, T.; Rosati, A. The Cost of Flowering in Olive (*Olea Europaea* L.). *Sci. Hortic.* **2019**, *252*, 268–273. [CrossRef]
36. *Aquaponics Food Production Systems: Combined Aquaculture and Hydroponic Production Technologies for the Future*; Life Sciences; Göddeke, S.; Joyce, A.; Kotzen, B.; Burnell, G.M. (Eds.) Springer: Cham, Switzerland, 2019; ISBN 978-3-030-15943-.
37. Tabatabai, M.A.; Bremner, J.M. A Simple Turbidimetric Method of Determining Total Sulfur in Plant Materials1. *Agron. J.* **1970**, *62*, 805–806. [CrossRef]
38. Hansen, T.H.; De Bang, T.C.; Laursen, K.H.; Pedas, P.; Husted, S.; Schjoerring, J.K. Multielement Plant Tissue Analysis Using ICP Spectrometry. *Methods Mol. Biol.* **2013**, *953*, 121–141.
39. Lenth, R.V. Least-Squares Means: The R Package Lsmeans. *J. Stat. Softw.* **2016**, *69*, 17496. [CrossRef]

40. Povero, G.; Mejia, J.F.; Di Tommaso, D.; Piaggesi, A.; Warrior, P. A Systematic Approach to Discover and Characterize Natural Plant Biostimulants. *Front. Plant Sci.* **2016**, *7*, 435. [CrossRef] [PubMed]
41. Bulgari, R.; Podetta, N.; Cocetta, G.; Piaggesi, A.; Ferrante, A. The Effect of a Complete Fertilizer for Leafy Vegetables Production in Family and Urban Gardens. *Bulg. J. Agric. Sci.* **2014**, *20*, 1361–1367.
42. Yakhin, O.I.; Lubyanov, A.A.; Yakhin, I.A.; Brown, P.H. Biostimulants in Plant Science: A Global Perspective. *Front. Plant Sci.* **2017**, *7*, 2049. [CrossRef]
43. Puglia, D.; Pezzolla, D.; Gigliotti, G.; Torre, L.; Bartucca, M.L.; Del Buono, D. The Opportunity of Valorizing Agricultural Waste, Through Its Conversion into Biostimulants, Biofertilizers, and Biopolymers. *Sustainability* **2021**, *13*, 2710. [CrossRef]
44. Bulgari, R.; Cocetta, G.; Trivellini, A.; Vernieri, P.; Ferrante, A. Biostimulants and Crop Responses: A Review. *Biol. Agric. Hortic.* **2015**, *31*, 1–17. [CrossRef]
45. Abdalla, M.M. The Potential of Moringa Oleifera Extract as a Biostimulant in Enhancing the Growth, Biochemical and Hormonal Contents in Rocket (Eruca Vesicaria Subsp. Sativa) Plants. *Int. J. Plant Physiol. Biochem.* **2013**, *5*, 42–49. [CrossRef]
46. Bulgari, R.; Morgutti, S.; Cocetta, G.; Negrini, N.; Farris, S.; Calcante, A.; Spinardi, A.; Ferrari, E.; Mignani, I.; Oberti, R.; et al. Evaluation of Borage Extracts as Potential Biostimulant Using a Phenomic, Agronomic, Physiological, and Biochemical Approach. *Front. Plant Sci.* **2017**, *8*, 935. [CrossRef] [PubMed]
47. Proietti, P.; Federici, E.; Fidati, L.; Scargetta, S.; Massaccesi, L.; Nasini, L.; Regni, L.; Ricci, A.; Cenci, G.; Gigliotti, G. Effects of Amendment with Oil Mill Waste and Its Derived-Compost on Soil Chemical and Microbiological Characteristics and Olive (Olea Europaea L.) Productivity. *Agric. Ecosyst. Environ.* **2015**, *207*, 51–60. [CrossRef]
48. Kałuzewicz, A.; Krzesiński, W.; Spizewski, T.; Zaworska, A. Effect of Biostimulants on Several Physiological Characteristics and Chlorophyll Content in Broccoli under Drought Stress and Re-Watering. *Not. Bot. Horti Agrobot. Cluj-Napoca* **2017**, *45*, 197–202. [CrossRef]
49. Popescu, G.C.; Popescu, M. Yield, Berry Quality and Physiological Response of Grapevine to Foliar Humic Acid Application. *Bragantia* **2018**, *77*, 273–282. [CrossRef]
50. Carillo, P.; Colla, G.; El-Nakhel, C.; Bonini, P.; D'Amelia, L.; Dell'Aversana, E.; Pannico, A.; Giordano, M.; Sifola, M.I.; Kyriacou, M.C.; et al. Biostimulant Application with a Tropical Plant Extract Enhances Corchorus Olitorius Adaptation to Sub-Optimal Nutrient Regimens by Improving Physiological Parameters. *Agronomy* **2019**, *9*, 249. [CrossRef]
51. Spinelli, F.; Fiori, G.; Noferini, M.; Sprocatti, M.; Costa, G. A Novel Type of Seaweed Extract as a Natural Alternative to the Use of Iron Chelates in Strawberry Production. *Sci. Hortic.* **2010**, *125*, 263–269. [CrossRef]
52. Massa, D.; Prisa, D.; Montoneri, E.; Battaglini, D.; Ginepro, M.; Negre, M.; Burchi, G. Application of Municipal Biowaste Derived Products in Hibiscus Cultivation: Effect on Leaf Gaseous Exchange Activity, and Plant Biomass Accumulation and Quality. *Sci. Hortic.* **2016**, *205*, 59–69. [CrossRef]
53. Fascella, G.; Montoneri, E.; Ginepro, M.; Francavilla, M. Effect of Urban Biowaste Derived Soluble Substances on Growth, Photosynthesis and Ornamental Value of Euphorbia x Lomi. *Sci. Hortic.* **2015**, *197*, 90–98. [CrossRef]
54. Khan, W.; Rayirath, U.P.; Subramanian, S.; Jithesh, M.N.; Rayorath, P.; Hodges, D.M.; Critchley, A.T.; Craigie, J.S.; Norrie, J.; Prithiviraj, B. Seaweed Extracts as Biostimulants of Plant Growth and Development. *J. Plant Growth Regul.* **2009**, *28*, 386–399. [CrossRef]
55. Fan, D.; Hodges, D.M.; Critchley, A.T.; Prithiviraj, B. A Commercial Extract of Brown Macroalga (*Ascophyllum Nodosum*) Affects Yield and the Nutritional Quality of Spinach In Vitro. *Commun. Soil Sci. Plant Anal.* **2013**, *44*, 1873–1884. [CrossRef]
56. Dudaš, S.; Šola, I.; Sladonja, B.; Erhatić, R.; Ban, D.; Poljuha, D. The Effect of Biostimulant and Fertilizer on "Low Input" Lettuce Production. *Acta Bot. Croat.* **2016**, *75*, 253–259. [CrossRef]
57. Regni, L.; Nasini, L.; Ilarioni, L.; Brunori, A.; Massaccesi, L.; Agnelli, A.; Proietti, P. Long Term Amendment with Fresh and Composted Solid Olive Mill Waste on Olive Grove Affects Carbon Sequestration by Prunings, Fruits, and Soil. *Front. Plant Sci.* **2017**, *7*, 2042. [CrossRef] [PubMed]
58. Du Jardin, P.; Xu, L.; Geelen, D. Agricultural Functions and Action Mechanisms of Plant Biostimulants (PBs). In *The Chemical Biology of Plant Biostimulants*; John Wiley & Sons, Ltd: Hoboken, NJ, USA, 2020; pp. 1–30. ISBN 978-1-119-35725-4.
59. Halpern, M.; Bar-Tal, A.; Ofek, M.; Minz, D.; Muller, T.; Yermiyahu, U. Chapter Two—The Use of Biostimulants for Enhancing Nutrient Uptake. In *Advances in Agronomy*; Sparks, D.L., Ed.; Academic Press: Cambridge, MA, USA, 2015; Volume 130, pp. 141–174. ISBN 0065-2113.
60. Rouphael, Y.; Giordano, M.; Cardarelli, M.; Cozzolino, E.; Mori, M.; Kyriacou, M.; Bonini, P.; Colla, G. Plant- and Seaweed-Based Extracts Increase Yield but Differentially Modulate Nutritional Quality of Greenhouse Spinach through Biostimulant Action. *Agronomy* **2018**, *8*, 126. [CrossRef]
61. Colla, G.; Rouphael, Y.; Canaguier, R.; Svecova, E.; Cardarelli, M. Biostimulant Action of a Plant-Derived Protein Hydrolysate Produced through Enzymatic Hydrolysis. *Front. Plant Sci.* **2014**, *5*, 448. [CrossRef] [PubMed]
62. Palumbo, G.; Schiavon, M.; Nardi, S.; Ertani, A.; Celano, G.; Colombo, C.M. Biostimulant Potential of Humic Acids Extracted from an Amendment Obtained via Combination of Olive Mill Wastewaters (OMW) and a Pre-Treated Organic Material Derived from Municipal Solid Waste (MSW). *Front. Plant Sci.* **2018**, *9*, 1028. [CrossRef] [PubMed]
63. Goñi, O.; Łangowski, Ł.; Feeney, E.; Quille, P.; O'Connell, S. Reducing Nitrogen Input in Barley Crops While Maintaining Yields Using an Engineered Biostimulant Derived from Ascophyllum Nodosum to Enhance Nitrogen Use Efficiency. *Front. Plant Sci.* **2021**, *12*, 789. [CrossRef]

64. Ertani, A.; Pizzeghello, D.; Altissimo, A.; Nardi, S. Use of Meat Hydrolyzate Derived from Tanning Residues as Plant Biostimulant for Hydroponically Grown Maize. *J. Plant Nutr. Soil Sci.* **2013**, *176*, 287–295. [CrossRef]
65. Zrig, A.; AbdElgawad, H.; Touneckti, T.; Mohamed, H.B.; Hamouda, F.; Khemira, H. Potassium and Calcium Improve Salt Tolerance of Thymus Vulgaris by Activating the Antioxidant Systems. *Sci. Hortic.* **2021**, *277*, 109812. [CrossRef]
66. Ahanger, M.A.; Agarwal, R.M. Potassium Up-Regulates Antioxidant Metabolism and Alleviates Growth Inhibition under Water and Osmotic Stress in Wheat (*Triticum Aestivum* L). *Protoplasma* **2017**, *254*, 1471–1486. [CrossRef]
67. Rehman, H.U.; Alharby, H.F.; Alzahrani, Y.; Rady, M.M. Magnesium and Organic Biostimulant Integrative Application Induces Physiological and Biochemical Changes in Sunflower Plants and Its Harvested Progeny on Sandy Soil. *Plant Physiol. Biochem.* **2018**, *126*, 97–105. [CrossRef]
68. Barroso, F.D.L.; Milagres, C.D.C.; Fontes, P.C.R.; Cecon, P.R. Magnesium-Influenced Seed Potato Development and Yield. *J. Plant Nutr.* **2021**, *44*, 296–308. [CrossRef]
69. Kroh, G.E.; Pilon, M. Regulation of Iron Homeostasis and Use in Chloroplasts. *Int. J. Mol. Sci.* **2020**, *21*, 3395. [CrossRef] [PubMed]
70. Salama, D.M.; Osman, S.A.; Abd El-Aziz, M.E.; Abd Elwahed, M.S.A.; Shaaban, E.A. Effect of Zinc Oxide Nanoparticles on the Growth, Genomic DNA, Production and the Quality of Common Dry Bean (*Phaseolus Vulgaris*). *Biocatal. Agric. Biotechnol.* **2019**, *18*, 101083. [CrossRef]
71. Szczepanek, M.; Siwik-Ziomek, A. P and K Accumulation by Rapeseed as Affected by Biostimulant under Different NPK and S Fertilization Doses. *Agronomy* **2019**, *9*, 477. [CrossRef]
72. Jannin, L.; Arkoun, M.; Etienne, P.; Laîné, P.; Goux, D.; Garnica, M.; Fuentes, M.; Francisco, S.S.; Baigorri, R.; Cruz, F.; et al. Brassica Napus Growth Is Promoted by *Ascophyllum Nodosum* (L.) Le Jol. Seaweed Extract: Microarray Analysis and Physiological Characterization of N, C, and S Metabolisms. *J. Plant Growth Regul.* **2013**, *32*, 31–52. [CrossRef]
73. Li, J.; Guan, Y.; Yuan, L.; Hou, J.; Wang, C.; Liu, F.; Yang, Y.; Lu, Z.; Chen, G.; Zhu, S. Effects of Exogenous IAA in Regulating Photosynthetic Capacity, Carbohydrate Metabolism and Yield of Zizania Latifolia. *Sci. Hortic.* **2019**, *253*, 276–285. [CrossRef]
74. Saini, S.; Sharma, I.; Kaur, N.; Pati, P.K. Auxin: A Master Regulator in Plant Root Development. *Plant Cell Rep.* **2013**, *32*, 741–757. [CrossRef]
75. Zhang, L.; Miras-Moreno, B.; Yildiztugay, E.; Ozfidan-Konakci, C.; Arikan, B.; Elbasan, F.; Ak, G.; Rouphael, Y.; Zengin, G.; Lucini, L. Metabolomics and Physiological Insights into the Ability of Exogenously Applied Chlorogenic Acid and Hesperidin to Modulate Salt Stress in Lettuce Distinctively. *Molecules* **2021**, *26*, 6291. [CrossRef]
76. Zhang, L.; Martinelli, E.; Senizza, B.; Miras-Moreno, B.; Yildiztugay, E.; Arikan, B.; Elbasan, F.; Ak, G.; Balci, M.; Zengin, G.; et al. The Combination of Mild Salinity Conditions and Exogenously Applied Phenolics Modulates Functional Traits in Lettuce. *Plants* **2021**, *10*, 1457. [CrossRef] [PubMed]
77. Ahmed, C.B.; Rouina, B.B.; Sensoy, S.; Boukhriss, M.; Abdullah, F.B. Exogenous Proline Effects on Photosynthetic Performance and Antioxidant Defense System of Young Olive Tree. *J. Agric. Food Chem.* **2010**, *58*, 4216–4222. [CrossRef] [PubMed]
78. Lucini, L.; Miras-Moreno, B.; Rouphael, Y.; Cardarelli, M.; Colla, G. Combining Molecular Weight Fractionation and Metabolomics to Elucidate the Bioactivity of Vegetal Protein Hydrolysates in Tomato Plants. *Front. Plant Sci.* **2020**, *11*, 976. [CrossRef] [PubMed]
79. Sohag, A.A.M.; Tahjib-Ul-Arif, M.; Polash, M.A.S.; Chowdhury, M.B.; Afrin, S.; Burritt, D.J.; Murata, Y.; Hossain, M.A.; Afzal Hossain, M. Exogenous Glutathione-Mediated Drought Stress Tolerance in Rice (*Oryza Sativa* L.) Is Associated with Lower Oxidative Damage and Favorable Ionic Homeostasis. *Iran. J. Sci. Technol. Trans. Sci.* **2020**, *44*, 955–971. [CrossRef]
80. Zhou, Y.; Diao, M.; Cui, J.-X.; Chen, X.-J.; Wen, Z.-L.; Zhang, J.-W.; Liu, H.-Y. Exogenous GSH Protects Tomatoes against Salt Stress by Modulating Photosystem II Efficiency, Absorbed Light Allocation and H2O2-Scavenging System in Chloroplasts. *J. Integr. Agric.* **2018**, *17*, 2257–2272. [CrossRef]

Article

Assessment of Drought Tolerance of *Miscanthus* Genotypes through Dry-Down Treatment and Fixed-Soil-Moisture-Content Techniques

Tzu-Ya Weng [1], Taiken Nakashima [2], Antonio Villanueva-Morales [3], J. Ryan Stewart [4,5,6], Erik J. Sacks [5,7] and Toshihiko Yamada [5,6,*]

1 Graduate School of Global Food Resources, Hokkaido University, Sapporo 0600809, Hokkaido, Japan; tzuyaweng@eis.hokudai.ac.jp
2 Research Faculty of Agriculture, Hokkaido University, Sapporo 0608589, Hokkaido, Japan; tnak005@res.agr.hokudai.ac.jp
3 Department of Statistics, Mathematics and Computing, Forest Sciences Division, Universidad Autónoma Chapingo, Texcoco 56230, Mexico; avillanuevam@chapingo.mx
4 Department of Plant and Wildlife Sciences, Brigham Young University, Provo, UT 84604, USA; ryan.stewart@byu.edu
5 Field Science Center for Northern Biosphere, Hokkaido University, Sapporo 0600811, Hokkaido, Japan; esacks@illinois.edu
6 Global Center for Food, Land and Water Resources, Hokkaido University, Sapporo 0608589, Hokkaido, Japan
7 Department of Crop Sciences, University of Illinois, Urbana-Champaign, Urbana, IL 61801, USA
* Correspondence: yamada@fsc.hokudai.ac.jp; Tel.: +81-(0)11-706-3644

Citation: Weng, T.-Y.; Nakashima, T.; Villanueva-Morales, A.; Stewart, J.R.; Sacks, E.J.; Yamada, T. Assessment of Drought Tolerance of *Miscanthus* Genotypes through Dry-Down Treatment and Fixed-Soil-Moisture-Content Techniques. *Agriculture* **2022**, *12*, 6. https://doi.org/10.3390/agriculture12010006

Academic Editors: Daniele Del Buono, Primo Proietti and Luca Regni

Received: 16 November 2021
Accepted: 14 December 2021
Published: 22 December 2021

Abstract: *Miscanthus*, a high-yielding, warm-season C_4 grass, shows promise as a potential bioenergy crop in temperate regions. However, drought may restrain productivity of most genotypes. In this study, total 29 *Miscanthus* genotypes of East-Asian origin were screened for drought tolerance with two methods, a dry-down treatment in two locations and a system where soil moisture content (SMC) was maintained at fixed levels using an automatic irrigation system in one location. One genotype, *Miscanthus sinensis* PMS-285, showed relatively high drought-tolerance capacity under moderate drought stress. *Miscanthus sinensis* PMS-285, aligned with the *M. sinensis* 'Yangtze-Qinling' genetic cluster, had relatively high principal component analysis ranking values in both two locations experiments, Hokkaido University and Brigham Young University. Genotypes derived from the 'Yangtze-Qinling' genetic cluster showed relatively greater photosynthetic performance than other genetic clusters, suggesting germplasm from this group could be a potential source of drought-tolerant plant material. Diploid genotypes showed stronger drought tolerance than tetraploid genotypes, suggesting ploidy could be an influential factor for this trait. Of the two methods, the dry-down treatment appears more suitable for selecting drought-tolerant genotypes given that it reflects water-stress conditions in the field. However, the fixed-SMC experiment may be good for understanding the physiological responses of plants to relatively constant water-stress levels.

Keywords: *Miscanthus* spp.; drought tolerance; photosynthetic parameters; bioenergy crops; automated irrigation control

1. Introduction

Drought stress limits plant growth and yield and acts as a barrier to the successful cultivation of bioenergy crops, such as sugarcane and maize, particularly in world arid and semi-arid regions [1]. Drought impairs plant metabolism, such that plants cannot provide sufficient photosynthetic energy for cell growth and maintenance, which sometimes results in death [2].

To adapt and survive under drought stress, mechanisms involving drought resistance and drought recovery are key aspects of adaptations [3]. Plants with drought tolerance generally express certain traits under stress, such as leaf area reduction to minimize transpirational water loss and maintenance of high chlorophyll content to enable high photosynthetic levels in order to produce enough energy for survival. Therefore, photosynthetic parameters, especially photosynthetic rate (Pn), are considered as an effective measure of drought tolerance in plants, such as in *Leucaena leucocephala* (Lam.) de Wit [4]. Liu et al. (2015) reported that the photosynthetic rate of drought-tolerant switchgrass (*Panicum virgatum* L.) genotypes was positively correlated with other physiological parameters, such as relative water content, transpiration rate (Tr), stomatal conductance (gs), and water-use efficiency (WUE) when subjected to low-water conditions [5]. It means that the performance of Pn can be regarded as the physiological response of plants under drought.

Extreme drought will likely increase in the future due to global warming [6]. Consequently, there is a strong need for identifying crop accessions with high recovery capacity to drought stress. Such a trait enables crops to access water from the soil from short-term rain events and to maintain physiological function to survive drought. Lauenroth et al. (1987) observed that the warm-season perennial grass species, *Bouteloua gracilis* H.B.K. Lag. ex Steud., in response to low soil moisture, generated new root growth after the root zone was replenished with water, which led to increased soil water uptake [7]. Also, lipid peroxidation and H_2O_2 content, which were generated in tea plants (*Camellia sinensis* (L) O. Kuntze) in response to drought, decreased after post-drought soil-water recharge [8]. In addition, the catalase activity of pea (*Pisum sativum* L., cv. Progress 9), which is involved in removing H_2O_2 molecules, increased during drought [9]. However, H_2O_2 molecules decreased to normal levels after re-watering. Moreover, Chen et al. (2016) reported drought adaptability of maize (*Zea mays* L.) seedlings was more associated with drought recovery ($r = 0.714$) than drought resistance ($r = 0.332$) in correlation analysis, suggesting recovery capacity is a key component of plant survival to drought stress [3]. They also used it as a screening criterion to identify drought-tolerant genotypes.

For evaluation of drought tolerance in plants under greenhouse studies, two types of techniques, the dry-down treatment [10] and fixed-soil moisture content (SMC) methods [11,12], have been used to apply low-water conditions in potted plants. In the dry-down method, water is withheld from plants, often for several days, after initially being well watered. As evapotranspiration occurs, SMC will continue to decline, which often leads to gradually increasing levels of plant drought stress. The SMC of plants in dry-down treatments usually decreases quickly over a short period of time. Advantages of the dry-down technique include cost-efficiency and ease of operation. However, the method affords little time for researchers to observe how plants respond to drought.

On the other hand, the fixed SMC technique is used to keep the SMC of target plants at fixed soil-moisture levels by regularly adding water through a computerized irrigation system to the rhizosphere of the potted plants based on the amount of water evapotranspired from the plant and medium [11]. In this technique, the rhizosphere of target plants can be maintained at a relatively constant SMC, thus allowing for the plants to experience continuous drought conditions. The disadvantages of the fixed-SMC method include the large amount of time and effort required for calculating evapotranspiration and for maintaining irrigation levels. However, comparison between both techniques is warranted given that each method offers distinct advantages in terms of characterizing plant responses to drought stress.

Miscanthus, a C_4 perennial rhizomatous grass, has high biomass productivity in marginal lands and expresses high CO_2 fixation in low-temperature conditions, underscoring its potential as a bioenergy crop [13,14]. Two major important *Miscanthus* species are *Miscanthus sinensis* Andersson and *Miscanthus sacchariflorus* (Maxim.) Bentham. A single sterile triploid clone of *Miscanthus* × *giganteus* Greef & Deuter ex Hodk. & Renvoize, a hybrid between *M. sacchariflorus* and *M. sinensis*, has been adapted for commercial biomass production in Europe and North America. Based on data generated from restriction site-associated DNA sequencing and Golden Gate technologies, *M. sinensis* is mainly comprised of 6 genetic clusters, which include 'South-eastern China plus tropical', 'Yangtze-Qinling', 'Sichuan Basin', 'Korea, North China', 'Southern Japan', and 'Northern Japan' [15]. On the other hand, *M. sacchariflorus* consists of 'Yangtze' diploids, 'Northern China' diploids, 'Korea/Northeast China/Russia' diploids, 'Northern China/Korea/Russia' tetraploids, 'Southern Japan' tetraploids, and 'Northern Japan' tetraploids [16]. Relative to *M. sinensis* species, the diploid and tetraploid clusters of *M. sacchariflorus*, possibly play a role in stress-tolerance expression in the species complex, when used to breed *M.* × *giganteus* new genotype by crossing *M. sacchariflorus* with *M. sinensis* species. In the present study, a core population of several *Miscanthus* species, which were characterized by Clark et al. (2014, 2019) [15,16], were included for evaluation of their response to drought.

Miscanthus species are considered to have stronger drought tolerance than another potential energy crop, switchgrass [17,18]. Under drought conditions, relative to maize and switchgrass, *Miscanthus* exhibited higher light-use efficiency, photosynthetic rate, and above-ground biomass [19]. However, as a potential energy crop, selection needs to be made of drought-tolerant *Miscanthus* accessions [20]. Many cultivated *Miscanthus* genotypes, including the widely cultivated, high-yielding *Miscanthus* × *giganteus*, lack strong drought tolerance [21]. Moreover, *M.* × *giganteus* uses more water than maize due to its longer growing season and higher productivity [21]. Selecting for drought tolerance of *Miscanthus* is essential for wherever it may be cultivated as a bioenergy crop because the ubiquity of drought also happens even in high-rainfall areas [20]. Selecting for and developing drought-tolerant *Miscanthus* genotypes as breeding material increases the versatility of *Miscanthus* as a sustainable bioenergy crop.

Little research appears to have been done to characterize drought tolerance of *Miscanthus*. Previous research on the impact of drought on *Miscanthus* mainly focused on *M.* × *giganteus* [22,23]. Most parameters, such as dry weight accumulation, leaf expansion chlorophyl content, decreased when *M.* × *giganteus* meet drought [22]. Moreover, there are many genetic resources of *Miscanthus* spp., which could be used as breeding stock to improve drought-adaptation capacity in high-yielding accessions. Consequently, there is a need to identify and evaluate drought-tolerant *Miscanthus* genotypes as breeding material from the core population.

Given that there are considerable genetic differences among *Miscanthus* genotypes, even under well-watered conditions, assessment of drought tolerance, based only on photosynthesis data collected during periods of low SMC, can be fraught with limitations. To avoid this problem, we employed the drought stress index (DSI) methodology of Liu et al. (2015) [5]. It shows promise in quantifying drought-induced effects in *Miscanthus* plants. Drought stress index values are calculated as follows:

$$\text{DSI} = (\text{value of traits under stress condition})/(\text{value of traits under well-watered condition}) \times 100 \quad (1)$$

The DSI can remove genetic differences among different genotypes and can be used as an indicator of drought tolerance throughout the *Miscanthus* genus.

The present study focused on two objectives to characterize the drought-tolerance capacity of *Miscanthus*. The first objective was to compare different techniques used to impose drought stress in plants in terms of their suitable applications. The second was to screen *Miscanthus* genotypes for drought tolerance with the express purpose of identifying germplasm to use as future breeding-stock material.

2. Materials and Methods

2.1. Screening Experiment for Dry-Down-Imposed Drought Stress

2.1.1. Experiment in Hokkaido University, Japan

A population of 23 *Miscanthus* genotypes, which included 10 *M. sinensis*, one *M. sinensis* var. *condensatus*, 11 *M. sacchariflorus*, and one *M. floridulus* genotypes, which were collected from across East Asia, served as the source of the selection materials for this study (Table 1; Table S1). The genotypes were divided into thirteen genetic clusters, based on analyses by Clark et al. (2014, 2019) [15,16]. Considerable genetic variation existed among the genotypes [15,16]. As such, we considered that even with only 23 genotypes, which had limited representation (i.e., between 1–6 genotypes) of each genetic cluster, there was sufficient genetic variation to draw broad-based inferences for the genus at large. The experiment was conducted in a semi-open rain-shelter greenhouse at Hokkaido University (HU) in Sapporo, Japan (43°4′43″N, 141°20′19″E). The dry-down experiment ran from July to August 2018. All 23 *Miscanthus* genotypes were propagated from rhizomes. Rhizome pieces of each genotype were cut into 10 cm lengths and grown in plastic pots (diameter = 19 cm, height = 27 cm). All plants were irrigated every day for 4 weeks before starting the experiment.

Table 1. List of *Miscanthus* genotypes included in screening experiments at Hokkaido University (HU), Sapporo, Japan and Brigham Young University (BYU), Provo, Utah, USA.

HU Screening Experiment (2017, 2018)			BYU Screening Experiment (2019)		
Species	**Accession**	**Type**	**Species**	**Accession**	**Type**
M. sacchariflorus	JM11-006	Wild	*M. sacchariflorus*	JM11-006	Wild
M. sacchariflorus	JPN-2011-004	Wild	*M. sacchariflorus*	JPN-2010-005	Wild
M. sacchariflorus	JPN-2011-006	Wild	*M. sacchariflorus*	JPN-2011-010	Wild
M. sacchariflorus	JPN-2011-010	Wild	*M. sacchariflorus*	UI11-00031	Wild
M. sacchariflorus	PMS-076	Wild	*M. sinensis*	PMS-007	Wild
M. sacchariflorus	RU2012-056.1WD (4x)	Wild	*M. sinensis*	PMS-014	Wild
M. sacchariflorus	RU2012-141	Wild	*M. sinensis*	PMS-164	Wild
M. sacchariflorus	RU2012-169	Wild	*M. sinensis*	PMS-285	Wild
M. sacchariflorus	RU2012-183	Wild	*M. sinensis*	PMS-347	Wild
M. sacchariflorus	UI10-00008	Cultivar	*M. sinensis*	PMS-586	Wild
M. sacchariflorus	UI11-00033	Wild	*M. sinensis*	UI10-00048	Cultivar
M. sinensis	PMS-164	Wild	*M. sinensis*	UI10-00088	Cultivar
M. sinensis	PMS-285	Wild	*M. sinensis*	UI10-00092	Wild
M. sinensis	PMS-347	Wild	*M. floridulus*	PI417947	Wild
M. sinensis	PMS-7	Wild			
M. sinensis var. *condensatus*	UI10-00015	Wild			
M. sinensis	UI10-00020	Wild			
M. sinensis	UI10-00024	Cultivar			
M. sinensis	UI10-00053	Cultivar			
M. sinensis	UI10-00080	Cultivar			
M. sinensis	UI10-00097	Cultivar			
M. sinensis	UI10-00100	Cultivar			
M. floridulus	PI417947	Wild			

The screening experiment was arranged as a randomized complete block design. There are three blocks and each block consisted of two pots of each of the 23 genotypes. One pot represented the well-watered treatment and one pot was assigned to the drought-stress treatment for each of the 23 genotypes in one block. The well-watered treatment involved daily irrigation to saturate the rhizosphere of each of the treated plants, while the dry-down treatment was applied by withholding water for 7 days. After 7 days, plants were irrigated to container capacity. The dry-down period was repeated four times.

The Soil Plant Analysis Development (SPAD) Chlorophyll meter (SPAD-502Plus, Konica Minolta, Osaka, Japan), used to measure chlorophyll content, is one of the simpler and quicker means to characterize drought stress due to its non-destructive nature and its close correlation with leaf-level photosynthesis [24,25]. Measurements of SPAD value were taken on all plants between 10:30 am to 2:00 pm on days 0, 7, 14, 21, and 28. Rhizosphere conditions after 28 days of the dry-down experiment could be equated with what occurs in the field in the spring and/or summer in temperate regions, such as the east-central U.S. [26].

To evaluate drought tolerance in 23 *Miscanthus* genotypes during the dry-down experiment, DSI of SPAD value was plotted against coefficient of variance (CV) of SPAD value (Figure 1). The DSI of the HU screening experiment was calculated as follows:

$$\text{DSI of SPAD value (HU screening experiment)} = (\text{value of traits on day 28 of drought})/(\text{value of traits on day 0 as well-watered treatment}) \times 100 \quad (2)$$

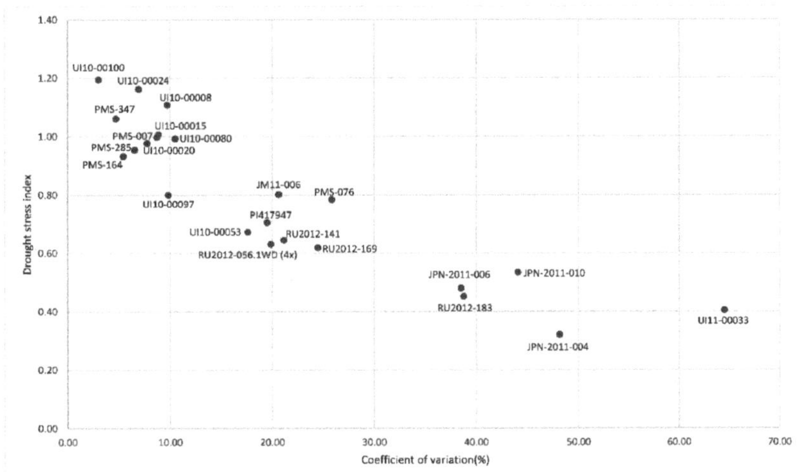

Figure 1. Scatter plot of coefficient of variation and drought stress index of the Soil Plant Analysis Development (SPAD) Chlorophyll meter value in screening experiment at Hokkaido University, Sapporo, Japan of drought-stress tolerance. Scatter points encircled by solid circles represent resistant genotypes. Scatter points encircled by dotted circles represent susceptible genotypes.

2.1.2. Experiment at Brigham Young University, USA

A population of 14 *Miscanthus* genotypes (Table 1), where each plant constituted the experimental units, were included in a drought-tolerance-evaluation experiment at Brigham Young University (BYU), Provo, Utah, USA (40°14′59″ N, 111°38′57″ W). The experiment was arranged in a completely randomized design. Due to independent *Miscanthus* genotype management at HU and BYU, six *Miscanthus* genotypes (JPN-2011-010, PMS-7, PMS-164, PMS-285, PMS-347, PI417947) were both evaluated in the HU screening experiment and BYU experiment. However, the remaining eight genotypes were only evaluated in the BYU experiment. The experiment was conducted under greenhouse conditions from 4 to 25 October 2019. Each genotype was replicated two times. All plants grown in plastic pots (diameter = 19 cm, height = 27 cm) were irrigated daily for one week prior to treatment initiation to keep them well watered before the dry-down experiment started. After measurements were collected on day 0 of the experiment, the dry-down treatment was applied by withholding water for 7 days. Plants were then irrigated to container capacity. The dry-down period was repeated three times.

The SPAD value was measured in all plants between 1:00 am to 3:30 pm on days 0, 7, 14, and 21 with a SPAD chlorophyll meter (MC-100 Chlorophyll Concentration Meter, Apogee Instruments, Inc., Logan, UT, USA). Photosynthesis parameters such as Pn, gs, Tr, intercellular CO_2 concentration (Ci), and leaf-level fluorescence (φPSII) were also measured for all genotypes with a portable photosynthesis system (LI-6400XT, LI-COR, Lincoln, NE, USA) with a 6400-40 leaf chamber fluorometer for use with the LI-6400 Portable Photosynthesis System (LI-COR, Lincoln, NE, USA). In addition, soil water potential was measured from collected soil samples on days 0, 7, 14, and 21 with a WP4C Dew Point Potentiometer (METER Group, Pullman, WA. USA).

Drought tolerance of *Miscanthus* genotypes was evaluated with the DSI data from the 14-day-dry-down data set from the BYU screening experiment, where the soil water potential (−2.6 MPa) led to slight levels of drought stress after the dry-down period.

$$\text{DSI (14-day dry-down data)} = \text{(value of traits from 14-day dry-down)}/ \\ \text{(value of traits of 0-day dry-down)} \times 100 \quad (3)$$

In order to comprehensively assess drought tolerance of the different genotypes, principal component analysis (PCA) ranking values, which were based on DSI values, were used to assess drought-tolerance capacity in each *Miscanthus* genotype based on the methodology of Liu et al. (2015) [5]. Liu et al. (2015) reported that the PCA based on the DSI of physiological parameters is considered to be a reliable method for evaluating drought tolerance among plants genotypes.

The 14 *Miscanthus* genotypes were ranked based on the PCA ranking values, which are based on DSI (14-day dry-down data) values. A significance test analysis done through SAS of the DSI data from the BYU screening experiment was used to complement the PCA results.

To understand the effect of different environments on *Miscanthus* genotype performance, SPAD value-based DSI values of the six *Miscanthus* genotypes were subjected to analysis of variance (ANOVA) in the HU screening and BYU experiments. As mentioned previously, six *Miscanthus* genotypes (JPN-2011-010, PMS-7, PMS-164, PMS-285, PMS-347, PI417947) were used in both the HU-screening and BYU screening experiments. Both experiments used the dry-down treatment to impose drought stress.

2.2. Precise-Comparison Experiment with Automated Irrigation System at HU

A total of ten *Miscanthus* genotypes, consisting of eight putatively drought-tolerant and two drought-sensitive *Miscanthus* genotypes, were selected based on preliminary results from the HU screening experiment. A scatterplot of SPAD value-based CV values and SPAD value-based DSI values in the HU screening experiment is shown in Figure 1. Relatively lower CV values and higher DSI values of some genotypes indicated that they had fewer variation between different drought levels and less differences between well-watered and drought conditions. Based on these results, eight putatively drought-tolerant genotypes (PMS-164, PMS-285, PMS-347, PMS-7, UI10-00008, UI10-00015, UI10-00020, UI10-00024) and two drought-sensitive genotypes (JPN-2011-004, UI11-00033) were selected to be included in the HU precise-comparison experiment for further analysis of their photosynthetic performance under specific drought levels through an automated irrigation system. Among the eight drought-tolerant genotypes, there was only one representative from the *M. sacchariflorus* species group, UI10-00008, while the other seven were *M. sinensis* genotypes. On the other hand, the most drought-sensitive genotypes, JPN-2011-004 and UI11-00033, were *M. sacchariflorus*. The genotypes were evaluated for drought tolerance in a precise-comparison experiment using an automated irrigation system following the methodology of Nemali and van Iersel (2006) [12]. A simplified diagram of the irrigation system can be seen in Figure S1. The experiment was conducted in a semi-open greenhouse at Hokkaido University from 10 September to 10 October 2018.

The precise-comparison experiment was arranged in a completely randomized design. Each genotype had three replicates. Soil moisture sensors (GS3, Meter Group, Pullman, WA) were inserted, along with drip emitters, into each of the potted plants (diameter = 19 cm, height = 27 cm). The sensors and emitters were connected to an automatic irrigation system, which regulated the amount of water applied to each plant. Soil-moisture treatments (20, 25, and 30% SMC) were arranged by setting the set-point of the system at pre-determined soil-moisture levels. The lowest SMC treatment (20%) was defined as the severe drought treatment and the highest SMC treatment (30%) was considered the well-watered treatment. After all potted plants achieved their SMC set points for 5 days, Pn, gs, Ci, and Tr were collected on the youngest, fully expanded leaf of each plant with a portable photosynthesis system (LI-6400XT). Leaf-level fluorescence (φPSII) and SPAD value, which were measured at the same time as photosynthesis, were measured with a fluorometer (Junior-PAM, Heinz Walz GmbH, Effeltrich, Germany) and a SPAD chlorophyll meter (SPAD-502Plus), respectively.

Soil moisture of all pots was controlled by the automated irrigation system. Average changes in SMC levels can be seen in Figure S2. The time taken for the potted media to reach the severe-drought-level set point required more time than media in the slight-drought-level treatment. For example, it only took 5 days for soil moisture to decrease from 30% to 25%, while it took 8 days for soil moisture to reduce from 25% to 20% (Figure S2).

Drought tolerance of these 10 *Miscanthus* genotypes was evaluated with the DSI data from the 25% SMC treatment in the HU precise-comparison experiment.

$$\text{DSI (25\% SMC treatment)} = \text{(value of traits of 25\% SMC)} / \text{(value of traits of 30\% SMC)} \times 100 \quad (4)$$

A PCA-ranking value based on DSI from the 25% SMC treatment was calculated for each genotype following the method of Liu et al. (2015) [5]. The 10 *Miscanthus* genotypes were ranked as relatively drought tolerant based on PCA ranking values. A significance test analysis done through SAS of the DSI data from the HU precise-comparison experiment was used to complement the PCA results.

To understand how different drought-treatment methods affected evaluation results of drought tolerance in *Miscanthus* spp., DSI of four photosynthetic parameters (Pn, gs, Tr, φPSII) of four *Miscanthus* genotypes (PMS-7, PMS-164, PMS-285, PMS-347), which were subjected to slight stress-level conditions (25% SMC in the HU precise-comparison experiment and −2.6 MPa of soil water potential on day 14 of the BYU experiment), were subjected to ANOVA. The fixed-SMC method was used as a drought-treatment method in the HU precise-comparison experiment, while in the BYU experiment, the dry-down method was used to subject plants to drought stress.

2.3. Post-Drought Recovery in the BYU Experiment

After the 21-day BYU dry-down screening experiment finished, a 7-day post-drought recovery experiment was conducted with the same plants in order to evaluate the drought-recovery capacity of the *Miscanthus* genotypes. A population of 14 *Miscanthus* genotypes (Table 1) was used in the 7-day post-drought-recovery experiment, which was the same material used in the BYU screening experiment. The recovery experiment was arranged in a completely randomized design and conducted under greenhouse conditions from 25 October to 1 November 2019, with two replicates of each genotype. Plants were watered daily over the 7-day experiment. Instrumentation and measurement parameters were the same as those used in the screening experiment. Measurements were made on the seventh day of the recovery experiment.

To understand the degree of recovery capacity from drought stress in *Miscanthus* genotypes, recovery DSI values were used to calculate PCA ranking values as assessment criteria.

$$\text{Recovery DSI} = \text{(value of traits of day 7 in BYU recovery experiment)} / \text{(value of traits of day 21 in BYU screening experiment)} \times 100 \quad (5)$$

Moreover, to comprehensively assess drought-recovery capacity of the different genotypes, the PCA ranking value based on recovery DSI values was calculated. The 14 *Miscanthus* genotypes were ranked according to their relative drought recovery capacity levels, which were based on the PCA-ranking-value results.

2.4. Drought Tolerance Evaluation and Statistical Analysis

The DSI values and PCA ranking values, which were based on DSI values, were used to assess the drought-tolerance capacity in *Miscanthus* genotypes which was based on the methodology of Liu et al. (2015) [5]. In order to quantify drought-induced effects in *Miscanthus* plants, the DSI value of each photosynthesis parameter was calculated using equation 1. Moreover, to comprehensively assess drought tolerance of the different genotypes, the PCA ranking value based on DSI values was calculated using the formula below:

$$\text{PCA ranking value} = \text{(contribution of the first principal components (PC1) (\%)} \times \text{PC1)} + \text{(contribution of the second principal components (PC2) (\%)} \times \text{PC2)} + \text{(contribution of the third principal components (PC3) (\%)} \times \text{PC3)} \quad (6)$$

In the BYU post-drought recovery experiment, recovery DSI values were used as an evaluation parameter of the recovery capacity of different genotypes. The formula used Equation (5). In the HU screening experiment, DSI of SPAD value and CV of SPAD value were used as drought-tolerance-evaluation parameters.

Microsoft Excel (Microsoft Office 2016, Microsoft Corporation, Redmond, WA, USA) was used to perform ANOVA. R statistical software (R3.5.1 by R Development Core Team, 2018) and ggplot2 package of R software were used to perform PCA of drought tolerance of *Miscanthus* genotypes. Statistical Analysis System (SAS Institute Inc., Cary, NC, USA) was used to respectively perform a significance test analysis of the DSI data from the HU precise-comparison and BYU screening experiments.

3. Results

3.1. *Comparison of* Miscanthus *Genotype Performance between HU and BYU Experiments*

Drought stress index values of 21-day dry-down of SPAD value of six *Miscanthus* genotypes (JPN-2011-010, PMS-007, PMS-164, PMS-285, PMS-347, PI417947) in the HU screening experiment and BYU experiment were subjected to ANOVA (Table 2). In the ANOVA results, there was no significant difference ($p > 0.05$) in DSI values of *Miscanthus* genotypes in the HU and BYU experiments.

Table 2. Analysis of Variance (ANOVA) result of six *Miscanthus* genotypes (JPN-2011-010, PMS-7, PMS-164, PMS-285, PMS-347, PI417947) between Hokkaido University screening experiment and Brigham Young University screening experiment using drought stress index of 21 days of SPAD value.

ANOVA					
Source of variation	SS	df	MS	F	*p*-value
Between Groups	0.0069	1	0.0069	0.1991	0.6650
Within Groups	0.3453	10	0.0345		
Total	0.3522	11			

Drought stress index values of four photosynthetic parameters (Pn, gs, Tr, φPSII) of four *Miscanthus* genotypes (PMS-7, PMS-164, PMS-285, PMS-347) under slight stress-level conditions (25% SMC in the HU precise-comparison experiment and soil water potential as −2.6 MPa on day 14 of the BYU experiment) were subjected to ANOVA (Table 3). The ANOVA results was significant ($p \leq 0.05$) in DSI values of *Miscanthus* genotypes in the HU precise-comparison and BYU experiments.

Table 3. Analysis of Variance (ANOVA) result of four *Miscanthus* genotypes (PMS-7, PMS-164, PMS-285, PMS-347) based on their drought stress index of four photosynthetic parameters (photosynthetic rate, stomatal conductance, transpiration rate, and chlorophyll fluorescence) under slight drought stress † of Hokkaido University (HU) precise-comparison and Brigham Young University (BYU) screening experiments.

ANOVA					
Source of variation	SS	df	MS	F	*p*-value
Between Groups	3.2866	1	3.2866	18.3031	0.0002
Within Groups	5.3870	30	0.1796		
Total	8.6736	31			

† Slight drought stress was set as 25% volumetric water content in soil of HU experiment and soil water potential as −2.6 MPa in BYU experiment.

The DSI φPSII mean values of *M. sinensis* genotype PMS-285 in the slight stress-level treatment in both the HU precise-comparison (94.1) and BYU (75.9) experiments significantly differed from that of the *M. sacchariflorus* genotype UI11-00033 (39.2) and *M. sinensis* genotype UI10-00015 (37.8) in the HU precise-comparison experiment (Table S2). In the BYU experiment, DSI φPSII of *M. sinensis* genotype PMS-007 (102.9) was higher than other genotypes and significantly differed from that of four *M. sinensis* genotypes (PMS-014, PMS-164, PMS-347, PMS-586) (Table S2-1). However, compared to its performance in the BYU experiment, the DSI φPSII of PMS-007 (71.5) was moderately high in the HU precise-comparison experiment (Table S2-2). The DSI Pn of *M. sinensis* genotype PMS-007 was relatively higher in the HU precise-comparison (77.3) and BYU (94.4) experiments than other genotypes subjected to slight stress levels, with the exception of *M. sinensis* genotype UI10-00088 in the BYU experiment (Table S3).

In addition, plants of *M. sinensis* genotype PMS-164 had higher DSI φPSII levels in the severe-stress-level treatment in the HU precise comparison (99.0) and BYU (67.8) experiments relative to those in slight stress-level treatment (HU: 42.9, BYU: 63.6) (Table S2). The DSI gs of *M. sinensis* genotype PMS-164 in the slight-stress-level treatment of the HU precise-comparison (824) and BYU (195) experiments statistically differed from 9 genotypes in the HU precise-comparison experiment (Table S4).

3.2. Changes in Soil Water Potential across Treatments in the BYU Experiment

Average changes in soil water potential of each genotype across treatments in the BYU experiment are shown in Table 4. Across treatments, soil water potential in the BYU experiment decreased, on average, from day 7 to 21 with the gradual exposure of plants to different levels of SMC. At first, soil water potential did not differ between days 0 and 7, but then considerably decreased on days 14 and 21 (Table 4). Soil water potential was around −0.1 MPa on days 0 and 7 and then decreased to −2.6 MPa on day 14 and −10.2 MPa on day 21 (Table 4). The soil water potential values on days 14 and 21 were more severe than those found at field capacity (−0.33 MPa) and permanent wilting (−1.5 MPa).

Table 4. Photosynthetic rate (Pn) of each *Miscanthus* genotype under each soil water potential in a screening experiment and a post-drought recovery experiment at Brigham Young University, Provo, Utah, USA.

Soil Water Potential (mPa)		**Day 0** −0.096	**Day 7** −0.14	**Day 14** −2.6025	**Day 21** −10.25	**Day 7 after Re-Watered** 0.04
Species	**Accession**	**Pn (μmol CO_2 m^{-2} s^{-1})**				
M. sacchariflorus	JM11-006	11.281	10.355	2.310	NA	NA
M. sacchariflorus	JPN-2010-005	7.673	9.899	2.744	NA	NA
M. sacchariflorus	JPN-2011-010	8.087	8.930	3.927	NA	NA
M. sacchariflorus	UI11-00031	12.044	12.292	7.707	5.372	8.145
M. floridulus	PI417947	3.961	5.595	3.192	3.006	2.774
M. sinensis	PMS-007	6.268	8.006	5.824	3.488	3.367
M. sinensis	PMS-014	10.724	12.436	6.394	1.655	11.899
M. sinensis	PMS-164	6.294	11.962	3.107	7.393	7.330
M. sinensis	PMS-285	7.613	8.364	6.810	6.484	5.121
M. sinensis	PMS-347	8.624	10.411	2.919	1.886	5.144
M. sinensis	PMS-586	5.148	9.832	2.051	1.438	6.569
M. sinensis	UI10-00048	5.034	15.136	0.777	NA	NA
M. sinensis	UI10-00088	4.418	5.081	4.312	1.160	3.275
M. sinensis	UI10-00092	5.334	13.335	5.049	NA	NA

3.3. Performance of Genotypes under Dry-Down Experiment in BYU Screening Experiment

After water-deficit treatments were initiated, photosynthetic levels of all *Miscanthus* genotypes decreased after day 7 as soil water potential decreased (Table 4). Most genotypes showed higher Pn on day 7 than on day 0, which corresponded to no changes in soil water potential (Table 4). After soil water potential values exhibited a large drop from day 7 to day 14 (−0.14 to −2.6 MPa), the Pn performance of all genotypes also showed a sharp decline, particularly going from a 15% decrease to a 77% decrease in Pn (Table 4). Moreover, five genotypes (JPN-2011-010, JM11-006, JPN-2010-005, UI10-00048, UI10-00092) died after day 14 due to serious drought. In addition, the Pn performance of the *M. sinensis* genotype, PMS-285, when experiencing low-water availability, showed almost no differences with conspecific genotypes in the well-watered treatment (Table 4). Although Pn of *M. sinensis* PMS-285 was at relatively moderate levels on days 0 and 7, it dropped when low soil-water conditions became more severe on days 14 and 21 (Table 4). However, the Pn of other genotypes experienced sharp decreases due to low soil-water availability during this time period, such as *M. sinensis* genotype UI10-00048 (Table 4).

In order to understand how photosynthetic traits contributed to drought tolerance of *Miscanthus* genotypes in the BYU experiment, we performed PCA using the DSI values (day 14) of six measured parameters (Pn, gs, Ci, Tr, φPSII, SPAD value) (Figure 2). The first (PC1) and second (PC2) principal components explained 76.6% of the variance among 14 *Miscanthus* genotypes. In addition, Pn and Tr had the largest contribution in PC1, suggesting Pn and Tr were the two most important photosynthesis parameters to the PCA results (Figure 2).

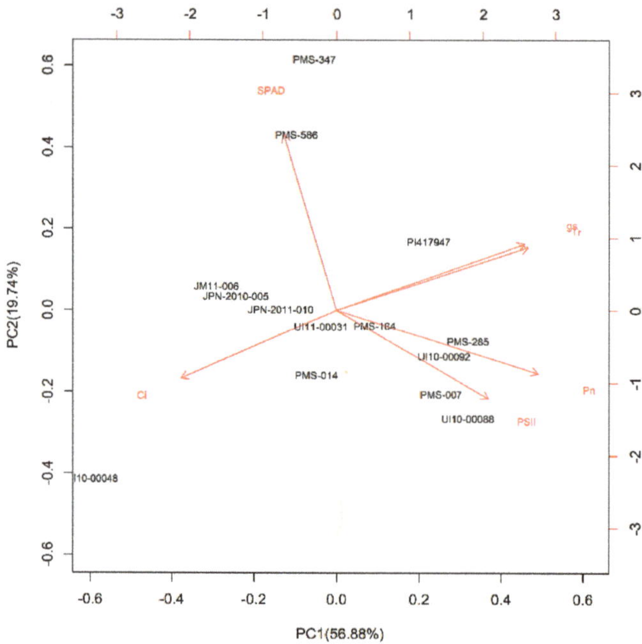

Figure 2. Principal component analysis (PCA) bi-plot of drought stress index (DSI) of six physiological traits (photosynthetic rate (Pn), stomatal conductance (gs), transpiration rate (Tr), intercellular CO_2 concentration (Ci), the Soil Plant Analysis Development (SPAD) Chlorophyll meter value, and chlorophyll fluorescence (PSII)) under drought over a 14-day period in screening experiment at Brigham Young University, Provo, UT, USA. Arrows represent physiological traits with various lengths, which were based on the impact of each trait on the separation of genotypes.

According to the PCA ranking value based on the DSI (day 14 of the BYU screening experiment) (Table 5), *M. sinensis* genotype PMS-285 and *M. floridulus* genotype PI417947 had relatively high ranking values compared to other genotypes, suggesting that they were more tolerant to drought stress among the 14 *Miscanthus* genotypes. In contrast, three of the *M. sacchariflorus* genotypes (JPN-2011-010, JPN-2010-005, JM11-006), which originated from Japan, showed relatively poor performance under low-water conditions while *M. sinensis* genotype UI10-00048 had the lowest PCA ranking relative to the other 13 *Miscanthus* genotypes in the BYU experiment (Table 5).

Table 5. Principal components analysis (PCA) ranking values † based on the drought stress index (Day 14) and the rank of drought-tolerance capacity of fourteen *Miscanthus* genotypes under slight drought stress ‡ in a screening experiment at Brigham Young University (BYU), Provo, Utah, USA.

Species	Accession	Origin	Genetic Clusters §	PC1	PC2	PC3	Ranking Value	Rank
M. sinensis	PMS-285	China	Yangtze-Qinling Msi	2.2033	−0.3101	0.4616	1.2647	1
M. floridulus	PI417947	Cultivar	SE China Msi	1.5610	0.6848	0.1983	1.0543	2
M. sinensis	UI10-00088	Cultivar	C Japan Msi	2.2217	−1.0798	0.0110	1.0521	3
M. sinensis	UI10-00092	Cultivar	C Japan Msi	1.8232	−0.4588	0.4144	1.0117	4
M. sinensis	PMS-007	China	Yangtze-Qinling Msi	1.7714	−0.8379	0.8903	0.9825	5
M. sinensis	PMS-347	China	SE China Msi	−0.3521	2.5066	1.3918	0.5140	6
M. sinensis	PMS-164	China	Yangtze-Qinling Msi	0.6587	−0.1539	−1.8284	0.0560	7
M. sinensis	PMS-586	China	Sichuan Msi	−0.6486	1.7557	0.0789	−0.0098	8
M. sacchariflorus	UI11-00031	China	Yangtze diploids (ssp. lutarioriparius) Msa	−0.2352	−0.1630	−0.0439	−0.1729	9
M. sinensis	PMS-014	China	Sichuan Msi	−0.3133	−0.6493	−0.3376	−0.3596	10
M. sacchariflorus	JPN-2011-010	Japan	N Japan 4x Msa	−0.9108	0.0087	−0.5782	−0.6075	11
M. sacchariflorus	JPN-2010-005	Japan	N Japan 4x Msa	−1.6686	0.1401	−1.2992	−1.1262	12
M. sacchariflorus	JM11-006	Japan	S Japan 4x Msa	−1.9928	0.2420	−0.9925	−1.2422	13
M. sinensis	UI10-00048	Cultivar	S Japan Msi	−4.1179	−1.6850	1.6335	−2.4174	14

† PCA ranking value was derived via calculation of first, second, and third principal components (PC1, PC2, and PC3). ‡ Slight drought stress was set as soil water potential as −2.6 MPa in the BYU experiment. § According to Clark et al. (2014) and Clark et al. (2019).

3.4. *Performance of Genotypes under Fixed Drought Level with Automated Irrigation System in the HU Precise-Comparison Experiment*

The PCA using the DSI (25% SMC) values of six parameters suggested that the PC1 and PC2 explained 76.9% of the variance among all 10 genotypes (Figure 3). Photosynthetic rate and Tr showed similar and strong influences on the PC1 axis. Stomatal conductance (gs), and Tr were the most important photosynthesis parameters to the PCA result of the HU precise-comparison experiment because they provided the largest contribution to PC1 (Figure 3).

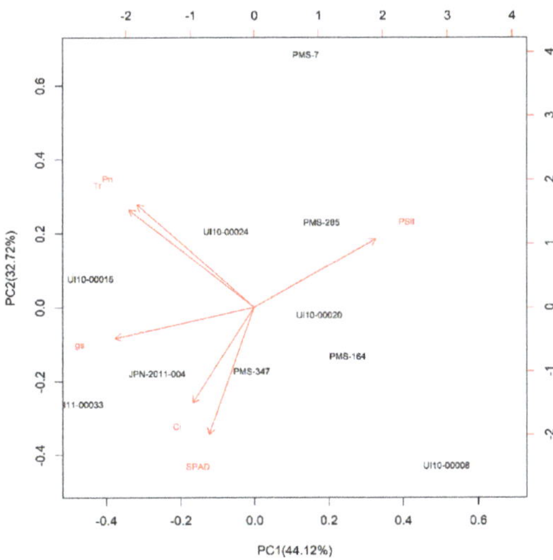

Figure 3. Principal component analysis (PCA) bi-plot of the drought stress index (DSI) of six physiological traits: photosynthetic rate (Pn), stomatal conductance (gs), transpiration rate (Tr), intercellular CO_2 concentration (Ci), the Soil Plant Analysis Development (SPAD) Chlorophyll meter value, chlorophyll fluorescence (PSII) under 25% soil moisture in a precise-comparison experiment at Hokkaido University, Sapporo, Japan. Arrows represent physiological traits with various lengths, which were based on the impact of each trait on the separation of genotypes.

According to the PCA ranking value based on the DSI (25% SMC) data (Table 6), *M. sinensis* genotypes, PMS-007 and PMS-285, had relatively high ranking values than the other genotypes, suggesting that they were more tolerant to drought stress while *M. sacchariflorus* genotypes, JPN-2011-004 and UI11-00033, had relatively lower ranking values than the other genotypes and were found to be more sensitive to drought stress. It is noteworthy that *M. sinensis* genotype PMS-285 also had a higher PCA ranking than other genotypes in the BYU screening experiment, while *M. sinensis* genotype PMS-007 did not have a high PCA ranking in the BYU screening experiment (Table 5). In contrast, *M. sacchariflorus* genotypes, with the exception of genotype UI10-00008, appeared to be more sensitive to drought than *M. sinensis* genotypes in the HU precise-comparison experiment (Table 6), which was also observed in the BYU screening experiment (Table 5).

Table 6. Principal components analysis (PCA) ranking values † based on the drought stress index (25% soil moisture content) and the rank of drought-tolerance capacity of ten *Miscanthus* genotypes under slight drought stress ‡ in a precise comparison experiment of a precise comparison experiment at Hokkaido University (HU), Sapporo, Japan.

Species	Accession	Origin	Genetic Clusters §	Leaf Width (cm)	Leaf Length (cm)	PC1	PC2	PC3	Ranking Value	Rank
M. sinensis	PMS-007	China	Yangtze-Qinling Msi	2.0	60	0.71	3.03	−0.57	1.232	1
M. sinensis	PMS-285	China	Yangtze-Qinling Msi	1.1	56	0.93	1.02	0.04	0.749	2
M. sacchariflorus	UI10-00008	Cultivar	NEChina/Korea/ Russia diploids Msa	0.8	44	2.65	−1.90	0.55	0.618	3
M. sinensis	UI10-00020	Cultivar	S Japan Msi	0.4	18	0.89	−0.09	0.68	0.450	4
M. sinensis	PMS-164	China	Yangtze-Qinling Msi	1.1	25	1.29	−0.58	−0.36	0.333	5
M. sinensis	UI10-00024	Cultivar	S Japan Msi	0.6	27	−0.40	0.90	0.47	0.178	6
M. sinensis	PMS-347	China	SE China Msi	1.8	48	−0.04	−0.76	−0.52	−0.333	7
M. sinensis var. *condensatus*	UI10-00015	Cultivar	C Japan Msi	1.6	40	−2.27	0.34	1.39	−0.712	8
M. sacchariflorus	JPN-2011-004	Japan	S Japan 4x Msa	1.8	55	−1.35	−0.79	−1.80	−1.085	9
M. sacchariflorus	UI11-00033	Japan	S Japan 4x Msa	2.0	61	−2.41	−1.16	0.14	−1.425	10

† PCA ranking value was derived via calculation of first, second, and third principal components (PC1, PC2, and PC3). ‡ Slight drought stress was set as 25% volumetric water content in the media of the HU experiment. § According to Clark et al. (2014) and Clark et al. (2019).

3.5. *Drought Recovery Capacity of* Miscanthus *Genotypes of Post-Drought Recovery Experiment in BYU*

Upon rewatering plants daily for 7 days after a 21-day dry-down treatment, the average soil water potential of all *Miscanthus* genotypes on day 7 of the BYU recovery experiment increased to 0.04 MPa, which was similar to that on day 0 of the BYU screening experiment (Table 4). This result suggests that the soil moisture level was high enough for plants to recover from drought (Table 4). Three *M. sacchariflorus* genotypes, JPN-2011-010, JM11-006, and JPN-2010-005, and two *M. sinensis* genotypes, UI10-00048 and UI10-00092, were nearly dead due to drought stress after a 21-day dry-down period in the BYU screening experiment (Table 4). Consequently, we were not able to characterize the recovery capacity of these genotypes.

On the other hand, the photosynthetic levels of *M. sinensis* genotypes PMS-014 and PMS-586 exhibited relatively quick recovery of Pn levels on day 7 in the BYU recovery experiment (Table 4). The Pn level of genotype PMS-014 on day 7 in the BYU recovery experiment was six times greater than its Pn performance on day 21 in the BYU screening experiment (Table 4). A similar pattern could be seen with genotype PMS-586, whose Pn level was four times greater than its Pn performance on day 21 in the BYU screening experiment (Table 4). In addition, these two genotypes had high recovery-PCA-ranking values, suggesting that they had the potential to recover from drought damage (Table 7). On the other hand, *M. sinensis* genotype PMS-285 had a relatively low recovery ranking value and was less capable of recovering from drought (Table 7), but it displayed higher Pn levels than other genotypes under low-water conditions in the BYU screening experiment (Table 4).

Table 7. Recovery principal components analysis (PCA) ranking values † based on the recovery drought stress index and the rank of recovery capacity from drought stress of fourteen *Miscanthus* genotypes in post-drought recovery experiment at Brigham Young University, Provo, Utah, USA.

Species	Accession	Origin	Genetic Clusters ‡	PC1	PC2	PC3	Ranking Value	Rank
M. sinensis	PMS-014	China	Sichuan Msi	4.3565	−1.4954	0.2379	2.8970	1
M. sinensis	PMS-586	China	Sichuan Msi	3.3939	0.4381	0.5461	2.5689	2
M. sinensis	PMS-347	China	SE China Msi	2.7456	1.2399	−1.0582	2.1484	3
M. sinensis	UI10-00088	Cultivar	C Japan Msi	1.2360	−0.6833	0.0990	0.7769	4
M. sacchariflorus	UI11-00031	China	Yangtze diploids (ssp. lutarioriparius) Msa	−0.2981	0.8091	0.6621	−0.0294	5
M. floridulus	PI417947	Cultivar	SE China Msi	−0.9572	2.3607	0.7977	−0.2173	6
M. sinensis	PMS-164	China	Yangtze-Qinling Msi	−0.5758	0.7901	−0.0801	−0.2786	7
M. sinensis	PMS-007	China	Yangtze-Qinling Msi	−0.5925	−0.3455	−0.4199	−0.5167	8
M. sinensis	PMS-285	China	Yangtze-Qinling Msi	−0.9592	−0.8056	0.0559	−0.8369	9

† PCA ranking value was derived via calculation of first, second, and third principal components (PC1, PC2, and PC3). ‡ According to Clark et al. (2014) and Clark et al. (2019).

4. Discussion

4.1. Comparison of Different Drought Treatment Methods for Evaluation

Some plants species show different physiological responses under rapidly-imposed and slowly-imposed drought-stress conditions [27]. As we mentioned earlier, two drought-treatment methods, the dry-down technique, and the fixed-SMC technique, imposed different patterns of drought stress on plants in the experiments. The dry-down technique made a quick and sizable decrease in SMC over a short period of time, while the fixed-SMC technique-controlled SMC at a relatively constant level at a slower rate and for a longer period of time. Both drought-imposition techniques were used in previous research for studying drought tolerance in plants [4,5,28–31].

Drought-tolerant genotypes, which were selected through the PCA ranking analysis, also showed different degrees of drought tolerance in the HU precise-comparison and BYU experiments. For example, *M. sinensis* PMS-007 showed high drought tolerance performance in the HU precise-comparison, but only medium-level performance in the BYU screening experiment. Environmental factors and methods of drought imposition could have been factors that influenced the results of the HU precise-comparison and BYU screening experiments. However, it appears that environmental factors did not influence the results of the two experiments. According to our results, there was no significant difference ($p > 0.5$) in DSI values of *Miscanthus* genotypes in the HU and BYU experiments (Table 2), suggesting that there was no effect of environment between the HU and BYU experiments when both experiments used the dry-down technique to impose drought on *Miscanthus* plants. Therefore, the different evaluation results between the HU precise-comparison and BYU experiments were likely due to differences in how drought was imposed.

Decreases in SMC showed different patterns in the two drought-treatment methods used in this study. As reflected in changes in soil water potential values, drought stress conditions caused by the dry-down technique became more severe (i.e., soil water potential went below the permanent wilting point (−1.5 MPa) over a 14-day period (day 7 to day 21 in the BYU screening experiment) with a quick and sizable decrease in SMC (Table 4). In this case, plants had little time to adjust low-water conditions. On the other hand, with the fixed-SMC method, SMC changed slowly and could be controlled at a relatively constant level for plants to respond low-water availability. In the HU precise-comparison experiment, soil moisture controlled by an automated irrigation system took around 30 days to change from slight stress to severe stress, and at each stress level plants had 3–5 days to adjust the stress before measurement (Figure S2). With the fixed-SMC method, plants had enough time to exhibit their responses to drought, presuming that there was some physiological regulation in their cells. Based on the different patterns we observed in decreases in SMC (Table 4; Figure S2), the dry-down method is suitable for selecting drought-tolerant genotypes for cultivar or breeding development. However, the fixed-SMC method can aid researchers in clarifying drought-induced response of plants, such as changes in cell-level osmotic potential changing or toxic ROS scavenging regulation [32].

Under field conditions, drought can be defined as a condition where plants cannot get take up enough water from dry soil for normal physiological function over an extended period of time [33]. Large decreases in soil moisture over a short period of time during a dry-down are more similar to drought in the field, which leads plants to perform all steps of drought-caused physiological regulation in a short time [27]. This aspect of the dry-down method leads plants to respond to low-water availability as if they were subjected to field conditions. However, rapidly decreasing soil moisture makes it difficult to capture and characterize ephemeral physiological changes in plants [27]. On the other hand, the fixed-SMC method is controlled by a computer, which can regulate irrigation and thereby control SMC to maintain continuous drought conditions [12]. Therefore, plants generally have enough time in this method to physiologically respond to drought due to being subjected to constant, low-SMC conditions. In addition, physiological responses of plants to different soil-moisture conditions (e.g., well-watered, moderate, severe) with this approach seem more straightforward than in the dry-down method [34]. However, such constant soil-moisture conditions, even when water levels are fairly low, differ from drought in the field such that genotypes identified as drought tolerant through the fixed-SMC method may not perform well when grown in the field.

4.2. *Characteristics of Drought Stress in* Miscanthus *spp.*

In general, *M. sinensis* appears to have stronger drought tolerance than *M. sacchariflorus* (Tables 5 and 6). Based on the PCA ranking results of the BYU screening experiment, four *M. sacchariflorus* genotypes (UI11-00031, JPN-2011-010, JPN-2010-005, and JM11-006) ranked relatively low in terms of drought-stress tolerance (Table 5). Similarly, based on the PCA ranking results of the HU precise-comparison experiment, two *M. sacchariflorus* genotypes, JPN-2011-004 and UI11-00033 ranked 9 and 10, suggesting they were sensitive to drought stress (Table 6). These results correspond to their native habitats. *Miscanthus sinensis* usually grows in dry, upland areas, while *M. sacchariflorus* occurs in mesic, lowland areas [35].

Miscanthus × *giganteus*, which is a triploid hybrid of tetraploid *M. sacchariflorus* and diploid *M. sinensis*, is considered as a potential high-yielding energy crop (29–38 Mg ha^{-1}) [13]. However, *M.* × *giganteus* expresses sensitivity to drought and needs more irrigation than maize under commercial cultivation conditions [21]. Genes inherited from *M. sacchariflorus* possibly influence the drought sensitivity of *M.* × *giganteus*.

Based on the PCA ranking results of the HU precise-comparison and BYU screening experiments, *M. sinensis* genotype PMS-285 had higher photosynthetic performance under drought in both experiments, suggesting that it can be used as germplasm in breeding programs (Tables 5 and 6). *Miscanthus sinensis* genotype PMS-007 showed relatively higher photosynthesis performance than other genotypes in the HU precise-comparison experiment (Table 6), but it exhibited only relatively moderate photosynthesis performance in the BYU screening experiment (Table 5). Considering the two drought-imposition methods used in our study, *M. sinensis* genotype PMS-007 appeared to maintain high photosynthesis performance for stable and consistent responses to low-water availability in the fixed-SMC method, but the photosynthesis performance was relatively lower at rapidly decreasing SMC conditions caused by the dry-down method (Table 4 and Table S5). Considering the different photosynthetic performance of *M. sinensis* genotypes PMS-285 and PMS-007 under dry-down and the fixed-SMC treatments, there should be some differences between the drought-response mechanisms of *M. sinensis* genotypes PMS-285 and PMS-007, which allowed for genotype PMS-285 to be tolerant of both rapidly and slowly decreasing soil-moisture availability, which needs to be clarified in the future. In addition, the genotypes, *M. sacchariflorus* UI10-00008 and *M. sinensis* UI10-00020, which had relatively narrow leaves and smaller leaf area than other genotypes, were more tolerant to drought than other genotypes in the HU precise-comparison experiment, with the exception of *M. sinensis* genotypes PMS-285 and PMS-007 (Table 6). A relatively small leaf area can lead to low transpiration levels, which could allow for plants to maintain photosynthetic rates at levels to sustain moderate growth despite having low soil-water availability [28,36].

The DSI φPSII of *Miscanthus sinensis* genotype PMS-285 in the slight-stress-level treatment in the HU precise-comparison (94.1) and BYU (75.9) experiments exceeded that of other genotypes in the study, except for *M. sinensis* genotype PMS007 and UI10-00088 in the BYU experiment (Table S2). On the other hand, the DSI Pn of *M. sinensis* genotype PMS-007 is relatively higher than other genotypes under slight stress levels in both the HU precise-comparison and BYU experiments (Table S3). The relatively high values of DSI φPSII of *M. sinensis* genotype PMS-285 and DSI Pn of *M. sinensis* genotype PMS-007 could help explain why these two genotypes showed stronger drought tolerance than other genotypes in this study. In addition, the DSI gs of *M. sinensis* genotype PMS-164 exceeded that of other genotypes in both experiments (Table S4).

Interestingly, the *Miscanthus* genotypes with strong drought-recovery capacity (PMS-014, PMS-586) did not exhibit high drought tolerance (Tables 5 and 7). On the other hand, genotypes with high drought tolerance may not have sufficient drought-recovery capacity. Based on the recovery PCA ranking results (Table 7), *M. sinensis* genotypes PMS-014 and PMS-586 ranked relatively higher than other genotypes, but they only displayed moderate levels of tolerance under 14 days of being subjected to the drought treatment in the BYU screening experiment (Table 5).

Miscanthus sinensis genotype PMS-285 had a higher photosynthetic performance of Pn and DSI φPSII than other genotypes under drought in both the HU precise comparison and BYU screening experiments (Table 4 and Table S2). In addition, this genotype had a higher Pn value on day 21 of the BYU screening experiment than its Pn value on day 7 of the BYU recovery experiment (Table 4). In addition, *M. sinensis* genotype PMS-285 did not have a high recovery PCA ranking value (Table 7), which suggests it did not recover from drought stress after being rewatered. This is surprising given that it had a high PCA ranking value under slight drought stress in both the HU precise comparison and BYU screening experiments (Tables 5 and 6).

Recovery capacity from drought is an important trait to help plants tide over from the effects of low SMC conditions [3]. Several plant species, whose photosynthetic machinery can often recover rapidly from drought stress, can absorb water when short-term rain events occur in the midst of a prolonged drought [3,7]. Lauenroth et al. (1987) reported that new root growth of *Bouteloua gracilis*, a warm-season perennial grass species, occurred nearly 40 h after being rewatered [7]. Such root growth has the capability of increasing water availability for plants. Another study, which focused on water relations of sugarcane (*Saccharum officinarum* L.), found that high WUE and deep root systems enable sugarcane to recover from drought damage [37]. Such traits may be a possible reason for the strong recovery performance of *M. sinensis* genotypes PMS-014 and PMS-586. These traits could be effective screening criteria for drought-tolerant genotypes of *Miscanthus*. The WUE and rooting depth were not measured in our experiments but should be focused on in future research.

Genetic clusters and ploidy levels may be factors that have considerable influence on drought tolerance in *Miscanthus* spp. [16]. Regarding the influence of genetic clusters on drought tolerance, genotypes in the *M. sinensis* 'Yangtze-Qinling' genetic clusters appear to have relatively stronger drought tolerance than other genetic clusters (Tables 5 and 6). In addition, genotypes in the *M. sinensis* 'Sichuan' genetic cluster can quickly recover from drought damage after being re-watered (Table 7). Both *M. sinensis* genotypes, PMS-007 and PMS-285, which align with the *M. sinensis* 'Yangtze-Qinling' genetic cluster (Tables 5 and 6), expressed relatively higher photosynthesis performance than other genotypes under stress in the HU precise comparison experiment (Table 6). In contrast, in the BYU screening experiment, *M. sinensis* genotypes PMS-014 and PMS-586, which are associated with the *M. sinensis* 'Sichuan' genetic cluster (Table 5), displayed low photosynthetic levels when subjected to drought (Table 5). However, both exhibited relatively high recovery of photosynthetic levels after re-irrigation in the BYU recovery experiment (Table 7).

For different ploidy-type accessions, *M. sacchariflorus* diploid genotype UI10-00008 (Table 6) showed much higher photosynthesis performance than two other *M. sacchariflorus* tetraploid genotypes, JPN-2011-004 and UI11-00033, in the HU precise-comparison experiment (Table 6). *Miscanthus sacchariflorus* UI10-00008 has, on average, a small leaf area with only 0.8 cm leaf width, while genotypes JPN-2011-004 and UI11-00033 have an average leaf width of 2 cm (Table 6). In the BYU screening experiment, three *M. sacchariflorus* tetraploid genotypes were dead after a dry-down period of 21 days, but *M. sacchariflorus* diploid genotype UI11-00031 survived despite prolonged exposure to severe drought stress (Table 4). In addition, *M. sacchariflorus* diploid genotype UI11-00031 showed considerable recovery of Pn on day 7 in the BYU recovery experiment (Table 4). Therefore, ploidy level may reflect how leaf area and transpiration rate of *Miscanthus* genotypes contribute to drought tolerance. Moreover, drought-tolerant diploid *M. sacchariflorus* genotypes could be used as breeding material to produce high-yielding *M.* × *giganteus* genotypes with strong drought tolerance. Using genetic clusters and ploidy levels for genotype evaluation will help to improve the efficiency of the selection and breeding of stress-tolerant crops.

Ornamental *Miscanthus* cultivars exhibited relatively higher drought tolerance than most wild-type accessions in both the HU precise-comparison and BYU screening experiments (Tables 5 and 6). Although wild-type *Miscanthus* accessions usually express stronger stress tolerance to drought, disease, and insect pests than ornamental cultivars [38], we found that some cultivars (*M. floridulus* PI417947, *M. sinensis* UI10-00088, *M. sinensis* UI10-00092, and *M. sacchariflorus* UI10-00008) showed relatively higher drought tolerance than wild-type accessions (Tables 5 and 6). *Miscanthus sinensis* cultivars can be found in gardens and yards throughout the U.S., Canada, and Europe [39]. For ornamental plants, drought tolerance ranks high as an important selection criteria because drought stress is commonly encountered in managed landscapes.

Further studies are needed to characterize drought-stress-response mechanisms in *Miscanthus*. Few information exists regarding the drought-stress physiology of *Miscanthus* [40–42]. Improvement of drought tolerance in *Miscanthus* spp. can enable them to survive when subjected to drought conditions caused by climate change. Such crops offer the opportunity to also generate biomass under low-soil-water conditions, which is important for developing *Miscanthus* as a sustainable energy crop.

Supplementary Materials: The following are available online at https://www.mdpi.com/article/10.3390/agriculture12010006/s1. Table S1: Detailed information of 29 *Miscanthus* genotypes used for the evaluation of low-water-adaptability capacity in *Miscanthus* spp., including entry number, species, origin location, and genetic groups background. Table S2: Least squares means of drought stress index (DSI) values of chlorophyll fluorescence (φPSII) of *Miscanthus* genotypes. Table S3: Least squares means of drought stress index (DSI) values of photosynthetic rate (Pn) of *Miscanthus* genotypes. Table S4: Least squares means of drought stress index (DSI) values of stomatal conductance (gs) of *Miscanthus* genotypes. Table S5: Photosynthetic rate (Pn) of each *Miscanthus* genotype under each soil water content level in a precise-comparison experiment at Hokkaido University, Sapporo, Japan. Figure S1: Simplified diagram showing various parts of the irrigation system. Figure S2: Average changes in soil moisture content controlled by the automated irrigation system in a precise-comparison experiment at Hokkaido University, Sapporo, Japan.

Author Contributions: Conceptualization, T.Y. and E.J.S.; methodology, T.-Y.W., T.N. and J.R.S.; data curation, T.-Y.W. and A.V.-M.; writing—original draft preparation, T.-Y.W.; writing—review and editing, J.R.S. and T.Y.; funding acquisition, T.Y. All authors have read and agreed to the published version of the manuscript.

Funding: This work was funded by Grants-in-Aid for Scientific Research (No. 17H04615 to T.Y.) from the Japanese Ministry of Education, Science, Sports and Culture.

Acknowledgments: The authors would like to thank the staff of the Field Science Center for Northern Biosphere of Hokkaido University and the Department of Plant and Wildlife Sciences of Brigham Young University for their kind support and cooperation during the study. All supports and assistance are sincerely appreciated.

Conflicts of Interest: The authors declare no conflict of interest.

References

1. Morrow, W.R., III; Gopal, A.; Fitts, G.; Lewis, S.; Dale, L.; Masanet, E. Feedstock loss from drought is a major economic risk for biofuel producers. *Biomass Bioenergy* **2014**, *69*, 135–143. [CrossRef]
2. Farooq, M.; Wahid, A.; Kobayashi, N.; Fujita, D.; Basra, S.M.A. Plant drought stress: Effects, mechanisms and management. *Agron. Sustain. Dev.* **2009**, *29*, 185–212. [CrossRef]
3. Chen, D.; Wang, S.; Cao, B.; Cao, D.; Leng, G.; Li, H.; Deng, X. Genotypic variation in growth and physiological response to drought stress and re-watering reveals the critical role of recovery in drought adaptation in maize seedlings. *Front. Plant Sci.* **2016**, *6*, 1241. [CrossRef]
4. Chen, Y.; Chen, F.; Liu, L.; Zhu, S. Physiological responses of *Leucaena leucocephala* seedlings to drought stress. *Procedia Eng.* **2012**, *28*, 110–116. [CrossRef]
5. Liu, Y.; Zhang, X.; Tran, H.; Shan, L.; Kim, J.; Childs, K.; Ervin, E.H.; Frazier, T.; Zhao, B. Assessment of drought tolerance of 49 switchgrass (*Panicum virgatum*) genotypes using physiological and morphological parameters. *Biotechnol. Biofuels* **2015**, *8*, 152. [CrossRef]
6. Trenberth, K.E.; Dai, A.; Schrier, G.; van der Jones, P.D.; Barichivich, J.; Briffa, K.R.; Sheffield, J. Global warming and changes in drought. *Nat. Clim. Chang.* **2014**, *4*, 17–22. [CrossRef]
7. Lauenroth, W.K.; Sala, O.E.; Milchunas, D.G.; Lathrop, R.W. Root dynamics of *Bouteloua gracilis* during short-term recovery from drought. *Funct. Ecol.* **1987**, *1*, 117–124. [CrossRef]
8. Upadhyaya, H.; Panda, S.K.; Dutta, B.K. $CaCl_2$ improves post-drought recovery potential in *Camellia sinensis* (L) O. Kuntze. *Plant Cell Rep.* **2011**, *30*, 495–503. [CrossRef] [PubMed]
9. Mittler, R.; Zilinskas, B.A. Regulation of pea cytosolic ascorbate peroxidase and other antioxidant enzymes during the progression of drought stress and following recovery from drought. *Plant J.* **1994**, *5*, 397–405. [CrossRef]
10. Blackman, C.J.; Li, X.; Choat, B.; Rymer, P.D.; DeKauwe, M.G.; Duursma, R.A.; Tissue, D.A.; Medlyn, B.E. Desiccation time during drought is highly predictable across species of *Eucalyptus* from contrasting climates. *New Phytol.* **2019**, *224*, 632–643. [CrossRef] [PubMed]
11. Bergsten, S.J.; Stewart, J.R. Measurement of the influence of low water availability on the productivity of *Agave weberi* cultivated under controlled irrigation. *Can. J. Plant Sci.* **2013**, *94*, 439–444. [CrossRef]
12. Nemali, K.S.; van Iersel, M.W. An automated system for controlling drought stress and irrigation in potted plants. *Sci. Hortic.* **2006**, *110*, 292–297. [CrossRef]
13. Heaton, E.A.; Dohleman, F.G.; Long, S.P. Meeting US biofuel goals with less land: The potential of *Miscanthus*. *Glob. Chang. Biol.* **2008**, *14*, 2000–2014. [CrossRef]
14. Toma, Y.; Fernandez, F.G.; Nishuwaki, A.; Yamada, T.; Bollero, G.; Stewart, J.R. Aboveground plant biomass, carbon, and nitrogen dynamics before and after burning in a seminatural grassland of *Miscanthus sinensis* in Kumamoto, Japan. *Glob. Chang. Biol. Bioenergy* **2010**, *2*, 52–62. [CrossRef]
15. Clark, L.V.; Brummer, J.E.; Głowacka, K.; Hall, M.C.; Heo, K.; Peng, J.; Yamada, T.; Yoo, J.H.; Yu, C.Y.; Zhao, H.; et al. A footprint of past climate change on the diversity and population structure of *Miscanthus sinensis*. *Ann. Bot.* **2014**, *114*, 97–107. [CrossRef]
16. Clark, L.V.; Jin, X.; Petersen, K.K.; Anzoua, K.G.; Bagmet, L.; Chebukin, P.; Deuter, M.; Dzyubenko, E.; Dzyubenko, N.; Heo, K.; et al. Population structure of *Miscanthus sacchariflorus* reveals two major polyploidization events, tetraploid-mediated unidirectional introgression from diploid *M. sinensis*, and diversity centred around the Yellow Sea. *Ann. Bot.* **2019**, *124*, 731–748. [CrossRef]
17. Heaton, E.; Voigt, T.; Long, S.P. A quantitative review comparing the yields of two candidate C_4 perennial biomass crops in relation to nitrogen, temperature and water. *Biomass Bioenergy* **2004**, *27*, 21–30. [CrossRef]
18. Mann, J.J.; Barney, J.N.; Kyser, G.B.; DiTomaso, J.M. Root system dynamics of *Miscanthus × giganteus* and *Panicum virgatum* in response to rainfed and irrigated conditions in California. *Bioenergy Res.* **2013**, *6*, 678–687. [CrossRef]
19. Joo, E.; Hussain, M.Z.; Zeri, M.; Masters, M.D.; Miller, J.N.; Gomez-Casanovas, N.; Bernacchi, C.J. The influence of drought and heat stress on long-term carbon fluxes of bioenergy crops grown in the Midwestern USA. *Plant Cell Environ.* **2016**, *39*, 1928–1940. [CrossRef]
20. Van der Weijde, T.; Huxley, L.M.; Hawkins, S.; Sembiring, E.H.; Farrar, K.; Dolstra, O.; Visser, R.G.F.; Trindade, L.M. Impact of drought stress on growth and quality of miscanthus for biofuel production. *Glob. Chang. Biol. Bioenergy* **2017**, *9*, 770–782. [CrossRef]

21. Vanloocke, A.; Bernacchi, C.J.; Twine, T.E. The impacts of *Miscanthus* × *giganteus* production on the Midwest US hydrologic cycle. *Glob. Chang. Biol. Bioenergy* **2010**, *2*, 180–191. [CrossRef]
22. Ings, J.; Mur, L.A.J.; Robson, P.R.H.; Bosch, M. Physiological and growth responses to water deficit in the bioenergy crop *Miscanthus* × *giganteus*. *Front. Plant Sci.* **2013**, *4*, 468. [CrossRef] [PubMed]
23. Emerson, R.; Hoover, A.; Ray, A.; Lacey, J.; Cortez, M.; Payne, C.; Karlen, D.; Birrell, S.; Laird, D.; Kallenbach, R.; et al. Drought effects on composition and yield for corn stover, mixed grasses, and *Miscanthus* as bioenergy feedstocks. *Biofuels* **2014**, *5*, 275–291. [CrossRef]
24. Kato, M.; Kobayashi, K.; Ogiso, E.; Yokoo, M. Photosynthesis and dry-matter production during ripening stage in a female-sterile line of rice. *Plant Prod. Sci.* **2004**, *7*, 184–188. [CrossRef]
25. Takai, T.; Kondo, M.; Yano, M.; Yamamoto, T. A quantitative trait locus for chlorophyll content and its association with leaf photosynthesis in rice. *Rice* **2010**, *3*, 172–180. [CrossRef]
26. Namias, J. Nature and possible causes of the northeastern United States drought during 1962–1965. *Mon. Weather. Rev.* **1966**, *94*, 543–554. [CrossRef]
27. Cornic, G.; Papgeorgiou, I.; Louason, G. Effect of a rapid and a slow drought cycle followed by rehydration on stomatal and non-stomatal components of leaf photosynthesis in *Phaseolus vulgaris* L. *J. Plant Physiol.* **1987**, *126*, 309–318. [CrossRef]
28. Ganjeali, A.; Porsa, H.; Bagheri, A. Assessment of Iranian chickpea (*Cicer arietinum* L.) germplasms for drought tolerance. *Agric. Water Manag.* **2011**, *98*, 1477–1484. [CrossRef]
29. Li, C.-N.; Yang, L.-T.; Srivastava, M.K.; Li, Y.-R. Foliar application of abscisic acid improves drought tolerance of sugarcane plant under severe water stress. *Int. J. Agric. Innov. Res.* **2014**, *3*, 101–107.
30. Nazari, L.; Pakniyat, H. Assessment of drought tolerance in barley genotypes. *J. Appl. Sci.* **2010**, *10*, 151–156. [CrossRef]
31. Percival, G.C.; Keary, I.P.; Al-Habsi, S. An assessment of the drought tolerance of *Fraxinus* genotypes for urban landscape plantings. *Urban For. Urban Green.* **2006**, *5*, 17–27. [CrossRef]
32. Rohollahi, I.; Khoshkholghsima, N.; Nagano, H.; Hoshino, Y.; Yamada, T. *Respiratory burst oxidase-D* Expression and Biochemical Responses in *Festuca arundinacea* under Drought Stress. *Crop Sci.* **2018**, *58*, 435–442. [CrossRef]
33. Dracup, J.A.; Lee, K.S.; Paulson, E.G. On the Definition of Droughts. *Water Resour. Res.* **1980**, *16*, 297–302. [CrossRef]
34. Kim, J.; van Iersel, M.W. Slowly developing drought stress increases photosynthetic acclimation of *Catharanthus roseus*. *Physiol. Plant.* **2011**, *143*, 166–177. [CrossRef]
35. Tamura, K.; Uwatoko, N.; Yamashita, H.; Fujimori, M.; Akiyama, Y.; Shoji, A.; Sanada, Y.; Okumura, K.; Gau, M. Discovery of natural interspecific hybrids between *Miscanthus sacchariflorus* and *Miscanthus sinensis* in southern Japan: Morphological characterization, genetic structure, and origin. *Bioenergy Res.* **2016**, *9*, 315–325. [CrossRef]
36. Smith, W.K. Temperatures of desert plants: Another perspective on the adaptability of leaf size. *Science* **1978**, *201*, 614–616. [CrossRef]
37. Jangpromma, N.; Thammasirirak, S.; Jaisil, P.; Songsri, P. Effects of drought and recovery from drought stress on above ground and root growth, and water use efficiency in sugarcane (*Saccharum officinarum* L.). *Aust. J. Crop Sci.* **2012**, *6*, 1298–1304.
38. Dougherty, R.F.; Quinn, L.D.; Voigt, T.B.; Barney, J.N. Response of naturalized and ornamental biotypes of *Miscanthus sinensis* to soil-moisture and shade stress. *Northeast. Nat.* **2015**, *22*, 372–386. [CrossRef]
39. Linde-Laursen, I. Cytogenetic analysis of *Miscanthus* 'Giganteus', an interspecific hybrid. *Hereditas* **1993**, *119*, 297–300. [CrossRef]
40. Alvarez, E.; Scheiber, S.M.; Beeson, R.C.; Sandrock, D.R. Drought tolerance responses of purple lovegrass and 'Adagio' maiden grass. *HortScience* **2007**, *42*, 1695–1699. [CrossRef]
41. Chen, M.; Hou, X.; Fan, X.; Wu, J.; Pan, Y. Drought tolerance analysis of *Miscanthus sinensis* 'Gracillimu' seedlings. *Acta Prataculturae Sin.* **2013**, *22*, 184–189. [CrossRef]
42. Stavridou, E.; Webster, R.J.; Robson, P.R.H. Novel *Miscanthus* genotypes selected for different drought tolerance phenotypes show enhanced tolerance across combinations of salinity and drought treatments. *Ann. Bot.* **2019**, *124*, 653–674. [CrossRef] [PubMed]

Article

Drought Stress Amelioration in Maize (*Zea mays* L.) by Inoculation of *Bacillus* spp. Strains under Sterile Soil Conditions

Muhammad Azeem [1,†], Muhammad Zulqurnain Haider [2], Sadia Javed [1,*], Muhammad Hamzah Saleem [3] and Aishah Alatawi [4]

1 Department of Biochemistry, Government College University, Faisalabad 38000, Pakistan; muhammadazeem134@gmail.com
2 Department of Botany, Government College University, Faisalabad 38000, Pakistan; drmzhaider@gcuf.edu.pk
3 College of Plant Science and Technology, Huazhong Agricultural University, Wuhan 430070, China; saleemhamza312@webmail.hzau.edu.cn
4 Biology Department, Faculty of Science, Tabuk University, Tabuk 71421, Saudi Arabia; Amm.alatawi@ut.edu.sa
* Correspondence: sadiajaved@gcuf.edu.pk
† This research work is the part of M.Phil. thesis of Muhammad Azeem.

Abstract: The aim of the present study was to promote plant growth characteristics including mineral uptake and various phytohormone production by indigenously isolated *Bacillus* spp. strains. Plants subjected to normal and water stress conditions were collected after 21 days to measure physiological parameters, photosynthetic pigment estimation, biochemical attributes, lipid peroxidation and antioxidant enzyme response modulation. Our results correlated with drought stress amelioration with the inoculation of *Bacillus* spp. strains BEB1, BEB2, BEB3 and BEB4 under sterile soil condition. Inoculated plants of both maize cultivars showed increases in fresh (56.12%) and dry (103.5%) biomass, plant length (42.48%), photosynthetic pigments (32.76%), and biochemical attributes with enhanced nutrient uptake. The overall maize antioxidant response to bacterial inoculation lowered the malonaldehyde level (59.14%), generation of hydrogen peroxide (45.75%) and accumulation of flavonoid contents in both control and water stress condition. Activity of antioxidant enzymes, catalase (62.96%), peroxidase (23.46%), ascorbate peroxidase (24.44%), and superoxide dismutase (55.69%) were also decreased with the application of bacterial treatment. Stress amelioration is dependent on a specific plant–strain interaction evident in the differences in the evaluated biochemical attributes, lipid peroxidation and antioxidant responses. Such bacteria could be used for enhancing the crop productivity and plant protection under biotic and abiotic stresses for sustainable agriculture.

Keywords: plant growth-promoting bacteria; biological control; abiotic stress; drought tolerance; antioxidants enzymes; plant microbiome

Citation: Azeem, M.; Haider, M.Z.; Javed, S.; Saleem, M.H.; Alatawi, A. Drought Stress Amelioration in Maize (*Zea mays* L.) by Inoculation of *Bacillus* spp. Strains under Sterile Soil Conditions. *Agriculture* **2022**, *12*, 50. https://doi.org/10.3390/agriculture12010050

Academic Editors: Daniele Del Buono, Primo Proietti and Luca Regni

Received: 16 November 2021
Accepted: 21 December 2021
Published: 1 January 2022

1. Introduction

Drought is the major problem for corn plants and has become a source of dietary threat to food security and human health [1]. Drought stress is directly proportional to climate change. Because of the catastrophic loss of arable land due to drought stress, meeting the need of an overpopulated globe for food, shelter, and clothing will be a worrisome concern in the future [2,3]. The water deficiencies lead to reducing the overall plant growth by lowering photosynthetic activity, hormone production and membrane stability [4,5]. Lack of water causes the production of reactive oxygen species (ROS) in plants and induces irreversible damage to a metabolic system that ultimately triggers the cell damage [6].

Under water deficit conditions, the ROS including hydroxyl, superoxide, hydrogen peroxide and many other free radicals are very damaging for the normal metabolic pathways of the plants. The excessive presence of ROS in plant leaves causes the chemical oxidation of important cellular biomolecules, including proteins, lipids, nucleic acids and chlorophylls, which leads to cell death [7]. The antioxidant system of plant enzymatic or non-enzymatic mechanisms work in coordination to normalize the oxidative stress by neutralizing toxic ROS [2]. Both mechanisms involve the removal of ROS to reduce the oxidative damage to the plant cell [8]. The low molecular weight organic compounds including many sugars, amino acids, proline help the plant cells to normalize the fluid level for balanced cellular metabolic pathways [9].

In Pakistan, maize is one of the top three cereal crops, along with wheat and rice. Total maize production was 7,800,000 tons in 2020. To meet the required food and nutrition for an ever-growing population all over the world, the cereal production needs to be upgraded in drought conditions. In this respect, several approaches and biofertilizers are used to enhance the drought tolerance and promote plant growth [10,11]. The current agricultural crop production approaches are costly and non-renewable; for example, improper fertilizer and pesticide inputs may result in the production of greenhouse gases and cause a variety of environmental and human health problems [12]. In this regard, the use of beneficial microbes could be a stress protecting agent for plants and lead to promising solutions for a sustainable and environment-friendly agriculture [2,13].

Beneficial soil microorganisms are being used to boost plant growth and production as part of these efforts. Plant growth-promoting rhizobacteria (PGPR) are beneficial rhizobacteria that can be employed to boost plant growth and productivity [9]. When PGPR is introduced to plant seeds or roots, it has the potential to colonize the entire root system, consuming amino acids and sugars contained in root exudates as a source of food and energy to activate plant growth-promotion activities, resulting in increased plant growth and yield [14]. On various crops, inoculating plants with PGPR can enhance growth by up to 500 percent and yield by up to 57 percent.

Plant growth promoting bacteria eliminate abiotic stress in plants and grasses including by inoculation to their rhizosphere [14]. The PGP bacteria in the rhizosphere promote the nutrient uptakes and modulate the production and metabolism of plant hormones including auxin, abscisic acid, cytokinin, ethylene and gibbereellins [13,15]. Bacteria having the ability to produce enzyme 1-aminocyclopropane-1-carboxylate (ACC) deaminase reduce the level of endogenous ethylene in plants to regulate the growth and development [13]. The endophytic bacteria comprising the capability of solubilizing phosphate and fixing nitrogen are considered very efficient biofertilizers for the accessibility of atmospheric inorganic elements to the plant [16]. Numerous microbes have been described that solubilize the phosphorus by converting it from an insoluble to soluble form through acidification, chelation, and redox reactions [2].

On the many taxonomic groups of PGPR, numerous reviews and research papers have been published [2,17]. Indigenous bacillus strains from the textile effluents are the focus of this study because bacillus PGPR gets a lot of attention around antioxidant stress relief. Because of their ability to withstand severe environments and mitigate plant drought stress, these strains may lead to the development of biotechnologies for plant growth promotion in arid and semiarid regions. Accordingly, the purpose of this study was to analyze effect of inoculation of four textile effluent isolated *Bacillus* spp. strains on the growth, physiology, biochemical attributes, lipid peroxidation and antioxidant enzyme response modulation of a drought-sensitive plant like maize under water deficit.

2. Materials and Methods

2.1. Isolation and Characterization of Bacterial Microflora

The textile industrial effluent samples were collected in 500 mL sterile bottles from Kamal textile (Pvt) Jaranwala Road, Faisalabad, Punjab, Pakistan (Google map location: 31.469006616012493, 73.30483972821574). Serially diluted (10^{-1}, 10^{-2}, 10^{-3}, 10^{-4} and 10^{-5}) effluent samples applied for microbial growth by a standard spread plate method [18] on nutrient agar medium and incubated at 37 °C for 24 h. The bacterial cell shape and color were examined under the microscope at 100× using oil emersion after staining with a gram staining method. A single colony mixed in a drop of 3% H_2O_2 (v/v) on a glass slide and observed for effervescence of O_2 bubbles [19]. A starch hydrolysis test was performed to check the ability of microbes to produce amylase enzyme using Lugol's solution (KI and I_2) [20]. Casein agar plates were utilized to check the production of protease enzymes by isolates [21]. The ability of bacterial isolates to produce exopolysaccharides was checked using LB broth, after incubation of 2 days at 37 ± 2 °C cultures were centrifuged, supernatant was mixed with 100% chilled ethanol in equal volume and precipitation was observed [22].

2.2. Taxonomic Identification and Phylogenetic Analysis

Genomic DNA was extracted from pure isolate by a GF-1 extraction kit for bacterial DNA. The 16S rRNA were replicated utilizing a universal primer: 27F (AGAGTTTGATC-CTGGCTCAG) and 1492R (TACGGYTACCTTGTTACGACTT). The PCR amplification was performed according to DNA Taq polymerase protocol [23]. The products of PCR reaction were analyzed on agarose gel to ensure successful 16S rRNA gene amplification. DNA sequencing was done at MACROGEN, the sequencing company, Seoul, Korea working with ABI 3100, the automated sequencer with Big Dye Terminator Kit v. 3.1 with Sanger's dideoxy method using the same set of primers. The phylogeny was described by BLASTn analysis and by constructing the phylogenetic tree with the type species. The phylogenetic tree was built by utilizing the neighbor-joining (NJ) technique [24] following multiple sequence alignment results of ClustalW in the MEGA X [25] software package. The clustering constancy of the tree was estimated by a bootstrap study [26] of 100 data sets.

2.3. In Vitro PGP Traits

Peptone water broth medium was incubated with an isolated strain for 72 h at 37 ± 2 °C and 120 rpm in a shaking incubator. After 48 and 72 h, the amount of ammonia produced was estimated by Nessler's reagent [27].

Salkowski reagent method [28] was used for IAA quantification of bacterial isolates in LB medium containing 1 g/L tryptophan for 3 days in both normal and water stress conditions. Indole acetic acid (IAA) was used for a standard curve.

Estimation of nitrogen fixation was done to check the ability of bacterial isolates to covert the nitrogen gas from atmosphere into nitrogen form that is usable by plants. Nitrogen free malate medium was inoculated with bacteria and incubated for 7 days at 37 ± 2 °C and 120 rpm in a shaking incubator, and plates were incubated for 7 days at 37 ± 2 °C in an incubator to check the efficiency of bacteria to fix nitrogen [29].

The inorganic phosphate solubilizing activity of bacterial isolates was observed by using NBRIP agar medium [30]. Freshly grown pure colony of isolates was spotted on the NBRIP agar plate and allowed to incubate it for 7 days at 37 ± 2 °C and observed daily for zone formation. Formation of a clear halo zone around the bacterial colony indicates the phosphate solubilizing activity. The potential of bacterial isolate to solubilize inorganic phosphate was estimated by calculating the solubilization index (SI) using formula: Solubilization Index (SI) = Colony diameter (mm) + Zone diameter (mm)/Colony diameter (mm) [31].

2.4. In Vitro Drought Tolerance

The effects of drought on the growth of isolates were studied using polyethylene glycol MW 6000 (PEG) at different concentrations ranging from 0 to 25%. The isolates were inoculated in LB broth containing different concentrations of PEG (5%, 10%, 15%, 20%, and 25%) and incubated at 120 rpm and 37 °C for days. The bacterial growth was measured spectrophotometrically at OD 600 nm [32].

2.5. Pod Experiment

Inoculum suspension for seed treatment was prepared by mixing a single colony of a bacterial strain into the nutrient agar broth medium and incubating the culture for 24 to 48 h at 37 °C. The bacterial cells were washed three times with sterile distilled water by vortex mixing and centrifugation at 6000 rpm for 5 min using a sterile 15 mL centrifuge falcon tube. Following vertexing, a spectrophotometer was used to measure the absorbance (600 nm) of the cell suspension, which was then diluted to 10^8 cfu/mL with sterile PB. The surface-sterilized seeds were placed in bacterial suspension on continuous shaking at 120 rpm for 8 h to insure colonization during seed germination.

Experiment was designed in a completely randomized way with three replicates comprising a total of 60 pots (Bacterial treatments (5) × water stress (2) × cultivars (2) × replicates (3) = 60). Five treatments were used for pot experiments and include (T0) control (sterile distilled water); (T1) *Bacillus cereus* strain BEB1; (T2) *Bacillus cereus* strain BEB2; (T3) *Bacillus tropicus* strain BEB3; (T4) *Bacillus thuringiensis* strain BEB4. Seeds of two maize (*Zea mays* L.) cultivars (V1: FH-1046; V2: YH-5427) were collected from the Maize and Millets Research Sub-section, Ayub Agriculture Research Institute, Faisalabad, Pakistan. Soil was collected from the Botanical Garden, Department of Botany, GC university, Faisalabad, Pakistan and autoclaved at 121 °C for 15 min before use. The physiochemical properties of the soil used for the experiment are given (Table 1). Ten seeds of corn were sown in 350 g of soil for each pod. Pots were arranged in a completely randomized design on a bench in a greenhouse where temperatures varied from 24 °C (night) to 31 °C (day). The experiment lasted for 21 days during which pots were watered once a day. After germination, the seedlings were thinned to five per pod and maintained in a light growth chamber. The drought stress was maintained by soil water content method, by measuring the weight of dry and watered soil of normal and stress. The normal (non-stress) and drought (stressed) plants were maintained by weighing at 15–18% and 5–8% water content in the soil, respectively.

Table 1. Physicochemical soil characteristics used in the plant growth promotion experiment.

Parameter	Values
Texture	Clay loam
pH	8.6
EC (dSm^{-1})	2.93
Sodium adsorption ratio (SAR) ($(mmol^{-1})^2$)	6.5
Nitrogen (%)	0.014
Phosphorus (ppm)	3.0
Potassium (ppm)	40
Organic matter	0.28
Saturation (%)	32
Sand (%)	45.5
Silt (%)	42.5
Clay (%)	12.5
HCO_3 ($mmol^{-1}$)	3.55
Cl^- ($mmol^{-1}$)	2.34
SO_4^{2-} ($mmol^{-1}$)	6.67
Ca^{2+} + Mg^{2+} ($mmol^{-1}$)	3.5
Na^+ ($mmol^{-1}$)	3.7
K^+ ($mmol^{-1}$)	0.06
Available Cu^{2+} (mg kg^{-1})	0.35
Available Zn^{2+} (mg kg^{-1})	0.85

2.6. Plant Physiological Parameters

After 21 days of growth, plants were harvested and root and shoot fresh weight were measured. Plant roots were washed with distilled water after harvesting. Root and shoot lengths were measured manually using a ruler. Root and shoot dry weight were recorded, which was subsequently oven dried.

2.7. Photosynthetic Pigment Estimation

About 0.5 g of fresh leaf sample from each applied treatment replicate was completely homogenized in about 10 mL methanol (80%). Samples were centrifuged at 12,000× *g* for 10 min and kept at 4 °C overnight [33]. Absorbance of extract was measured using a UV visible spectrophotometer at 663, 645 and 480 for the Chlorophyll a, b and total carotenoid contents:

$$\text{Chlorophyll a (mg/g F·Wt)} = [12.7(\text{OD } 663) - 2.69\ (\text{OD } 645)] \times \text{V}/1000 \times \text{W} \quad (1)$$

$$\text{Chlorophyll b (mg/g F·Wt)} = [22.9(\text{OD } 663) - 4.68\ (\text{OD } 645)] \times \text{V}/1000 \times \text{W} \quad (2)$$

$$\text{Carotenoid (mg/g F·Wt)} = [(\text{OD } 480) - 0.114(\text{OD } 663) - 0.638\ (\text{OD } 645)] \times 1000/2500 \quad (3)$$

2.8. Plant Biochemical Attributes

About 0.5 g of fresh leaf sample from each applied treatment replicate was completely homogenized in about 10 mL methanol (80%). Samples were centrifuged at 12,000× *g* for 10 min. Flavonoids were assayed by the method of Zhishen [34]. Total soluble sugars were analyzed by Anthrone's reagent method [35]. About 0.1 mL sample was mixed with 1 mL Anthrone's reagent. After boiling for 15 min, the mixture could cool to room temperature. Absorbances of all treated samples were measured at 625 nm.

Estimation of total soluble protein was assayed using the Bradford method [36]. About 50 µL of sample was mixed with 1 mL of Bradford reagent, and absorbance was measured at 595 nm. Protein content was calculated by comparing with the BSA standard curve. For ascorbic acid estimation, a fresh leaf sample (0.5 g) was completely homogenized in 10 mL of TCA (6%). After that, 4 mL of the extract was mixed with 2 mL of dinitrophenyl hydrazine, followed by 1 drop of thiourea. After boiling for 15 min, the mixture could cool to room temperature. To the mixture, five milliliters of 80 percent H_2SO_4 were added. Following [37], absorbances of all treated samples were measured at 530 nm and compared to a standard curve drawn using ascorbic acid concentrations ranging from 10 to 100 mg/L.

Velikova's [38] method was used to determine total hydrogen peroxide (H_2O_2) levels. After filtration, 1.0 mL supernatant was combined with 0.5 mL phosphate buffer and 1 mL of 1 M potassium iodide in a mixture of 0.5 mL phosphate buffer and 1 mL of 1 M potassium iodide. The sample mixtures were thoroughly vortexed, and their absorbance was measured using a spectrophotometer at 390 nm. H_2O_2 was calculated using a standard curve created with tannic acid as the reference.

The method of Cakmak [39] was used to determine MDA levels. In a chilled mortar and pestle containing 5 mL of 1 percent (*w*/*v*) TCA, a fresh leaf sample (0.5 g) was ground. For 10 min, the mixture was centrifuged at 15,000 rpm. About 1 mL of 0.5 percent thiobarbituric acid (TBA) was added to 0.5 mL of the supernatant. The mixture was boiled and cooled at 95 °C. A spectrophotometer was used to measure the absorbance of all treated samples at 532 and 600 nm. The absorption co-efficient, 155 mmol/cm, was used to calculate the level of TBA:

$$\text{MDA} = \Delta\ (\text{OD532} - \text{OD600})/1.56 \times 10^5 \quad (4)$$

2.9. Antioxidant Enzyme Activities

For antioxidant enzyme activities, the sample of fresh leaf (0.5 g) was homogenized in 10 mL potassium phosphate buffer for enzyme extraction (pH 7.8). After centrifugation for 15 min at 15,000 rpm, the extract supernatant was frozen at −20 °C in an ultra-low freezer.

The method proposed by Chance and Maehly [40] was used to determine the activity of the CAT enzyme. We combined 0.1 mL of plant extract with 1 mL of 5.9 mM H_2O_2 and 1.9 mL of 50 mM phosphate buffer in a 50 mL flask (7.0 pH). At 240 nm, the absorbance was measured at 20-s intervals for two minutes. One unit of CAT activity was equal to a change of 0.01 A240 Units/min. The activities of CAT were then calculated and expressed in milligrams per milligram of total soluble protein.

The activity of POD enzyme was determined using a method proposed by [41]. A reaction mixture (750 µL phosphate buffers (7.0 pH) + 100 µL H_2O_2 (5.9 mM) + 100 µL guicol (0.5%) + 50 µL enzyme extract) was prepared to measure the activity of POD enzyme. The absorbance was then measured using a spectrophotometer at 470 nm for 2 min at 20-s intervals. The activities of POD were then calculated and expressed in milligrams per milligram of total soluble protein.

The activity of APX enzyme was determined using a method proposed by [40,41]. A reaction mixture (700 µL phosphate buffers (7.0 pH) + 100 µL H_2O_2 (5.9 mM) + 100 µL ascorbate (0.5 mM) + 100 µL enzyme extract) was prepared to measure the activity of APX enzyme. The absorbance was then measured using a spectrophotometer at 290 nm for 2 min at 20-s intervals. The activities of APX were then calculated and expressed in milligrams per milligram of total soluble protein.

A reaction mixture (400 µL H_2O + 350 µL phosphate buffer + 100 µL methionine + 50 µL NBT + 50 µL enzyme extract + 50 µL riboflavin) was prepared to measure the activity of the SOD enzyme [42]. The mixture was then exposed to light for 15 min, with the decrease in absorbance measured at 560 nm. A blank was made by omitting the enzyme extract. The activities of SOD were then calculated and expressed in milligrams per milligram of total soluble protein.

2.10. Statistical Analysis

A three-way fully randomized analysis of variance (ANOVA) with replication was carried out using CoStat V6.4 to test the influence of plant growth promoting *Bacillus* spp. strains isolated from textile wastewater on maize (*Zea mays* L.) under water deficit conditions. The Principle Component Analysis (PCA) and Pearson coefficient correlation among studied attributes were computed by using the IBM SPSS Statistics software windows version 25 (IBM Corp, Armonk, NY, USA).

3. Results

3.1. Screening of Plant Beneficial Bacillus spp. Strains

Bacterial strains were appeared as smooth, scatter, curved, and glittering colonies on the nutrient agar plates. Colonies isolated from mixed populations were obtained by characterizing and sub-culturing. Twelve isolates in total were obtained, and each isolate was used for further analysis. Four of those isolates with positive activity of important biochemical enzymes and efficient in plant growth promoting characteristics were selected for further studies (Table 2). The strains were identified as bacillus species based on the sequence similarity search in the NCBI database. The four isolated strains BEB1, BEB2, BEB3 and BEB4 showed 96.58%, 97.99%, 98.61% and 98.20% sequence identity with *Bacillus cereus* strain A22 (MG598445.1), *Bacillus cereus* strain YLB-P5 (KF376341.1), *Bacillus tropicus* strain SA31 (MK467555.1) and *Bacillus thuringiensis* strain a57 (KX057537.1), respectively. In phylogenetic trees, the 16S rRNA gene sequence is clustered with Bacillus species (Figure 1). The 16S rRNA sequences were submitted in the GenBank public database under the NCBI and received accession numbers (Table 2).

3.2. In Vitro Plant Growth Promoting Characteristics of Strains

Production of ammonia by isolated strains was examined by missing the cultured supernatant with Nessler's reagent. The highest level of ammonia (5.24 µmol/mL) was produced by strain BEB3 within 48 h, but, when estimated after a 72 h strain, BEB1 showed a maximum amount (6.44 µmol/mL) of ammonia produced in the culture medium. In terms of in vitro, plant-beneficial traits under normal and water stress (15% PEG6000) followed the somewhat similar trend for IAA synthesis in culture medium. In this assay, four selected isolates were able to produce IAA in the culture medium only in the presence of substrate L-Tryptophan (1 g/L). However, isolates BEB1 demonstrated IAA production 54.09 and 33.63 µg/mL under normal and the stress of PEG6000, respectively. However, varying nitrogenase activity, examined on a nitrogen-free malate medium, was detected by isolated strains in both liquid and solid media. In liquid media strain, BEB1 showed maximum atmospheric nitrogen fixation in terms of positive bacterial cell growth (OD 600 nm). On the other hand, on nitrogen-free malate, the agar plate BEB4 strain showed the highest blue halo zone diameter by nitrogenase enzyme activity (Table 2). For in-vitro phosphate solubilization assay, bacterial isolates formed halo-zones of varying diameters on NBRIP media plates; however, the maximum physiological competence solubilization index (2.54) and halo-zone formation was found in strain BEB3. Strains BEB1, BEB2 and BEB4 were able to demonstrate solubilization indexes 2.23, 2.33 and 2.47, respectively.

Table 2. Characterization of bacterial strain for enzyme production and in vitro PGP traits.

Characteristics	BEB1	BEB2	BEB3	BEB4
Cell Morphology				
Gram stain	+	+	+	+
Shape	Rod	Rod	Rod	Rod
Biochemical tests				
Catalase	+++	+++	+++	+++
Amylase	+++	++	+++	+++
Exopolysaccharides	+++	++	+++	++
Indole test	+	−	−	−
Proteases (Zone mm)	18 ± 1	15 ± 1	19 ± 1	17 ± 1
PGP Traits				
Ammonia Production 48 h (µmol/mL)	5.22	5.20	5.24	5.21
Ammonia Production 72 h (µmol/mL)	6.44	6.41	6.35	6.29
IAA Production (Without Trp)	−	−	−	−
IAA Production (Trp) (µg/mL)	54.09	20.70	22.13	21.41
IAA Production (Trp + 15%PEG) (µg/mL)	33.63	12.76	13.65	13.21
Nitrogen Fixation (mm ± SD)	26.33 ± 1.52	20.66 ± 1.52	19.66 ± 1.52	29.33 ± 1.52
Nitrogen Fixation (OD 600 nm)	0.975	0.453	0.541	0.424
Phosphate solubilization (SI ± SD)	2.23 ± 0.082	2.33 ± 0.082	2.52 ± 0.082	2.47 ± 0.082
Molecular				
BLAST Comparison (16S rDNA)	*B. cereus*	*B. cereus*	*B. tropicus*	*B. thuringiensis*
Accession Number	MW350048	MW350049	MW350050	MW350051

SI = solubilization index; SD = standard deviation; Trp = tryptophan; PEG; polyethylene glycol; OD = optical density; +, ++, and +++ = positive, high positive and strong positive, respectively.

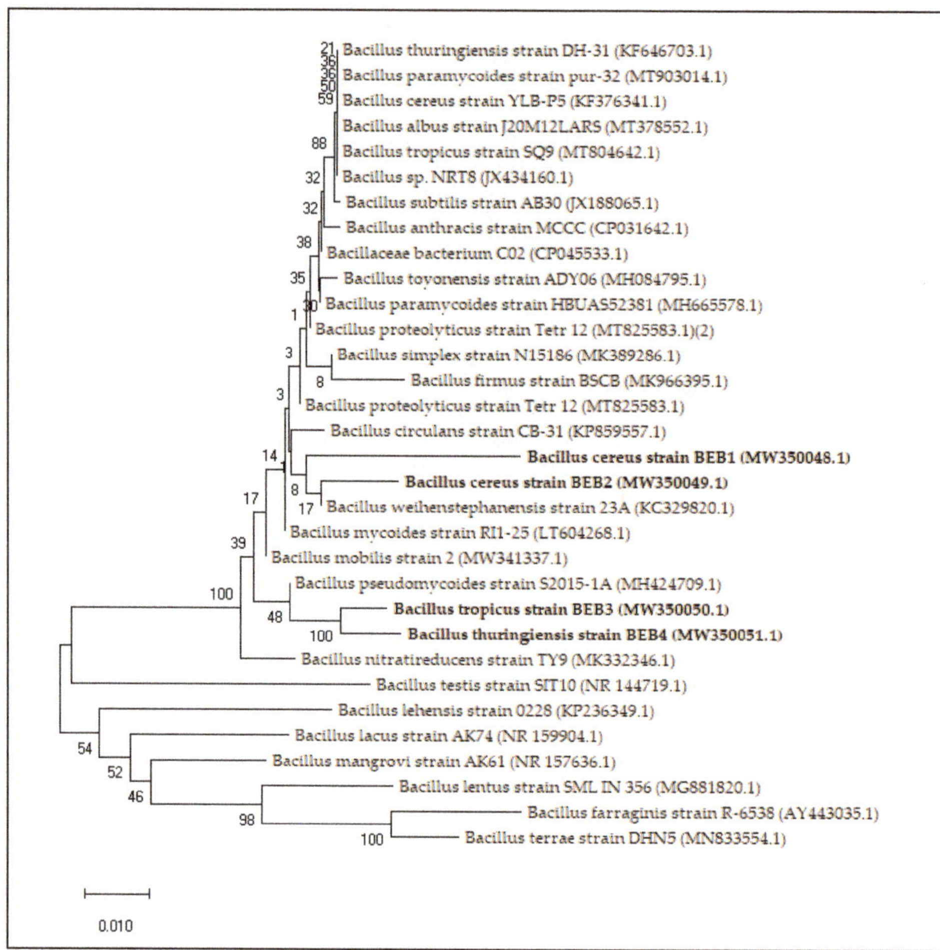

Figure 1. Phylogenetic tree based on a 16S rRNA nucleotide sequences showing the position of isolated strains among the most relevant species by BLAST comparative analysis. The tree was constructed using the Neighbor-Joining method and bootstrap values (expressed as percentage of 100 replications) are shown at branching points.

3.3. *In Vitro Drought Tolerance Testing of Bacterial Strains*

The four isolated bacterial strains showing best plant growth promoting properties were also analyzed for in vitro drought tolerance by using polyethylene glycol (PEG 6000). The growth of tested strains (OD 600 nm) on varying water potential by polyethylene glycol was shown in (Figure 2). Different concentrations (0, 5%, 10%, 15%, 20%, and 25%) of PEG 6000 were used in LB broth medium to control water potential. Strain BEB1 grew well and showed a maximum tolerance to the drought stress even at the lowest water potential.

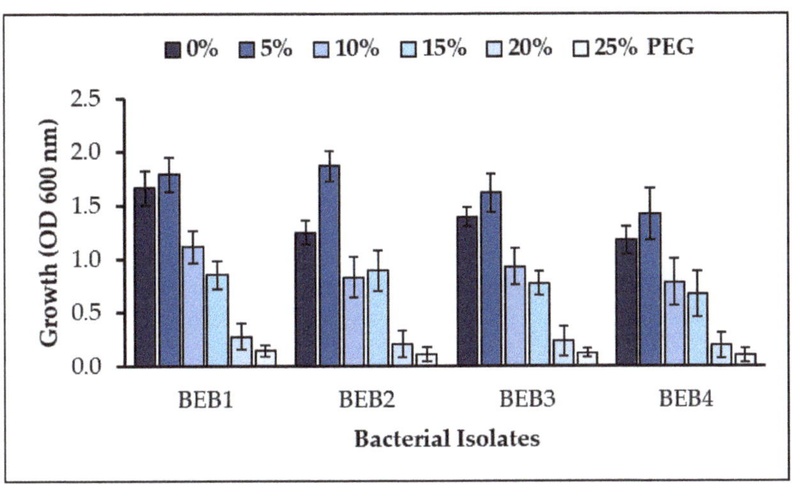

Figure 2. Relative growth of wastewater isolated bacterial strains in different concentrations of polyethylene glycol (PEG 6000) in culture medium (values are the mean of three replicates and the bar represents the standard error); LSD value ≤ 0.05.

3.4. *Plant Physiological Parameters Influenced by Bacterial Inoculation*

Plant growth decreased significantly under water stress (5–8% water content in the soil); reductions in plant shoot length of V1: FH-1016 (13.89%), V2: YH-5427 (9.75%), and root length V1 (15.09%), V2 (7.23%) were observed in uninoculated plants (Table 3, Figure 3A,B). However, bacterial inoculation improved the shoot as well as root lengths of both cultivars of maize (*Zea mays* L.) under water deficit regimes. The inoculation with *Bacillus* spp. BEB1 (V1:33.8, V2: 42.48%), BEB2 (39.6, 22.34%), BEB3 (33.1, 37.52%) and BEB4 (26.87, 33.14%) increased shoot length on stressed plants of both maize cultivars when compared to the uninoculated stressed control (Table 3, Figure 3A). Plant root length increased significantly by inoculation with *Bacillus* spp. BEB1 (17.7, 21.16%), BEB2 (24.64, 14.03%), BEB3 (22.99, 39.87%) and BEB4 (31.28, 13.81%) in both cultivars when compared with the uninoculated drought control (Table 3, Figure 3B).

Bacillus spp. inoculation increased fresh and dry biomass in both cultivars of maize plants under drought stress. Plants treated with PGPB inoculation recorded the improved fresh biomass, as shoot fresh weight followed by *Bacillus* sp. BEB1 (47.96, 48.0%), BEB2 (56.12, 29.0%), BEB3 (35.20, 52.5%), BEB4 (28.57, 39.5%) and root fresh weight *Bacillus* sp. BEB1 (36.73, 14.66%), BEB2 (21.64, 8.62%), BEB3 (17.14, 23.28%), BEB4 (13.88, 9.48%) when compared with the uninoculated drought control (Table 3, Figure 3C). Furthermore, plant shoot dry biomass accumulation increased significantly in stressed plants treated with *Bacillus* sp. BEB1 (103.44, 103.5%), BEB2 (71.73, 41.9%), BEB3 (65.25, 86.39%) and BEB4 (57.14, 70.5%) when compared with the uninoculated drought control. Furthermore, the inoculation with *Bacillus* sp. BEB1 (88.01, 57.65%), BEB2 (33.80, 19.48%), BEB3 (43.17, 50.67%) and BEB4 (39.18, 33.81%) significantly increased plant root dry weight in both maize cultivars when compared to the uninoculated drought control (Table 3, Figure 3F).

Table 3. Mean squares from three-way analysis of variance of data for different physiological, biochemical and antioxidant parameters of maize (*Zea mays* L.) plants inoculated with *Bacillus* spp. strains under well-watered (control, 15–18% water content in soil) and water deficit (stress, 5–8% water content in soil) conditions.

Source of Variation	**df**	**Shoot Length**	**Root Length**	**Shoot FW**	**Root FW**	**Shoot DW**
Cultivars (C)	1	40.83 *	0.020 ns	1.666 ns	0.158 ***	6.666 ns
Drought (D)	1	2.281 ns	33.00 **	0.078 *	0.152 ***	6.666 ns
Bacillus spp. (B)	4	85.87 ***	26.16 ***	0.115 ***	0.073 ***	0.005 ns
C × D		10.00 ns	9.680 ns	0.005 ns		6×10^{-5} ***
C × B	1	9.781 ns	6.978 ns	0.016 ns	0.008 ns	1.733 ns
D × B	4	13.24 ns	2.365 ns	0.015 ns	0.003 ns	1.333 ns
C × D × B	4	6.141 ns	2.681 ns	0.004 ns	0.011 ns	6×10^{-5} ns
Source of Variation	df	Root DW	Chl a	Chl b	Total Chl	Carotenoids
Cultivars (C)	1	0.001 ***	3.174 ns	1.815 ns	4.873 ns	0.080 ns
Drought (D)	1	0.001 ***	0.077 ***	0.014 ***	0.024 ***	10.41 ***
Bacillus spp. (B)	4	0.005 ***	0.027 ***	6.518 ns	0.020 ***	1.437 ***
C × D		2.2 ns	0.001 ns	0.002 ns	3.601 ns	0.864 ns
C × B	1	1.3 ns	2.144 ns	2.184 ns	4.35 ns	0.111 ns
D × B	4	2.5 ns	0.002 ***	2.123 ns	0.003 **	0.201 ns
C × D × B	4	1.6 ns	1.298 ns	1.596 ns	1.768 ns	0.058 ns
Source of Variation	df	Flavonoids	TSS	TSP	AsA	MDA
Cultivars (C)	1	403.0 ***	85,126 ***	6242 ***	93.50 ***	0.140 ns
Drought (D)	1	67.20 ***	80,227 ***	12,973 ***	3749 ***	13.72 ***
Bacillus spp. (B)	4	61.23 ***	33,266 ***	17,168 ***	361.7 ***	9.156 ***
C × D		4.873 **	4968 ***	11.26 ns	117.3 ***	0.121 ns
C × B	1	3.821 ***	19,325 ***	780.7 ***	160.2 ***	0.465 ns
D × B	4	10.23 ***	3273 ***	2614 ***	59.83 ***	0.369 ns
C × D × B	4	30.26 ***	2013 ***	2550 ***	84.21 ***	0.120 ns
Source of Variation	df	H_2O_2	CAT	POD	APX	SOD
Cultivars (C)	1	3.504 ***	0.596 ***	1.134 **	3.243 ***	1.666 ns
Drought (D)	1	82.83 ***	25.53 ***	522.3 ***	62.07 ***	219.6 ***
Bacillus spp. (B)	4	11.78 ***	11.01 ***	36.43 ***	14.33 ***	99.05 ***
C × D		0.060 ns	0.233 **	15.39 ***	0.019 ns	5.985 ns
C × B	1	0.072 **	0.406 ***	4.037 ***	0.298 ***	5.614 ns
D × B	4	3.186 ***	0.414 ***	0.768 **	0.742 ***	7.023 ns
C × D × B	4	0.268 ***	0.186 ***	1.530 ***	0.833 ***	5.334 ns

ns = non-significant; *, ** and *** = significant at 0.05, 0.01 and 0.001 levels, respectively.

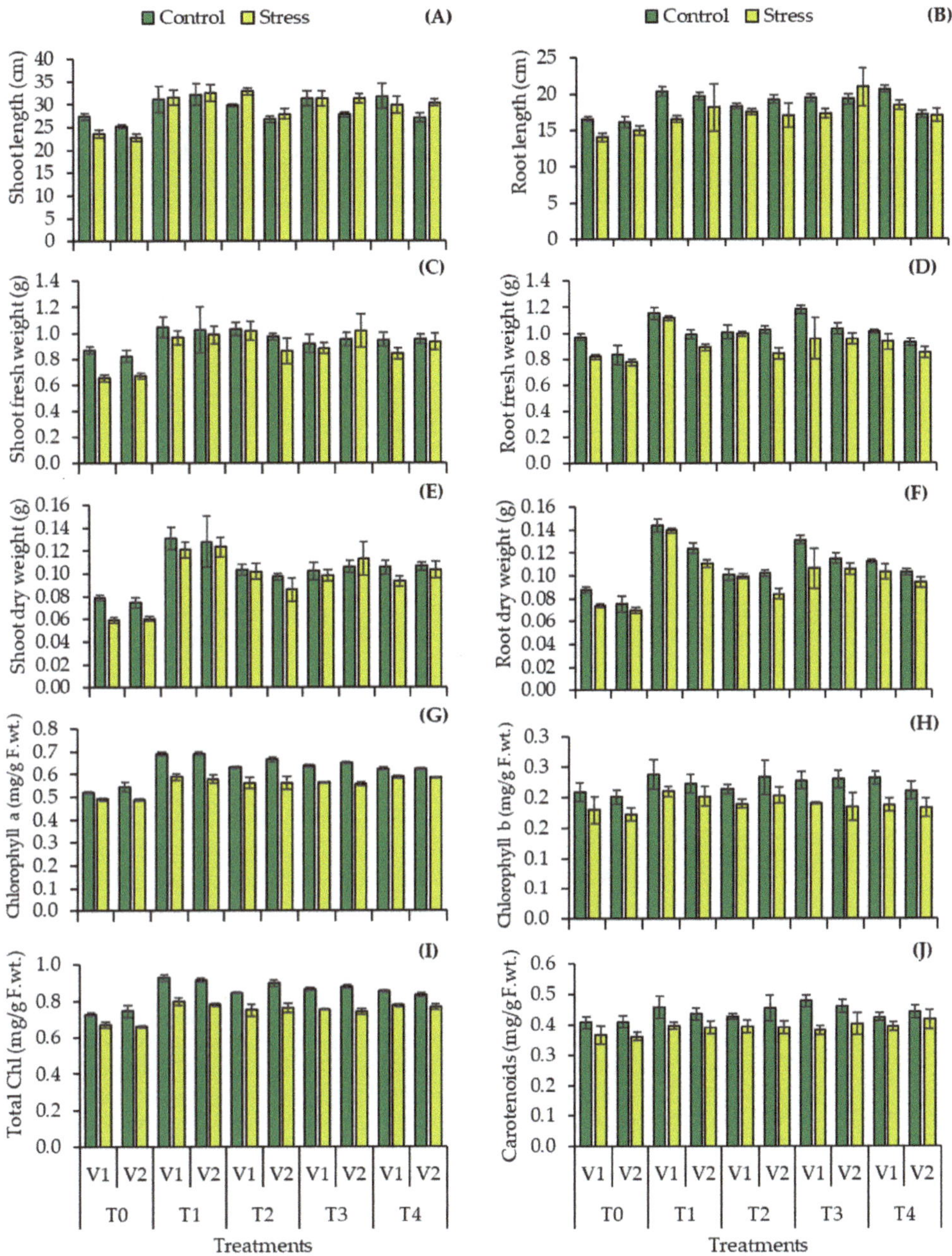

Figure 3. (**A**) shoot length, (**B**) root length, (**C**) shoot fresh weight, (**D**) root fresh weight, (**E**) shoot dry weight, (**F**) root dry weight, (**G**) Chlorophyll a, (**H**) chlorophyll b, (**I**) total chlorophyll, and (**J**) carotenoid contents of two cultivars of maize (*Zea mays* L.) inoculated with *Bacillus* spp. strains subjected to water stress condition (Mean ± S.E). Here T0 = uninoculated control, T1 = *Bacillus cereus* strain BEB1, T2 = *Bacillus cereus* strain BEB2, T3 = *Bacillus tropicus* strain BEB3, T4 = *Bacillus thuringiensis* strain BEB4. LSD value ≤ 0.05.

3.5. Estimation of Plant Photosynthetic Pigments

Photosynthetic pigment synthesis decreased in uninoculated drought-stressed followed by significant reduction in chlorophyll a, chlorophyll b, total chlorophyll and carotenoid contents in both maize cultivars when compared to the uninoculated control (Table 3, Figure 3). PGPR strains significantly ($p \leq 0.001$) improved chlorophyll (a, b & total) and carotenoid contents under both water regimes. Both maize cultivars had higher chlorophyll a, b, total chlorophyll and carotenoid contents as compared to the uninoculated plants under water stress conditions. The maximum improvement in chlorophyll a with *Bacillus* sp. BEB1 in V1 (32.76%), chlorophyll b with *Bacillus* sp. BEB2 in V2 (15.8%) and total chlorophyll contents was observed in plants inoculated with *Bacillus* sp. BEB1 (27.36, 22.82%) when compared to the uninoculated control (Table 3, Figure 3G–I). In contrast, plants inoculated with *Bacillus* sp. BEB3 (17.14, 12.12%) showed improved carotenoid contents when compared to the uninoculated control (Table 3, Figure 3J).

3.6. Biochemical Attributes Influenced by Bacterial Inoculation

A significant reduction was observed in flavonoid contents in both maize cultivars when inoculated with *Bacillus* spp. BEB1 (48.79, 22.25%), BEB2 (38.33, 29.62%), BEB3 (35.15, 31.57%) and BEB4 (39.55, 22.67%) under water deficit conditions when compared to the uninoculated drought control which was accumulated (46.67, 28.85%) in uninoculated stress plants (Table 3, Figure 4A). Plant biochemical parameters decreased significantly under water stress (5–8% water content in the soil); reductions in plant total soluble sugars (51.91, 31.07%), total soluble protein (32.42, 46.24%) and ascorbic acid contents (22.91, 34.94%) were observed in uninoculated plants (Table 3, Figure 4B–D). The inoculation with *Bacillus* spp. BEB1 (150.79, 123.88%), BEB2 (86.67, 31.06%), BEB3 (59.68, 39.40%) and BEB4 (253.02, 17.31%) increased total soluble sugars in stressed plants of both maize cultivars when compared to the uninoculated stressed control (Table 3, Figure 4B). An increase in total soluble protein contents was observed under both well water and water deficit condition with the inoculation of *Bacillus* spp. BEB1 (39.60, 89.64%), BEB2 (10.40, 41.45%), BEB3 (34.68, 42.55%) and BEB4 (52.17, 70.55%) when compared to the uninoculated drought control (Table 3, Figure 4C). Plant ascorbic acid contents increased significantly by inoculation with *Bacillus* sp. BEB1 (26.61, 43.29%), BEB2 (35.27, 26.57%), BEB3 (18.22, 51.29%) and BEB4 (23.26, 39.57%) in both cultivars (V1, V2) when compared with the uninoculated drought control (Table 3, Figure 4D). A significant difference was observed among all treatments, and among both cultivars in biochemical attributes estimated under a well-watered regime and under water deficit stress.

Malonaldehyde and hydrogen peroxide contents accumulation increased significantly in uninoculated drought-stressed followed by MDA (23.83, 19.89%) and H_2O_2 (94.76, 100%) in both maize cultivars when compared to the uninoculated control (Table 3, Figure 4). Due to inoculation of isolated PGPR bacterial species, MDA contents and H_2O_2 were lowered under both well-watered and water deficit conditions. Plants treated with *Bacillus* sp. BEB1 (50.78, 59.14%), BEB2 (50.26, 45.16%), BEB3 (48.19, 40.32%) and BEB4 (39.38, 30.65%) lowered the malonaldehyde contents when compared to the uninoculated control (Table 3, Figure 4E). In contrast, plants inoculated with *Bacillus* sp. BEB1 (37.04, 44.71%), BEB2 (32.44, 31.94%), BEB3 (45.75, 41.07%) and BEB4 (39.24, 36.18%) showed decreased hydrogen peroxide generation when compared to the uninoculated drought control (Table 3, Figure 4F). A significant difference was observed among maize cultivars, and bacterial treatments were found to be varying significantly in both cultivars specifically under water deficit conditions.

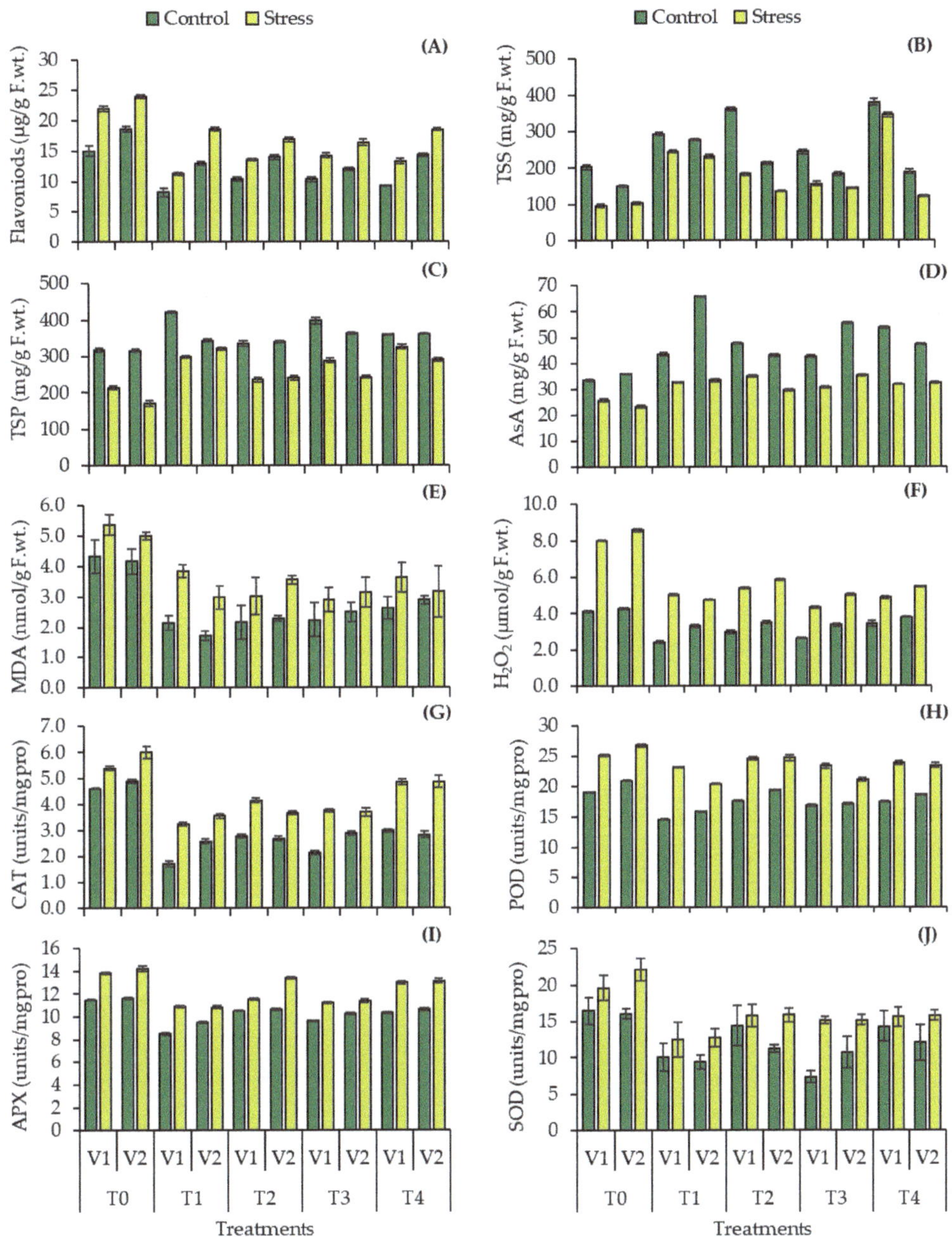

Figure 4. (**A**) Flavonoids, (**B**) total soluble sugars, (**C**) total soluble proteins and (**D**) ascorbic acid (**E**) malonaldehyde, (**F**) hydrogen peroxide contents (**G**) catalase, (**H**) peroxidase, (**I**) ascorbate peroxidase, and (**J**) superoxide dismutase enzyme activity of two cultivars of maize (*Zea mays* L.) inoculated with *Bacillus* spp. strains subjected to water stress condition (Mean ± S.E). Here, T0 = uninoculated control, T1 = *Bacillus cereus* strain BEB1, T2 = *Bacillus cereus* strain BEB2, T3 = *Bacillus tropicus* strain BEB3, T4 = *Bacillus thuringiensis* strain BEB4. LSD ≤ 0.05.

3.7. Antioxidant Enzyme Activities of Plant Leaf Extract

The antioxidant response to drought stress was stimulated in uninoculated plants, increasing CAT (16.85, 23.01%), POD (32.21, 27.86%), APX (20.36, 22.45%) and SOD (19.16, 37.42%) activities (Table 3, Figure 4). The trend in increase in enzyme activities in both cultivars was significantly different. Moreover, *Bacillus* spp. inoculation showed the mitigation of the plant enzymatic antioxidant responses under drought stress. The reductions in CAT activity were recorded in plants inoculated with *Bacillus* sp. BEB1 (62.96, 47.34%), BEB2 (39.67, 44.96%), BEB3 (53.51, 40.72%) and BEB4 (35.13, 42.02%) (Table 3, Figure 4G) when compared to the uninoculated control. The higher reductions in POD and APX activity were recorded in plants inoculated with *Bacillus* sp. BEB1, which was (23.46, 24.44%) and (25.95, 18.10%), respectively, as compared to the uninoculated control (Table 3, Figure 4H,I). In addition, a significant reduction in SOD activity was observed in plants inoculated with *Bacillus* sp. BEB1 (38.92, 41.72%), BEB2 (12.57, 30.06%), BEB3 (55.69, 33.74%) and BEB4 (13.17, 25.15%) when compared to the drought control (Table 3, Figure 4J).

3.8. Principal Component Analysis and Pearson Coefficient Correlation

The values of Pearson correlation of all the studied parameters with their significance at probability levels (p) < 0.05, 0.01, 0.001 are presented (Table 4). It clearly shows that the plant biomass has a positive significant correlation with studied physiological attributes like shoot and root length, chlorophyll a, total chlorophyll as well as biochemical attributes such as total soluble sugars, total soluble proteins and ascorbic acid contents. However, shoot biomass was negatively correlated with the MDA, H_2O_2 and flavonoid contents. Moreover, the shoot dry weight also has a negative significant correlation with antioxidant enzyme activities CAT, POD, APX and SOD. Correlation studies presented in (Figure 5) generated through PCA show that the studied attributes are categorized into two major groups. The first component of the PCA explained 58.3%, while the second component explained 10.4% of the variance.

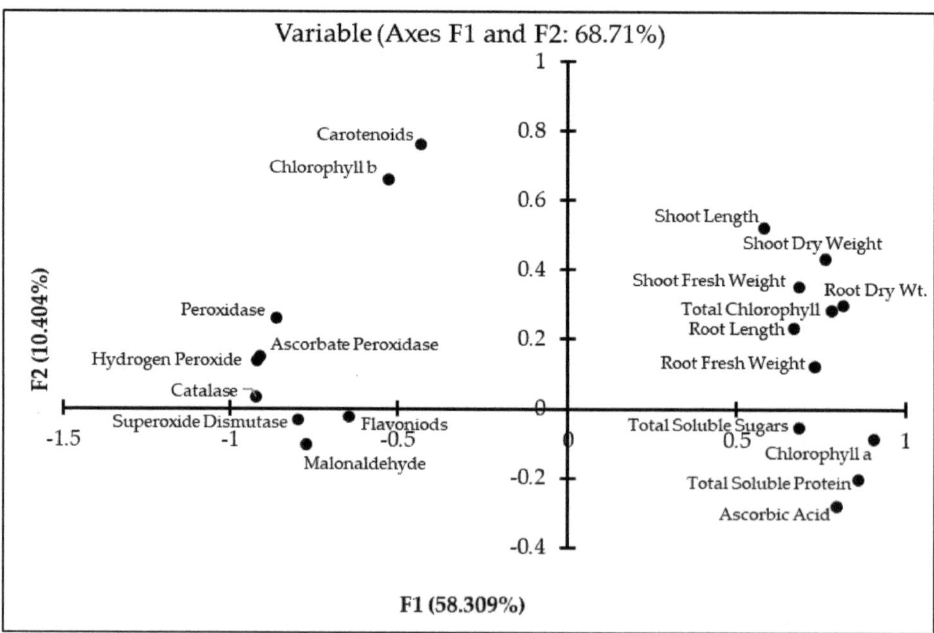

Figure 5. Principal component analysis (PCA) of studied attributes of maize inoculated with *Bacillus* spp. strains under drought stress.

Table 4. Pearson coefficient correlation values of estimated attributes of maize (*Zea mays* L.) inoculated with *Bacillus* spp. strains showing significance differences under drought stress.

	RFW	SFW	RDW	SDW	RL	SL	Chla	Chlb	TChl	Caro	Flav	TSS	TSP	H_2O_2	MDA	AsA	CAT	APX	POD	SOD
RFW	1.000																			
SFW	0.504 ***	1.000																		
RDW	0.863 ***	0.571 ***	1.000																	
SDW	0.531 ***	0.900 ***	0.765 ***	1.000																
RL	0.449 ***	0.602 ***	0.519 ***	0.607 ***	1.000															
SL	0.480 ***	0.688 ***	0.632 ***	0.762 ***	0.576 ***	1.000														
Chla	0.561 ***	0.542 ***	0.683 **	0.634 ***	0.594 ***	0.401 ***	1.000													
Chlb	−0.327 ***	−0.313 ***	−0.291 ***	−0.281 ***	−0.305 ***	−0.158 ns	−0.556 ***	1.000												
TChl	0.482 ***	0.466 ***	0.649 ***	0.596 ***	0.534 ***	0.390 ***	0.874 ***	−0.082 ns	1.000											
Caro	−0.312 **	−0.216 *	−0.195 ns	−0.143 ns	−0.178 **	−0.063 ns	−0.402 ***	0.936 ***	0.065 ns	1.000										
Flav	−0.588 **	−0.319 ***	−0.588 ***	−0.407 ***	−0.350 **	−0.417 ***	−0.519 ***	0.274 **	−0.463 **	0.283 *	1.000									
TSS	0.466 **	0.369 ***	0.492 ***	0.430 ***	0.439 ***	0.412 ***	0.587 ***	−0.304 ***	0.526 ***	−0.310 **	−0.648 ***	1.000								
TSP	0.577 ***	0.423 ***	0.616 ***	0.509 ***	0.457 ***	0.300 **	0.783 ***	−0.449 ***	0.676 ***	−0.400 ***	−0.543 ***	0.651 ***	1.000							
H_2O_2	−0.623 ***	−0.548 ***	−0.618 ***	−0.564 ***	−0.556 ***	−0.441 ***	−0.799 ***	0.485 ***	−0.675 ***	0.428 ***	0.537 ***	−0.651 ***	−0.911 ***	1.000						
MDA	−0.480 ***	−0.463 ***	−0.573 ***	−0.538 ***	−0.463 ***	−0.381 ***	−0.754 ***	0.264 *	−0.750 ***	0.126 ns	0.402 ***	−0.474 ***	−0.622 ***	0.712 ***	1.000					
AsA	0.411 ***	0.449 ***	0.474 ***	0.487 ***	0.496 ***	0.290 **	0.791 ***	−0.527 ***	0.639 ***	−0.489 ***	−0.401 ***	0.558 ***	0.682 ***	−0.738 ***	−0.647 ***	1.000				
CAT	−0.671 ***	−0.567 ***	−0.750 ***	−0.657 ***	−0.535 ***	−0.443 ***	−0.845 ***	0.467 ***	−0.740 ***	0.372 **	0.632 ***	−0.547 ***	−0.773 ***	0.838 ***	0.745 ***	−0.713 ***	1.000			
APX	−0.670 ***	−0.544 ***	−0.736 ***	−0.635 ***	−0.535 ***	−0.434 ***	−0.784 ***	0.500 ***	−0.648 ***	0.467 ***	0.560 ***	−0.570 ***	−0.827 ***	0.891 ***	0.694 ***	−0.748 ***	0.890 ***	1.000		
POD	−0.530 ***	−0.463 ***	−0.554 ***	−0.505 ***	−0.524 ***	−0.272 **	−0.763 ***	0.466 ***	−0.642 ***	0.445 ***	0.453 ***	−0.585 ***	−0.875 ***	0.883 ***	0.633 ***	−0.831 ***	0.814 ***	0.896 ***	1.000	
SOD	−0.559 ***	−0.511 ***	−0.658 ***	−0.599 ***	−0.459 ***	−0.424 ***	−0.729 ***	0.383 ***	−0.651 ***	0.279 **	0.424 ***	−0.396 ***	−0.713 ***	0.703 ***	0.638 ***	−0.589 ***	0.740 ***	0.716 ***	0.657 ***	1.000

ns = non-significant; *, ** and *** = significant at 0.05, 0.01 and 0.001 levels, respectively. RFW = root fresh weight; SFW = shoot fresh weight; RDW = root dry weight; SDW = shoot dry weight; RL = root length; SL = shoot length; Chla = chlorophyll a; Chlb = Chlorophyll b; TChl = total chlorophyll; Caro = Carotenoids; Flav = flavonoids; TSS = total soluble sugars; TSP = total soluble proteins; H_2O_2 = hydrogen peroxide; MDA = malonaldehyde; AsA = ascorbic acid; CAT = catalase; APX = ascorbate peroxidase; POD = peroxidase; SOD = superoxide dismutase.

4. Discussion

In the current study, twelve different *Bacillus* spp. strains were isolated from textile effluent of Kamal textile, Faisalabad, Punjab, Pakistan. The selection criteria of Bacillus species out of 12 isolates were superior in drought tolerance and better plant PGP characteristics. Comparative analysis based on the bacterial 16S rRNA gene sequence has arisen as a preferable molecular technique for the identification [43]. The gene sequences of the smallest subunit (16S) were commonly used to study the phylogenetic relationships, but, later, it has been extensively used as a molecular marker for the identification of an unknown bacterium to the genus and species level [44]. In the present study, 16S rRNA nucleotides sequences of bacterial isolate strain (accession number) BEB1 (MW350048), BEB2 (MW350049) showed 96.58 and 97.99 percent identity with *Bacillus cereus*, respectively. The 16S rRNA sequence of isolated strain BEB3 (MW350050) showed a 98.61 percent identity with *Bacillus tropicus* keeping 100% query coverage. Isolated strain BEB4 (MW350051) showed a 98.20 percent identity with *Bacillus thuringiensis*. Kadam [45] also revealed the presence of indigenous bacillus microflora in textile wastewaters. The previous studies on the tannery wastewater confirmed the presence and isolation of *B. tropicus* and *B. thuringiensis* species [46,47]. Bacteria isolated from different textile effluents were closely related to genera bacillus recently also identified from the hospital, domestic and industrial wastewater samples [48,49]. Studies by [45] also revealed the presence of bacillus microflora in textile wastewaters. These observations corroborate with a previous study which reported the isolation of *Bacillus*, *Pseudomonas*, *Serattia* and other bacterial species from textile dye effluent [50].

Selected PGPR strains were also subjected to characterize their potential for in-vitro plant growth promoting traits including nitrogen fixation, ammonia production, IAA synthesis and phosphate solubilization (Table 1). The appearance of a blue halo zone diameter on nitrogen free malate medium was observed as a qualitative indication of atmospheric nitrogen fixation and results correlate with previous studies [51,52]. In a previous study, the bacillus strain FSS2C was reported as an ammonia producing bacteria [27]. In another study, the plant growth promoting bacillus species were described as positive for the production of ammonia [53]. The production of IAA is highly promoted by tryptophan as a substrate for many bacterial species; such bacterial strains follow the biochemical pathway of IAA synthesis dependent upon tryptophan as a source substrate [3]. Our results are analogous to the previously isolated bacterial strains from the soil and wastewater that were only able to produce IAA when supplemented with substrate [31]. The decrease in IAA production under stress might be due to the low bacterial growth by osmotic pressure caused by the PEG [54]. Many of the bacterial strains previously isolated from the water and wastewater are also plant beneficial and able to solubilize inorganic phosphate for the plants [2,3]. Results are also in correlation with previous findings on the phosphate solubilization by bacterial species [31]. Strain BEB1 grows well and shows a maximum tolerance to the drought stress even at the lowest water potential. The growth of tested strains (OD 600 nm) on varying water potential by PEG concentration is also shown (Figure 2). Our findings were demonstrated with previously isolated bacteria from semi-arid conditions [32]. The growth of bacterial strains decreases as the concentration of the PEG increases in the LB broth medium, and the concentration of PEG tolerated by a bacterial strain is marked as the tolerance level of that isolate [54].

Overall plant growth and yield reduction in numerous cereal crops have been reported due to the water deficient condition [7]. Drought stress was lessened in maize plants that were inoculated with *Bacillus* spp. strains in this study. There were increases in nutrient uptake and plant height, as well as root and leaf biomass for the inoculated plants under drought stress. All these effects have been related to PGPB's ability to reduce drought stress in maize [55]. Many bacillus bacterial strains have been reported as efficient PGPR in recent years [56]. In the present investigation, the water deficit (5–8% water in soil) significantly ($p \leq 0.001$) lowered the shoot and root fresh and dry weight of maize cultivars (V1: FH-1046; V2: YH-5427). The reduction in plant growth and biomass might be due to increased levels of lipid peroxidation, H_2O_2 and disruption in nutrient contents of non-inoculated maize seedlings. Drought stress is well known for dramatically reducing photosynthetic pigments, associated with decreased plant growth and yield [57]. Furthermore, IAA production by PGPB increases plant growth, root elongation and development to substantially improve nutrient and water acquisition [58]. In the present study, increased plant biomasses were envisioned owing to contribution of nitrogen fixation, ammonia production IAA synthesis and phosphate-solubilization capability of bacillus species as had been earlier reported in drought stressed maize grown with application of dry-Caribbean *Bacillus* spp. strains [59].

Drought stress also reduces the synthesis of green pigments (chlorophyll contents), resulting in a slower rate of photosynthesis [60]. Chlorophyll a, b and total chlorophyll content decreased significantly in both maize cultivars examined in this study. Higher photosynthetic pigments of drought stressed maize plants under bacillus species inoculation probably results from the activation of enzymatic pathways for chlorophyll biosynthesis, limits in ROS production or increased solubilization and bioavailability of organic minerals i.e., Mg [3]. Increased chlorophyll contents are likely to boost the plant's photosynthetic activity as obvious from higher fresh and dry weights of inoculated maize plants, particularly the bacterial inoculation. Our results coincide with the findings that Bacillus strain application boosted the growth and chlorophyll content of maize due to an increase in chlorophyll biosynthesis and nutrient balance [59]. An interesting observation was that PGPB influenced the biomass and chlorophyll contents not only when maize plants were exposed to drought, which was consistent with our hypothesis.

In the absence of bacillus species strains, drought stress significantly increased in plants H_2O_2 levels with concomitant rise in lipid peroxidation and relative membrane permeability in both maize cultivars (Figure 4). Under water-stress conditions, high lipid peroxidation occurs because of elevated levels of ROS damaging plant ultra-structures [61]. In the current study, maize plants subjected to water deficit conditions accumulated a high amount of H_2O_2 and MDA. However, exogenous PGPR strains are said to significantly reduce H_2O_2 and MDA levels [3]. Under abiotic stress, such as drought, H_2O_2 is believed to be produced due to high oxidative stress [62]. Both water-stressed and non-stressed maize cultivars benefited from bio seed primed bacterial treatments, which significantly increased the ascorbic acid content. An increase in ascorbic acid level has a significant impact on plant fresh and dry biomass of both maize cultivars grown under water-limited conditions, as can be inferred from the data. An increase in ascorbic acid was observed in different crops, such as wheat [63], tomato [64], and canola [65], which are becoming more resistant to water stress.

Antioxidant enzymes are indicators of plant defense against stress, which scavenge ROS under drought stressed conditions. Alterations in plant protein contents upon drought exposure might be owing to protein degradation or enhanced antioxidative activities which disrupt the plant growth [31]. Antioxidant enzymes such as catalase (CAT), peroxidase (POD), ascorbate peroxidase (APX), and superoxide dismutase (SOD) can protect plants from oxidative stress [66]. Under water stress, the activities of CAT, POD, APX, and SOD enzymes increased in the current study (Figure 4). Under both water regimes, various applied bacterial PGPR strains resulted in deceased activity of CAT, POD, APX, and SOD in both cultivars. Inoculation with bacillus specie strains significantly decrease in ROS generation and lipid peroxidation contents as well as the activities of CAT, POD, APX and SOD in both of the cultivars. These findings corroborate previous reports of increased antioxidant enzyme activity under stress in maize treated with bacteria [31]. The fact that both the cultivars had different responses confirmed that the activities of enzymatic antioxidants could be due to genetic differences between them [2,7]. The correlation between all physiological and biochemical parameters studied and their placement in a component chart by factor analysis resembled previous studies.

The Bacillus genera have proven to be an effective ally in the management of drought and other abiotic stresses in a variety of crops [4,56]. Previous research [59] demonstrated the ability of co-inoculation to alleviate drought stress under non-sterile soil conditions.

Nonetheless, our findings in maize indicate that these four *Bacillus* spp. strains individually induce a similar drought stress amelioration effect in plant growth under sterile soil conditions, which is dependent on a specific plant–strain interaction evident in the differences in the evaluated antioxidant responses.

5. Conclusions

It was concluded that isolated bacterial strains BEB1–4 are capable of tolerating drought stress levels though efficient nitrogen fixation, phosphate solubilization, ammonia and IAA production. Our results confirmed that drought stress inhibited plant biomass and nutrient contents in maize cultivars in a cultivar specific manner. Application of bacterial species confers more advantageous growth as reflected from higher root (36.73%) and shoot (56.12%) biomasses, photosynthetic pigments (32.76%), biochemical attributes while maintaining lower levels of lipid peroxidation (59.14%), and antioxidants. It was concluded that these bacillus species may be used as bio-inoculant and bioremediation tools synergistically with crop plants in water deficit environments for sustainable food production. To determine the use of these PGPR candidates as microbial agents under drought stress, corresponding assays may be necessary to evaluate the performance of the four selected isolates in field conditions.

Author Contributions: Conceptualization, M.Z.H. and S.J.; methodology, M.Z.H. and S.J.; software, M.Z.H. and M.A.; validation, M.Z.H., S.J. and M.A.; formal analysis, M.A.; investigation, M.A.; resources, S.J., M.Z.H., M.H.S. and A.A.; data curation, M.A.; writing—original draft preparation, M.A.; writing—review and editing, S.J. and M.A.; visualization, M.A.; supervision, S.J.; project administration, S.J. and M.Z.H.; funding acquisition, M.H.S. and A.A. All authors have read and agreed to the published version of the manuscript.

Funding: A part of this research was funded by National Research Program for Universities (NRPU), Higher Education Commission, Islamabad (grant number 5567/Punjab/NRPU/R&D/HEC/2016).

Institutional Review Board Statement: Not applicable.

Informed Consent Statement: Not applicable.

Data Availability Statement: The dataset generated during and/or analyzed during the current study are available from the corresponding author on reasonable request.

Acknowledgments: The authors are thankful to the Ayub Agricultural Research Institute, Faisalabad, Pakistan for providing the seeds of maize varieties. The authors would like to express their deepest gratitude to the University of Tabuk for the technical support for this study.

Conflicts of Interest: The authors declare no conflict of interest.

References

1. Naveed, M.; Mitter, B.; Reichenauer, T.G.; Wieczorek, K.; Sessitsch, A. Increased drought stress resilience of maize through endophytic colonization by Burkholderia phytofirmans PsJN and *Enterobacter* sp. FD17. *Environ. Exp. Bot.* **2014**, *97*, 30–39. [CrossRef]
2. Ullah, A.; Akbar, A.; Luo, Q.; Khan, A.H.; Manghwar, H.; Shaban, M.; Yang, X. Microbiome diversity in cotton rhizosphere under normal and drought conditions. *Microb. Ecol.* **2019**, *77*, 429–439. [CrossRef] [PubMed]
3. Khan, Z.; Rho, H.; Firrincieli, A.; Hung, S.H.; Luna, V.; Masciarelli, O.; Kim, S.-H.; Doty, S.L. Growth enhancement and drought tolerance of hybrid poplar upon inoculation with endophyte consortia. *Curr. Plant Biol.* **2016**, *6*, 38–47. [CrossRef]
4. Ali, L.; Khalid, M.; Asghar, H.N.; Asgher, M. Scrutinizing of rhizobacterial isolates for improving drought resilience in maize (*Zea mays*). *Int. J. Agric. Biol.* **2017**, *19*, 1054–1064. [CrossRef]
5. Daffonchio, D.; Hirt, H.; Berg, G. Plant-microbe interactions and water management in arid and saline soils. In *Principles of Plant-Microbe Interactions: Microbes for Sustainable Agriculture*; Springer: Cham, Switzerland, 2015; pp. 265–276. [CrossRef]
6. Gepstein, S.; Glick, B.R. Strategies to ameliorate abiotic stress-induced plant senescence. *Plant Mol. Biol.* **2013**, *82*, 623–633. [CrossRef] [PubMed]
7. Moreno-Galván, A.; Romero-Perdomo, F.A.; Estrada-Bonilla, G.; Meneses, C.H.S.G.; Bonilla, R.R.J.M. Dry-caribbean *Bacillus* spp. strains ameliorate drought stress in maize by a strain-specific antioxidant response modulation. *Microorganism* **2020**, *8*, 823. [CrossRef]

8. Allah, E.F.; Alqarawi, A.A. Alleviation of adverse impact of salt in *Phaseolus vulgaris* L. by arbuscular mycorrhizal fungi. *Pak. J. Bot.* **2015**, *47*, 1167–1176.
9. Agami, R.A.; Medani, R.A.; Abd El-Mola, I.A.; Taha, R.S. Exogenous application with plant growth promoting rhizobacteria (PGPR) or proline induces stress tolerance in basil plants (*Ocimum basilicum* L.) exposed to water stress. *Int. J. Environ. Agric. Res.* **2016**, *2*, 78.
10. Ullah, A.; Sun, H.; Yang, X.; Zhang, X. Drought coping strategies in cotton: Increased crop per drop. *Plant Biotechnol. J.* **2017**, *15*, 271–284. [CrossRef]
11. Zulfiqar, U.; Ishfaq, M.; Yasin, M.U.; Ali, N.; Ahmad, M.; Ullah, A.; Hameed, W. Performance of maize yield and quality under different irrigation regimes and nitrogen levels. *J. Glob. Innov. Agric. Sci.* **2017**, *5*, 159–164.
12. Shams, S.; Sahu, J.; Rahman, S.S.; Ahsan, A.J. Sustainable waste management policy in Bangladesh for reduction of greenhouse gases. *Sustain. Cities Soc.* **2017**, *33*, 18–26. [CrossRef]
13. Glick, B.R. Bacteria with ACC deaminase can promote plant growth and help to feed the world. *Microbiol. Res.* **2014**, *169*, 30–39. [CrossRef] [PubMed]
14. Sandhya, V.; Shrivastava, M.; Ali, S.Z.; Prasad, V.S.S.K.J. Endophytes from maize with plant growth promotion and biocontrol activity under drought stress. *Russ. Agric. Sci.* **2017**, *43*, 22–34. [CrossRef]
15. Shahzad, R.; Khan, A.L.; Bilal, S.; Waqas, M.; Kang, S.-M.; Lee, I.-J. Inoculation of abscisic acid-producing endophytic bacteria enhances salinity stress tolerance in *Oryza sativa*. *Environ. Exp. Bot.* **2017**, *136*, 68–77. [CrossRef]
16. Rashid, M.I.; Mujawar, L.H.; Shahzad, T.; Almeelbi, T.; Ismail, I.M.I.; Oves, M. Bacteria and fungi can contribute to nutrients bioavailability and aggregate formation in degraded soils. *Microbiol. Res.* **2016**, *183*, 26–41. [CrossRef] [PubMed]
17. Adesemoye, A.O.; Yuen, G.; Watts, D.B. Microbial inoculants for optimized plant nutrient use in integrated pest and input management systems. In *Probiotics and Plant Health*; Springer: Singapore, 2017; pp. 21–40.
18. Prescott, H. *Laboratory Exercises in Microbiology*, 5th ed.; John Wiley & Sons: Hoboken, NJ, USA, 2002.
19. Barrow, G.I.; Feltham, R.K.A. *Manual for the Identification of Medical Bacteria, Cowan and Steels*; Cambridge University Press: Cambridge, UK, 1993; Volume 1, p. 331.
20. Seibold, G.; Auchter, M.; Berens, S.; Kalinowski, J.; Eikmanns, B.J. Utilization of soluble starch by a recombinant Corynebacterium glutamicum strain: Growth and lysine production. *J. Biotechnol.* **2006**, *124*, 381–391. [CrossRef]
21. Vijayaraghavan, P.; Vincent, S.G.P. A simple method for the detection of protease activity on agar plates using bromocresolgreen dye. *J. Biochem. Technol.* **2013**, *4*, 628–630.
22. Paulo, E.M.; Vasconcelos, M.P.; Oliveira, I.S.; Affe, H.M.d.J.; Nascimento, R.; Melo, I.S.d.; Roque, M.R.d.A.; Assis, S.A.d. An alternative method for screening lactic acid bacteria for the production of exopolysaccharides with rapid confirmation. *Food Sci. Technol.* **2012**, *32*, 710–714. [CrossRef]
23. Weisburg, W.G.; Barns, S.M.; Pelletier, D.A.; Lane, D.J. 16S ribosomal DNA amplification for phylogenetic study. *J. Bacteriol.* **1991**, *173*, 697–703. [CrossRef]
24. Saitou, N.; Nei, M. The neighbor-joining method: A new method for reconstructing phylogenetic trees. *Mol. Biol. Evol.* **1987**, *4*, 406–425. [CrossRef]
25. Kumar, S.; Stecher, G.; Li, M.; Knyaz, C.; Tamura, K. MEGA X: Molecular evolutionary genetics analysis across computing platforms. *Mol. Biol. Evol.* **2018**, *35*, 1547–1549. [CrossRef] [PubMed]
26. Felsenstein, J. Confidence limits on phylogenies: An approach using the bootstrap. *Evolution* **1985**, *39*, 783–791. [CrossRef] [PubMed]
27. Mahmood, F.; Shahid, M.; Hussain, S.; Haider, M.Z.; Shahzad, T.; Ahmed, T.; Noman, M.; Rasheed, F.; Khan, M.B. Bacillus firmus strain FSS2C ameliorated oxidative stress in wheat plants induced by azo dye (reactive black-5). *3 Biotech* **2020**, *10*, 40. [CrossRef]
28. Penrose, D.M.; Glick, B.R. Methods for isolating and characterizing ACC deaminase-containing plant growth-promoting rhizobacteria. *Physiol. Plant* **2003**, *118*, 10–15. [CrossRef]
29. Bilal, R.; Rasul, G.; Qureshi, J.A.; Malik, K.A. Characterization of Azospirillum and related diazotrophs associated with roots of plants growing in saline soils. *World J. Microbiol. Biotechnol.* **1990**, *6*, 46–52. [CrossRef]
30. Nautiyal, C.S. An efficient microbiological growth medium for screening phosphate solubilizing microorganisms. *FEMS Microbiol. Lett.* **1999**, *170*, 265–270. [CrossRef]
31. Abbas, S.; Javed, M.T.; Shahid, M.; Hussain, I.; Haider, M.Z.; Chaudhary, H.J.; Tanwir, K.; Maqsood, A. Acinetobacter sp. SG-5 inoculation alleviates cadmium toxicity in differentially Cd tolerant maize cultivars as deciphered by improved physio-biochemical attributes, antioxidants and nutrient physiology. *Plant. Physiol. Biochem.* **2020**, *155*, 815–827. [CrossRef] [PubMed]
32. Rashid, U.; Yasmin, H.; Hassan, M.N.; Naz, R.; Nosheen, A.; Sajjad, M.; Ilyas, N.; Keyani, R.; Jabeen, Z.; Mumtaz, S. Drought-tolerant *Bacillus megaterium* isolated from semi-arid conditions induces systemic tolerance of wheat under drought conditions. *Plant Cell Rep.* **2021**. (Online ahead of print). [CrossRef]
33. Arnon, D.I. Copper enzymes in isolated chloroplasts. Polyphenoloxidase in Beta vulgaris. *Plant. Physiol.* **1949**, *24*, 1–15. [CrossRef]
34. Zhishen, J.; Mengcheng, T.; Jianming, W. The determination of flavonoid contents in mulberry and their scavenging effects on superoxide radicals. *Food Chem.* **1999**, *64*, 555–559. [CrossRef]
35. Irigoyen, J.; Einerich, D.; Sánchez-Díaz, M. Water stress induced changes in concentrations of proline and total soluble sugars in nodulated alfalfa (*Medicago sativd*) plants. *Physiol. Plant.* **1992**, *84*, 55–60. [CrossRef]

36. Bradford, M.M. A rapid and sensitive method for the quantitation of microgram quantities of protein utilizing the principle of protein-dye binding. *Anal. Biochem.* **1976**, *72*, 248–254. [CrossRef]
37. Mukherjee, S.; Choudhuri, M. Implications of water stress-induced changes in the levels of endogenous ascorbic acid and hydrogen peroxide in Vigna seedlings. *Physiol. Plant.* **1983**, *58*, 166–170. [CrossRef]
38. Velikova, V.; Yordanov, I.; Edreva, A. Oxidative stress and some antioxidant systems in acid rain-treated bean plants: Protective role of exogenous polyamines. *Plant Sci.* **2000**, *151*, 59–66. [CrossRef]
39. Cakmak, I.; Horst, W.J. Effect of aluminium on lipid peroxidation, superoxide dismutase, catalase, and peroxidase activities in root tips of soybean (*Glycine max*). *Physiol. Plant.* **1991**, *83*, 463–468. [CrossRef]
40. Chance, B.; Maehly, A. Assay of catalases and peroxidases. In *Methods in Enzymology*; Colowick, S.P., Kaplan, N.O., Eds.; Academic Press: Cambridge, MA, USA, 1955.
41. Zhang, J.; Kirkham, M.B. Drought-stress-induced changes in activities of superoxide dismutase, catalase, and peroxidase in wheat species. *Plant Cell Physiol.* **1994**, *35*, 785–791. [CrossRef]
42. Giannopolitis, C.N.; Ries, S.K. Superoxide dismutases: II. Purification and quantitative relationship with water-soluble protein in seedlings. *Plant. Physiol.* **1977**, *59*, 315–318. [CrossRef]
43. Fuks, G.; Elgart, M.; Amir, A.; Zeisel, A.; Turnbaugh, P.J.; Soen, Y.; Shental, N.J.M. Combining 16S rRNA gene variable regions enables high-resolution microbial community profiling. *Microbiome* **2018**, *6*, 17. [CrossRef]
44. Kai, S.; Matsuo, Y.; Nakagawa, S.; Kryukov, K.; Matsukawa, S.; Tanaka, H.; Iwai, T.; Imanishi, T.; Hirota, K.J. Rapid bacterial identification by direct PCR amplification of 16S rRNA genes using the MinION™ nanopore sequencer. *FEBS Open Bio* **2019**, *9*, 548–557. [CrossRef]
45. Kadam, A.A.; Kamatkar, J.D.; Khandare, R.V.; Jadhav, J.P.; Govindwar, S.P. Solid-state fermentation: Tool for bioremediation of adsorbed textile dyestuff on distillery industry waste-yeast biomass using isolated Bacillus cereus strain EBT1. *Environ. Sci. Pollut. Res.* **2013**, *20*, 1009–1020. [CrossRef]
46. Mohanty, S.; Bose, S.; Jain, K.G.; Bhargava, B.; Airan, B. TGFβ1 contributes to cardiomyogenic-like differentiation of human bone marrow mesenchymal stem cells. *Int. J. Cardiol.* **2013**, *163*, 93–99. [CrossRef] [PubMed]
47. Samanta, S.; Datta, D.; Halder, G. Biodegradation efficacy of soil inherent novel sp. Bacillus tropicus (MK318648) onto low density polyethylene matrix. *J. Polym. Res.* **2020**, *27*, 324. [CrossRef]
48. Chowdhary, P.; More, N.; Raj, A.; Bharagava, R.N. Characterization and identification of bacterial pathogens from treated tannery wastewater. *Microbiol. Res. Int.* **2017**, *5*, 30–36. [CrossRef]
49. Greay, T.L.; Gofton, A.W.; Zahedi, A.; Paparini, A.; Linge, K.L.; Joll, C.A.; Ryan, U.M. Evaluation of 16S next-generation sequencing of hypervariable region 4 in wastewater samples: An unsuitable approach for bacterial enteric pathogen identification. *Sci. Total Environ.* **2019**, *670*, 1111–1124. [CrossRef]
50. Jayaseelan, T.; Damodaran, R.; Ganesan, S.; Mani, P. Biochemical characterization and 16s rRNA sequencing of different bacteria from textile dye effluents. *J. Drug Deliv. Ther.* **2019**, *8*, 35–40. [CrossRef]
51. Li, X.; Geng, X.; Xie, R.; Fu, L.; Jiang, J.; Gao, L.; Sun, J. The endophytic bacteria isolated from elephant grass (*Pennisetum purpureum* Schumach) promote plant growth and enhance salt tolerance of Hybrid Pennisetum. *Biotechnol. Biofuels* **2016**, *9*, 190. [CrossRef]
52. Wen, P.; Han, Y.; Wu, Z.; He, Y.; Ye, B.-C.; Wang, J. Rapid synthesis of a corncob-based semi-interpenetrating polymer network slow-release nitrogen fertilizer by microwave irradiation to control water and nutrient losses. *Arab. J. Chem.* **2017**, *10*, 922–934. [CrossRef]
53. Karthika, S.; Midhun, S.J.; Jisha, M. A potential antifungal and growth-promoting bacterium *Bacillus* sp. KTMA4 from tomato rhizosphere. *Microb. Pathog.* **2020**, *142*, 104049. [CrossRef] [PubMed]
54. Kumar, A.S.; Sridar, R.; Uthandi, S. Mitigation of drought in rice by a phyllosphere bacterium *Bacillus altitudinis* FD48. *Afr. J. Microbiol. Res.* **2017**, *11*, 1614–1625.
55. Shirinbayan, S.; Khosravi, H.; Malakouti, M.J. Alleviation of drought stress in maize (*Zea mays*) by inoculation with Azotobacter strains isolated from semi-arid regions. *Appl. Soil Ecol.* **2019**, *133*, 138–145. [CrossRef]
56. AbdAllah, E.F.; Alqarawi, A.A.; Hashem, A.; Radhakrishnan, R.; Al-Huqail, A.A.; Al-Otibi, F.O.N.; Malik, J.A.; Alharbi, R.I.; Egamberdieva, D. Endophytic bacterium *Bacillus subtilis* (BERA 71) improves salt tolerance in chickpea plants by regulating the plant defense mechanisms. *J. Plant Interact.* **2018**, *13*, 37–44. [CrossRef]
57. Ashraf, M.; Harris, P.J. Photosynthesis under stressful environments: An overview. *Photosynthetica* **2013**, *51*, 163–190. [CrossRef]
58. Kamaruzzaman, M.; Abdullah, S.; Hasan, H.A.; Hassan, M.; Othman, A.; Idris, M. Characterisation of Pb-resistant plant growth-promoting rhizobacteria (PGPR) from Scirpus grossus. *Biocatal. Agric. Biotechnol.* **2020**, *23*, 101456. [CrossRef]
59. Moreno-Galván, A.E.; Cortés-Patiño, S.; Romero-Perdomo, F.; Uribe-Vélez, D.; Bashan, Y.; Bonilla, R.R. Proline accumulation and glutathione reductase activity induced by drought-tolerant rhizobacteria as potential mechanisms to alleviate drought stress in Guinea grass. *Appl. Soil Ecol.* **2020**, *147*, 103367. [CrossRef]
60. Mibei, E.K.; Ambuko, J.; Giovannoni, J.J.; Onyango, A.N.; Owino, W.O. Carotenoid profiling of the leaves of selected African eggplant accessions subjected to drought stress. *Food Sci. Nutr.* **2017**, *5*, 113–122. [CrossRef] [PubMed]
61. Lum, M.; Hanafi, M.; Rafii, Y.; Akmar, A. Effect of drought stress on growth, proline and antioxidant enzyme activities of upland rice. *J. Anim. Plant Sci.* **2014**, *24*, 1487–1493.

62. Alam, M.M.; Nahar, K.; Hasanuzzaman, M.; Fujita, M. Exogenous jasmonic acid modulates the physiology, antioxidant defense and glyoxalase systems in imparting drought stress tolerance in different Brassica species. *Plant Biotechnol. Rep.* **2014**, *8*, 279–293. [CrossRef]
63. Smirnoff, N. Ascorbic acid metabolism and functions: A comparison of plants and mammals. *Free Radic. Biol. Med.* **2018**, *122*, 116–129. [CrossRef]
64. Amirjani, M.R.; Mahdiyeh, M. Antioxidative and biochemical responses of wheat to drought stress. *J. Agric. Biol. Sci.* **2013**, *8*, 291–301.
65. Shafiq, S.; Akram, N.A.; Ashraf, M.; Arshad, A. Synergistic effects of drought and ascorbic acid on growth, mineral nutrients and oxidative defense system in canola (*Brassica napus* L.) plants. *Acta Physiol. Plant.* **2014**, *36*, 1539–1553. [CrossRef]
66. Ghorbanli, M.; Gafarabad, M.; Amirkian, T.; Allahverdi, M.B. Investigation of proline, total protein, chlorophyll, ascorbate and dehydroascorbate changes under drought stress in Akria and Mobil tomato cultivars. *Iran. J. Plant. Physiol.* **2013**, *3*, 651–658.

Article

Individual and Interactive Effects of Multiple Abiotic Stress Treatments on Early-Season Growth and Development of Two *Brassica* Species

Akanksha Sehgal [1], Kambham Raja Reddy [1,*], Charles Hunt Walne [1], T. Casey Barickman [2], Skyler Brazel [2], Daryl Chastain [3] and Wei Gao [4]

1. Department of Plant and Soil Sciences, Mississippi State University, P.O. Box 9555, Mississippi State, MS 39762, USA; as5002@msstate.edu (A.S.); chw148@msstate.edu (C.H.W.)
2. North Mississippi Research and Extension Center, Mississippi State University, P.O. Box 1690, Verona, MS 38879, USA; t.c.barickman@msstate.edu (T.C.B.); s.r.brazel@msstate.edu (S.B.)
3. Delta Research and Extension Center, Mississippi State University, P.O. Box 197, Stoneville, MS 38776, USA; drc373@msstate.edu
4. USDA UVB Monitoring and Research Program, Natural Resource Ecology Laboratory, Department of Ecosystem Science and Sustainability, Colorado State University, Fort Collins, CO 80523, USA; wei.gao@colostate.edu

* Correspondence: krreddy@pss.msstate.edu

Abstract: Potential global climate change-related impacts on crop production have emerged as a major research priority and societal concern during the past decade. Future changes, natural and human-induced, projected in the climate have implications for regional and global crop production. The simultaneous occurrence of several abiotic stresses instead of stress conditions is most detrimental to crops, and this has been long known by farmers and breeders. The green leafy vegetables of the Brassicaceae family have especially gained attention due to their many health benefits. However, little information is available about abiotic stress's effects on *Brassica* vegetables' growth and development. An experiment was conducted on two *Brassica* species: *B. oleracea* L. var. acephala WINTERBOR F1 (hybrid kale) and *B. juncea* var. GREEN WAVE OG (mustard greens). Seven treatments were imposed on the two brassica species in soil–plant–atmosphere–research (SPAR) units under optimum moisture and nutrient conditions, including a control treatment (optimal temperature and UV-B conditions at ambient CO_2 levels), and six treatments where stresses were elevated: CO_2, UV-B, temperature (T), CO_2+UV-B, CO_2+T, and CO_2+UV-B+T. Above- and below-ground growth parameters were assessed at 26 d after sowing. Several shoot and root morphological and developmental traits were evaluated under all the treatments. The measured growth and development traits declined significantly under individual stresses and under the interaction of these stresses in both the species, except under elevated CO_2 treatment. All the traits showed maximum reductions under high IV-B levels in both species. Leaf area showed 78% and 72% reductions, and stem dry weight decreased by 73% and 81% in kale and mustard, respectively, under high UV-B levels. The increased CO_2 concentrations alleviated some deleterious impacts of high temperature and UV-B stresses. The results of our current study will improve our understanding of the adverse effects of environmental stresses on the early-season growth and development of two *Brassica* species.

Keywords: temperature stress; elevated CO_2; UV-B; *Brassica oleracea*; *Brassica juncea*

Citation: Sehgal, A.; Reddy, K.R.; Walne, C.H.; Barickman, T.C.; Brazel, S.; Chastain, D.; Gao, W. Individual and Interactive Effects of Multiple Abiotic Stress Treatments on Early-Season Growth and Development of Two *Brassica* Species. *Agriculture* **2022**, *12*, 453. https://doi.org/10.3390/agriculture12040453

Academic Editors: Daniele Del Buono, Primo Proietti and Luca Regni

Received: 25 January 2022
Accepted: 21 March 2022
Published: 23 March 2022

1. Introduction

Many scientific and intergovernmental reports warn of the dangerous consequences of climate change for various aspects of human life, with considerable threats to plant productivity [1–7]. Current temperatures are approximately 1 °C above pre-industrial levels, and a further increase in global temperatures by 0.5 °C would elevate the related risks [6]. Elevated atmospheric CO_2 is the most eminent cause of global warming. At present, the global atmospheric CO_2 concentration is 417 ppm (recorded in March 2021 by Mauna Loa observatory, Waimea, HI, USA), which was only 270 ppm during the pre-industrial era. Over the last two centuries, such an unprecedented rise in atmospheric CO_2 has occurred due to massive anthropogenic activities such as deforestation, fossil-based fuel combustion, rapid urbanization, and industrialization [8]. Climate change is a cumulative effect of multiple factors, such as changes in temperatures, radiation, precipitation, and CO_2 levels [9]. Thus, it is imperative to understand the combined effects of multiple factors and elevated CO_2 to mimic real-world situations.

There has been an increasing trend in consuming more green leafy vegetables in the human diet. Among the variety of green leafy vegetables available for human consumption, kale (*Brassica oleracea* L.) and mustard (*Brassica juncea* L.) are among the most consumed vegetables [10]. The vegetables of the *Brassicaceae* family gained attention due to their sulfur-containing phytonutrients, known as glucosinolates, that are known to promote health. Glucosinolates, flavonoids, and phenolic compounds are responsible for antioxidant and free radical scavenging properties [11,12]. Kale leaves are generally consumed fresh and unprocessed as a salad or cooked and used as a garnish, and they are usually sold in fresh, canned, and frozen forms [13]. Kale is reported to have much higher protein than other *Brassica* family vegetables [14] and other green leafy vegetables such as spinach (2.9% on a fresh weight basis). Both kale and mustard are excellent sources of vitamin A and β-carotenes, and flavonoids [15]. Research studies have reported other health-beneficial activities of kale and a mustard-like protective role in coronary artery diseases, anti-inflammatory activities, antigenotoxic ability, and gastroprotective activity [10]. The presence of compounds such as polyphenols, glucosinolates, carotenoids, and vitamins E and C in kale and mustard is associated with cardiovascular protection [16], and mustard is also beneficial in the treatment of diabetes and cataracts [17].

Both kale and mustard are considered cool-season crops, generally thriving at daytime temperatures of 18 to 24 °C and nighttime temperatures of 4 to 7 °C [18]. Kale and mustard are sensitive to high temperatures [19]. Thus, an early-season planting can help prevent high temperatures during the seedling stage and mitigate the losses due to stress. Since abiotic stresses are interconnected, their concurrent occurrence and combined effects have been shown to be more destructive to plant growth, productivity, and yield, and will be essential in devising management and breeding decisions in the coming years.

Several previous studies on elevated CO_2 concentration in crops have revealed a significant direct impact on plant growth and crop yield that can compensate for a potentially hotter climate. Even though it has been noticed that an increase in CO_2 concentration leads to a significant yield increase in C_3 plants [20–22], few direct effects have been recorded on kale and mustard plants. However, the impact of elevated CO_2 on these two *Brassica* species, much less the interaction of elevated CO_2, temperature, and UV-B, is not adequately understood to allow accurate predictions of future crop production.

The projected higher doses of incoming UV-B radiation have been reported to stimulate various responses of higher plants [23,24]. Some of the deleterious effects of UV-B radiation on plants include DNA damage, the disintegration of cellular membranes, photooxidation of leaf pigments and phytohormones, and inhibition of photosynthesis [25–27]. Moreover, higher UV-B radiation can affect whole-plant photosynthesis via alterations in leaf thickness, anatomy, and canopy morphology [28]. Therefore, to optimize the production with suitable management and breeding strategies for green leafy vegetables in the future, it is crucial to understand the effects of UV-B radiation and the other stresses on these crops. Although little data exists on the growth, development, and productivity of crops in response to CO_2, temperature, or UV-B applied alone, to the best of our knowledge no data is available on the interactive effects of multiple factors on the growth and development of kale and mustard.

Root systems are challenging to study because of their highly structured underground distribution, the complexity of vigorous interactions with the environment, and their diversity of functions. The root system can be more affected than the aerial parts by multiple abiotic stresses. Despite this, the influence of abiotic stresses on root development has been considerably less studied than on shoots because of the limited accessibility of root observations [29]. Different methodologies have been developed to study root growth under both field and controlled environmental conditions. Root scanning based on the WinRHIZO optical scanner [30] is an efficient method that allows image analysis and examination of the root morphological traits. This technique provides data that, using established software protocols, enables quick analysis and rapid, straightforward, and accurate screening of root characteristics. Therefore, this method is most suited for screening the root traits of kale and mustard plants grown under controlled environmental conditions.

Previous studies on sunlit, controlled environment chambers demonstrated the effects of multiple abiotic stress interactions on plant growth, development, physiology, and reproduction in cotton, *Gossypium hirsutum* L. [31], soybean, *Glycine max* (L.) Merr. [32,33], and some other crops. Still, none of these studies have been conducted on kale and mustard. One of the studies performed on soybean revealed that plant height and leaf area development were the most sensitive processes in responses to multiple stresses, leading to a profound loss in biomass production [33]. However, it was observed that elevated CO_2 concentration ameliorated the damaging effects caused by various abiotic stresses on most of the plant growth and physiological parameters. According to Reddy et al. [34], numerous environmental stresses affect crop growth, development, and physiological processes multiplicatively, not additively. Hence, rather than a particular stress condition, the simultaneous occurrence of multiple abiotic stresses is most harmful to any crop. Therefore, the interactive effects of various environmental stresses on kale and mustard must be sufficiently understood to allow accurate predictions of future crop production. The objectives of this study were to characterize the changes in vegetative growth and developmental traits in kale and mustard in their response to multiple environmental factors of (CO_2) [400 and 720 μmol mol^{-1} (+(CO_2))], temperature treatments [25/17 °C and 35/27 °C (day/night) (+T)], and B radiation [0 and 10 kJ m^{-2} d^{-1} (+UV-B)] during early-season growth.

2. Materials and Methods

2.1. Seed Material and Experiment Conditions

Brassica species: *B. oleracea* var. WINTERBOR F1 (hybrid kale) and *B. juncea* var. GREEN WAVE OG (mustard greens) were used for this study. The experiment was conducted in August 2019 in sunlit soil–plant–atmosphere–research (SPAR) chambers located at the Rodney Foil Plant Science Research facility of Mississippi State University, Mississippi State, MS (lat. 33°28′ N, long. 88°47′ W). Each SPAR chamber consists of a steel soil bin and a 1.27 cm thick Plexiglas chamber to accommodate root and aerial plant parts. The Plexiglas allows 97% of the visible solar radiation to pass without spectral variability in absorption (wavelength 400–700 nm). During the experiment, the incoming daily solar radiation measured with a pyranometer (Model 4–8; The Eppley Laboratory Inc.) outside the SPAR units ranged from 11.3 to 31.3 MJ m^2 d^{-1} with an average value of 25.10 ± 0.82 MJ m^2 d^{-1}. More details of the SPAR chamber operations and control have been described by Reddy et al. [34]. Briefly, air ducts on each SPAR unit's northern side were connected to the heating and cooling devices. Conditioned air was passed through the plant canopy with sufficient velocity to cause leaf flutter (4.7 km h^{-1}) and was returned to the air-handling unit just above the soil level. Two electrical resistance heaters provided short heat pulses as needed to fine-tune the air temperature control. Chamber air temperature, CO_2 concentration, soil watering in each SPAR unit, and continuous monitoring of environmental and plant gas exchange variables were controlled by a dedicated computer system [35] (Table 1). The vapor pressure deficits in the units were estimated from these measurements as per Murray [36] (Table 1).

Table 1. The set treatments and measured day, night, and average temperatures, chamber carbon dioxide concentration (CO_2), daytime and nighttime vapor pressure deficit (VPD), and evapotranspiration (ET) during the experimental period of each treatment in kale and mustard.

Treatments	Measured Temperature (°C)			CO_2 (μmol mol^{-1})	VPD (kPa)		Mean ET (L H_2O d^{-1})
	Day	Night	Day/Night	Day	Day	Night	Day/Night
Control	24.8 ± 0.03	17.5 ± 0.04	21.6 ± 0.03	434.3 ± 1.77	1.4 ± 0.01	0.98 ± 0.01	8.6 ± 0.45
+CO_2	25 ± 0.03	17.6 ± 0.03	21.8 ± 0.02	723.6 ± 0.33	1.4 ± 0.01	1 ± 0.01	7.4 ± 0.67
+T	30 ± 0.88	22.4 ± 0.88	26.7± 0.87	435.1 ± 1.76	2.1 ± 0.15	1.5 ± 0.11	8.5 ± 0.63
+UV-B	24.7 ±0.04	17.4 ± 0.04	21.5 ± 0.03	436.9 ± 2.44	1.4 ± 0.01	1 ± 0.01	6.6 ± 0.24
+T+CO_2	30.3 ± 0.94	22.6 ± 0.93	27 ± 0.92	724 ± 0.36	2.6± 0.15	1.7 ± 0.12	8.7 ± 0.81
+UV-B+CO_2	24.8 ± 0.03	17.4 ± 0.03	21.5 ± 0.03	720.5 ± 0.55	1.3 ± 0.02	0.93 ± 0.01	5.2 ± 0.47
+UV-B+CO_2+T	30 ± 0.89	22.4 ± 0.88	26.7 ± 0.87	733.2 ±0.49	2.6 ± 0.19	1.8 ± 0.14	6.91 ± 0.59

During the experiment, the incoming daily solar radiation measured with a pyranometer (Model 4–8; The Eppley Laboratory Inc., Newport, RI, USA) outside the SPAR units ranged from 11.3 to 30.9 MJ m^2 d^{-1} with an average value of 24.12 ± 1.14 MJ m^2 d^{-1}.

Seeds were sown in 210 polyvinyl-chloride pots (15.2 cm diameter and 30.5 cm height) filled with the soil medium consisting of 3:1 sand/topsoil classified as a sandy loam (87% sand, 2% clay, and 11% silt) with 500 g of gravel at the bottom of each pot. Initially, three seeds were sown in each pot, and 7 d after emergence the plants were thinned to one per pot. Pots were arranged in 10 rows with three pots per row in each SPAR chamber with alternating kale and mustard plants. Plants were irrigated three times a day through an automated, computer-controlled drip system with full-strength Hoagland's nutrient solution [37], delivered at 0700, 1200, and 1700 h, based on treatment-based evapotranspiration values. Evapotranspiration rates expressed on a ground area basis (L d^{-1}) throughout the treatment period were measured in each SPAR unit as the rate at which the cooling coils removed the condensate at 900 s intervals [35,38,39]. They were obtained by measuring the mass of water in collecting devices connected to a calibrated pressure transducer. Average evapotranspiration values for each treatment during the experimental period are provided in Table 1.

2.2. Treatments

The treatments included combinations of two [CO_2], [400 and 720 μmol mol^{-1} (+CO_2)], two different temperatures, [25/17 °C and 35/27 °C (+T) (day/night)], and two daily biologically effective UV-B radiation intensities, [0 and 10 kJ m^{-2} d^{-1} (+UV-B)].

The control treatment was 400 μmol mol^{-1} [CO_2], 25/17 °C (day/night) temperatures, and 0 kJ m^{-2} d^{-1} UV-B treatment. All SPAR chambers were maintained at control conditions until 12 days after sowing (DAS). Subsequently, each chamber was set at one of the seven treatments until the final harvest (26 DAS; 14 DAT): (1) a control treatment with optimum temperature, ambient CO_2 levels, and no UV-B; (2) optimum temperature with elevated CO_2 levels and no UV-B (+CO_2); (3) elevated temperature with ambient CO_2 levels and no UV-B (+T); (4) optimum temperature and ambient CO_2 levels with 10 kJ UV-B (+UV-B); (5) elevated temperature, and CO_2 levels with no UV-B (+T+CO_2); (6) optimum temperature with elevated CO_2 levels and 10 kJ UV-B (+CO_2+UV-B); (7) elevated temperature and elevated CO_2 levels with 10 kJ UV-B (+UV-B+CO_2+T). For each treatment, fifteen replications were maintained per species per SPAR unit.

A humidity and temperature sensor (HMV 70Y, Vaisala, Inc., St. Louis, MO, USA) was used to monitor the relative humidity of each chamber. The monitor was installed in the returning path of airline ducts. A chilled mixture of ethylene glycol and water was injected through the cooling coils located outside the air handler of each chamber via several parallel solenoid valves. The valves opened or closed depending on the cooling requirement to maintain a constant humidity [38].

Pure CO_2 supply was maintained through a compressed gas cylinder using a system that included a pressure regulator, solenoid and needle valves, and a calibrated flow meter [35]. The chamber CO_2 was measured and maintained either at 400 or 720 μmol mol^{-1} with a dedicated infrared gas analyzer (LI-COR, model LI-6252, Lincoln, NE, USA); the drawn gas sample through the lines run underground from SPAR units to the field laboratory building. The sample lines were run through refrigerated water (4 °C) that was automatically drained and through a column of $Mg(ClO_4)_2$ to remove moisture from the gas sample.

The desired elevated UV-B treatment, 10 kJ m^{-2} d^{-1}, was imposed from 12 DAS to the end of the experiment. The square-wave UV-B supplementation systems were used under near-ambient PAR to provide anticipated UV-B radiation dosage. The UV-B radiation was delivered from 0.5 m above the plant canopy for 8 h each day, from 08:00 to 16:00, by eight fluorescent UV-313 lamps (Q-Panel Company, Cleveland, OH, USA) attached horizontally on a metal frame inside each chamber, powered by 40 W variable dimming ballasts. The individual UV lamp was wrapped with solarized 0.07 mm diacetate film to filter UV-C (<280 nm) radiation. The UV-B radiation supplied at the top of the plant canopy was monitored daily at 08:00 with a UVX digital radiometer (UVP Inc., San Gabriel, CA, USA) calibrated against an Optronic Laboratory (Orlando, FL, USA) Model 754 Spectroradiometer, which was used initially to quantify lamp output. The lamp output was adjusted, and the cellulose diacetate films were replaced as needed to maintain the individual UV-B radiation level. The actual biologically effective UV-B radiation was measured in each SPAR chamber at three different locations (in the middle and two corners) to ensure the plants received the exact UV-B dosage of 10 ± 0.18 kJ m^{-2} d^{-1} during the crop growth period.

2.3. Measurements

2.3.1. Phenology and Growth

The total number of leaves (LN) was counted, and plant height (PH) and marketable fresh weight (MFW) were measured on all plants at harvest (26 DAS). Leaf area was measured using the LI-3100 leaf-area meter (LI-COR, Inc., Lincoln, NE, USA). Plant component total dry weights (TD) were measured after oven drying at 75 °C until a constant weight was reached.

2.3.2. Root Image Acquisition and Analysis

Aboveground plant parts were cut and separated from the root systems. Roots were gently washed free of all soil media. The longest root length (LRL) was determined using a ruler. The cleaned individual root systems were floated in 5 mm of water in a 0.4 by 0.3 m Plexiglas tray. A plastic paintbrush was used to untangle and separate roots to minimize root overlap. The tray was placed on top of a specialized dual-scan optical scanner [30] linked to a computer. Gray-scale root images were acquired by setting the parameters to high accuracy (resolution 800 × 800 dpi). Acquired images were analyzed for the total root length (TRL), root surface area (RSA), average root diameter (RAD), root length per volume (RLPV), root volume (RV), number of tips (RT), number of forks (RF), and number of crossings (RC) using WinRHIZO Pro software 2009c [Regent Instruments, Inc., Québec, QC, Canada] [30].

2.4. Data Analysis

2.4.1. Combined Stress Response Index (CSRI)

Based on the summation of relative individual stress responses at each treatment and similar to the cumulative response index quoted in other UV-B studies [32], the combined stress response index (CSRI) was calculated to evaluate the interactive effects of six treatments (+CO_2, +T, +UV-B, +CO_2+T, +CO_2+UV-B, and +CO_2+T+UV-B) in comparison to the control treatment. The CSRI was calculated as the value of a parameter under control (c), subtracted from the value of the parameter under treatment (t), and then by dividing by the value of a parameter under control (c) as follows:

$$\text{CSRI} = \frac{(\text{PHt}-\text{PHc})}{(\text{PHc})} + \frac{(\text{LNt}-\text{LNc})}{(\text{LNc})} + \frac{(\text{LAt}-\text{LAc})}{(\text{LAc})} + \frac{(\text{MFWt}-\text{MFWc})}{(\text{MFWc})} + \frac{(\text{ADWt}-\text{ADWc})}{(\text{ADWc})} + \frac{(\text{RDWt}-\text{RDWc}}{(\text{RDWc})} + \frac{(\text{TDWt}-\text{TDWc})}{(\text{TDWc})} + \frac{(\text{LRLt}-\text{LRLc})}{(\text{LRLc})} + \frac{(\text{TRLt}-\text{TRLc})}{(\text{TRLc})} + \frac{(\text{RSAt}-\text{RSAc})}{(\text{RSAc})} + \frac{(\text{RADt}-\text{RADc})}{(\text{RADc})} + \frac{(\text{RLPVt}-\text{RLPVc})}{(\text{RLPVc})} + \frac{(\text{RVt}-\text{RVc})}{(\text{RVc})} + \frac{(\text{RTt}-\text{RTc})}{(\text{RTc})} + \frac{(\text{RFt}-\text{RFc})}{(\text{RFc})} + \frac{(\text{RCt}-\text{RCc})}{(\text{RCc})}$$

where CSRI is the combined stress response index, PH—the plant height, LN—the leaf number, LA—the leaf area of the plant, MFW—the marketable fresh weight, ADW—aboveground dry weight, RDW—the root dry weight, TDW—Total dry weight, LRL—the longest root length, TRL—the total root length, RSA—the root surface area, RAD—the root average diameter, RLPV—the root length per volume, RV—the root volume, RT—the number of root tips, RF—the number of root forks, RC—the number of root crossings under t (treatment) and c (control).

2.4.2. Statistical Analysis

Data were subjected to analysis of variance [40] with a split-plot design considering species and treatment as sources of variance. Replicated values for LN, PH, LA, MFW, ADW, RDW, TDW, LRL, TRL, RSA, RAD, RLPV, RV, RT, RF, and RC were analyzed using one-way ANOVA of the general linear model, PROC GLM, in SAS [40] to determine the effect of multi-stress treatments on the morphological and developmental parameters of kale and mustard. Fisher-protected least significant difference tests at $p = 0.05$ were employed to test the differences among treatments for measured parameters. The standard errors of the mean were calculated and are presented in the figures as error bars.

3. Results

This is the first study providing data for the effects of abiotic multi-stress on the growth and development of roots and shoots of green leafy vegetables of the Brassica family.

3.1. Shoot Growth and Developmental Attributes

3.1.1. Plant Height

Interactive effects of increased CO_2, temperature, and UV-B radiation led to significant differences in plant height (Table 2). Compared to the control treatment, the plants were significantly taller under elevated CO_2 (+CO_2) (8% and 12% in kale and mustard, respectively). Plants grown under elevated temperature conditions along with CO_2 treatment had minimal adverse effects on plant height, showing owing only 12% and 11.7% (+T), and 3.3% and 1.8% (+CO_2+T), reductions in average plant height in kale and mustard, respectively, compared to the control. Plants grown under UV-B conditions alone and elevated temperature and CO_2 produced significantly shorter plants in both Brassica sp., as evident in Figure 1. The greatest plant height reduction was observed under UV-B (53% for kale and 39% for mustard) treatment (Table 3). Among the two Brassica sp., mustard plants were taller than kale plants under all the treatments and showed lesser reductions.

Table 2. The analysis of variance across the treatments of carbon dioxide concentration CO_2, temperature., UV-B radiation, and two crops (kale and mustard), and their interactions on kale and mustard root and shoot growth and developmental traits, plant height (PH), mainstem leaves (LN), whole plant leaf area (LA), marketable fresh weight (MFW), aboveground dry weight (ADW), root dry weight (RDW), total plant dry weight (TDW), the longest root length (LRL), total root length (TRL), root surface area (RSA), average root diameter (RAD), root length per volume (RLPV), root volume (RV), root tips (RT), root forks (RF), and root crossings (RC).

Source of Variance	PH	LN	LA	MFW	ADW	RDW	TDW	LRL	TRL	RSA	RAD	RLPV	RV	RT	RF	RC
Treatment	***	***	***	***	***	***	***	*	***	***	***	***	***	***	***	***
Crop	**	***	***	***	***	***	***	***	***	***	***	***	***	***	***	***
Trt × Crop	NS	***	***	***	**	**	**	*	**	**	***	**	**	***	**	**

*** indicates significance levels, **, *, and NS, representing $p < 0.001$, $p < 0.01$, $p < 0.05$ and $p > 0.05$, respectively.

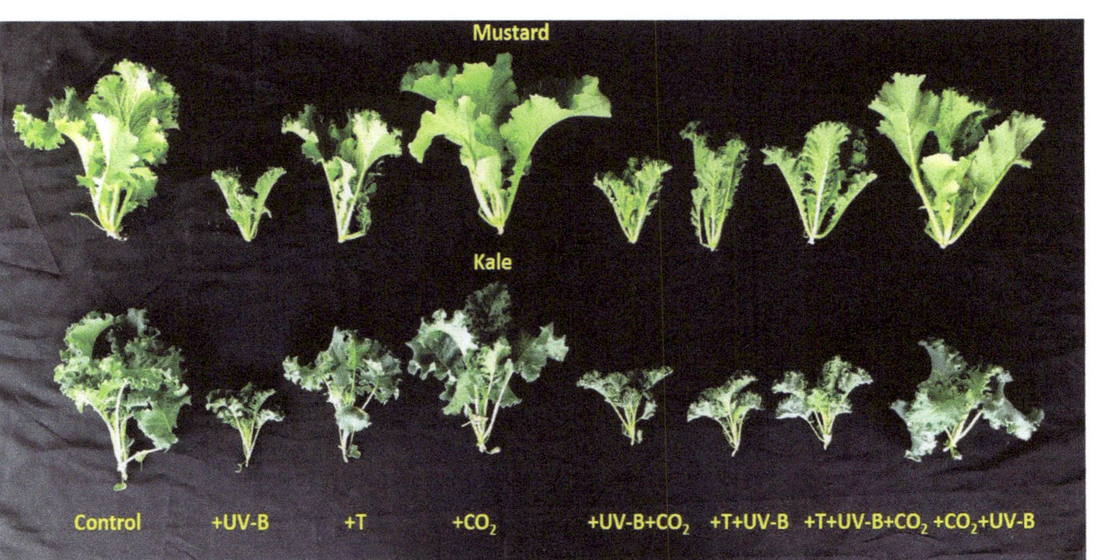

Figure 1. A pictorial presentation of the impact of CO_2 concentration (control, 400 µmol mol^{-1} and +CO_2, 720 µmol mol^{-1}), elevated temperatures (25/17 °C and 35/27 °C (day/night)), and radiation (control, 0 and +UV-B, 10 kJ m^{-2} d^{-1}), and their interactions on the morphological growth of kale and mustard.

Table 3. Mean values and percent change for plant height (PH), leaf number (LN), leaf area (LA), marketable fresh weight (MFW), aboveground dry weight (ADW), root dry weight (RDW), and total dry weight (TDW) measured under CO_2 concentration (control, 400 µmol mol^{-1} and +CO_2, 720 µmol mol^{-1}), elevated temperatures (25/17 °C and 35/27 °C (day/night)), and UV-B radiation (control, 0 kJ m^{-2} d^{-1}, and +UV-B, 10 kJ m^{-2} d^{-1}), and their interactions for kale and mustard at 26 DAS.

	Traits	Crop	Treatments						
			Control	0	+T	+UV-B	+T+CO_2	+UV-B+CO_2	+UV-B+CO_2+T
Shoot Traits	PH (cm plant^{-1})	Kale	30.8	33.3 (+8%)	27.2 (−11.7%)	14.3 (−53%)	29.8 (−3.3%)	17.3 (−43.8%)	18.5 (−40%)
		Mustard	31.9	35.8 (+12%)	28 (−12%)	19.3 (−39%)	31.3 (−1.8%)	19.5 (−38.8%)	22.3 (−30%)
	LN (plant^{-1})	Kale	9.33	9.7 (+3.9%)	8.7 (−8.8%)	9.5 (+1.8%)	9 (−3.5%)	17.3 (−43.8%)	18.5 (−40%)
		Mustard	11.7	12.7 (+8.5%)	9.7 (+17%)	10.8 (−7.6%)	15.8 (+35%)	19.5 (−38.8%)	22.3 (−30%)
	LA (cm^2 plant^{-1})	Kale	600.5	779.3 (+29.8%)	424.5 (−29.3%)	131.7 (−78%)	565.7 (−5.7%)	226.3 (−62.3%)	292.9 (−51.2%)
		Mustard	1305.1	1629.6 (+24.8%)	983.9 (−24.6%)	363.5 (−72%)	1495.5 (−8.2%)	380.5 (−70.8%)	653.5 (−50%)
	MFW (g plant^{-1})	Kale	42	51.5 (+22.7%)	24 (−42.7%)	12.7 (−69.7%)	35 (−16.9%)	20.9 (−50%)	22.5 (−46.3%)
		Mustard	76.3	116.2 (+52%)	61.3 (−19.7%)	30.4 (−60.2%)	96.9 (26.8%)	30.6 (−60%)	50 (−34.5%)
Dry weight traits	ADW (g plant^{-1})	Kale	3.69	5.7 (+55%)	2.8 (−24.6%)	1.3 (−64%)	4.2 (+13.7%)	2.6 (−28.4%)	2.7 (−27.6%)
		Mustard	5.93	8.8 (+48%)	5.4 (−9.5%)	2.6 (−56.8%)	8.6 (+45.7%)	2.7 (−54.5%)	4.6 (−22.4%)
	RDW (g plant^{-1})	Kale	0.37	0.6 (+55.2%)	0.3 (−30%)	0.1 (−63%)	0.4 (0%)	0.3 (−65.6%)	0.3 (−20%)
		Mustard	0.79	1 (+32.8%)	0.7 (−5.9%)	0.4 (−52.7%)	1.1 (+43.4%)	0.3 (−60%)	0.7 (−7%)
	TDW (g plant^{-1})	Kale	4.05	6.3 (+55.3%)	3 (−25%)	1.5 (−64%)	4.6 (+12.5%)	2.9 (−28%)	3 (−27%)
		Mustard	6.71	9.9 (+47%)	6.1 (−9%)	2.9 (−60%)	9.8 (+45.6%)	3 (−55%)	5.3 (−20%)

3.1.2. Leaf Number

The number of leaves produced in the plants under increased CO_2, UV-B, and temperature treatments was significant in the present study (Table 2). The crops showed different responses under different treatments for the number of leaves produced during the treatment (Figure 1). Fewer leaves in mustard were observed under +UV-B+CO_2 treatment, where the reduction was 25% compared to the control (Table 3; Figure 1). In the case of kale, the maximum deduction was observed under high-temperature treatment. More leaves were observed under +CO_2 and high temperatures and UV-B (+CO_2+T+UV-B) in both the crops (4% and 8.5% at +CO_2; 9% and 4.2% at +CO_2+T+UV-B in kale and mustard, respectively). A maximum increase in the number of leaves was recorded at high UV-B and CO_2 (+CO_2+UV-B) for kale (12.5%) and high temperature and CO_2 (+T+CO_2) for mustard (35%).

3.1.3. Leaf Area

The leaf area exhibited significant differences under all the treatments. Higher leaf area was recorded under +CO_2 treatment, as clearly noticeable from Figure 1 (30% in kale and 25% in mustard), compared to the control (Table 3; Figure 2A.). The highest reduction in leaf area in both crops was observed under the UV-B treatment alone (+UV-B), 78% (kale), and 72% (mustard) compared to their respective controls. Elevated CO_2 (+CO_2) seemed to alleviate the adverse effects of high temperature (+T) on leaf area leading to the most negligible reduction in both crops (5.7% and 8.2% in kale and mustard, respectively).

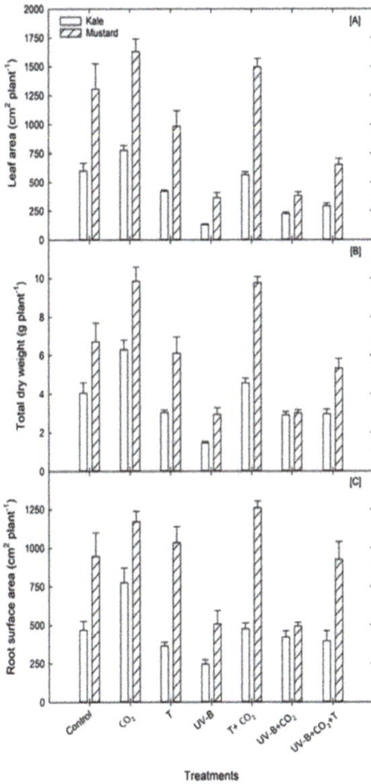

Figure 2. Impact of CO_2 concentration (control, 400 μmol mol^{-1} and +CO_2, 720 μmol mol^{-1}), elevated temperatures (25/17 °C and 35/27 °C (day/night)), and UV-B radiation (control, 0 and +UV-B, 10 kJ m^{-2} d^{-1}), and their interactions on (**A**) leaf area, (**B**) total dry weight and (**C**) root surface area for kale and mustard. Bars indicate standard errors of the mean.

3.1.4. Marketable Fresh Weight

Marketable fresh weight, which determines the economic value of green leafy vegetables, decreased significantly under all the treatments except for the $+CO_2$ treatment. The average marketable fresh weight ranged from 12.7 to 116.2 g plant^{-1}, with the lowest value under +UV-B treatment and the highest under $+CO_2$ treatment alone (Figures 1 and 3; Table 3). Accordingly, the marketable fresh weight decreased by 46.3% and 34.5% in kale and mustard under the +UV-B+CO_2+T treatment. The marketable fresh weight doubled in mustard under the CO_2 treatment compared to its control counterparts, whereas an increase of 23% was recorded in kale.

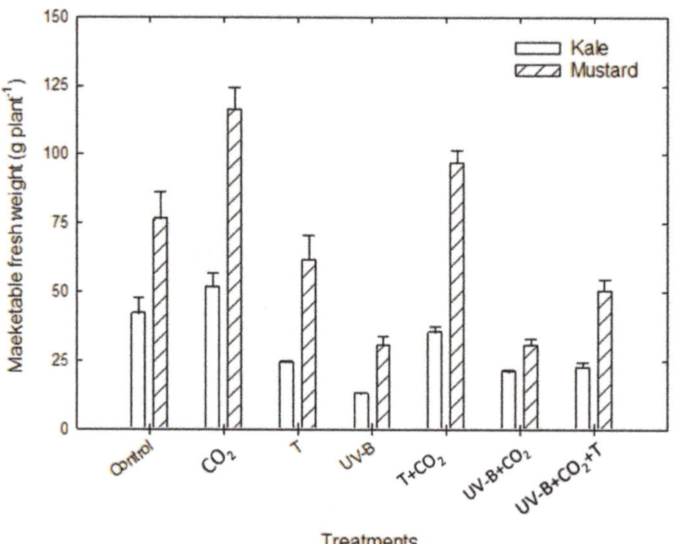

Figure 3. Impact of CO_2 concentration (control, 400 µmolmol^{-1} and $+CO_2$, 720 µmol mol^{-1}), elevated temperatures (25/17 °C and 35/27 °C (day/night)), and UV-B radiation (control, 0 and +UV-B, 10 kJ m^{-2} d^{-1}), and their interactions on marketable fresh weight for kale and mustard. Bars indicate standard errors of the mean.

3.1.5. Dry Weight Components

Like marketable fresh weight, significant reductions in aboveground dry weight, accounting for 64% (kale) and 57% (mustard) of the decrease compared to their control, were observed under +UV-B treatments. The CO_2 treatment alone ($+CO_2$) and in combination with elevated temperature (+T+CO_2) recorded an increase of 55% (kale) and 48% (mustard), and of 14% (kale) and 46% (mustard), in aboveground dry weight, respectively, compared to their control treatments (Table 3). Unlike other shoot parameters, the highest reduction in root dry weight was observed under the elevated UV-B and CO_2 treatment (+UV-B+CO_2), which was 66% and 60% in kale and mustard, respectively, compared to their respective controls.

The minimum adverse effect on root dry weight in kale was observed under the +UV-B+T+CO_2 treatment with an average decrease of 20%. In mustard, the minimum decrease was observed under the temperature treatment alone (6%) in comparison to the control. Under the CO_2 treatment and elevated temperature (+T+CO_2), there was no change in the root dry weight of kale, whereas an increase of 43.4% was observed in mustard.

The total dry weight produced during the experimental period ranged from 1.46 to 9.87 g plant^{-1} (Table 3, Figure 2B); plants grown under +UV-B registered the lowest while plants grown under $+CO_2$ treatment alone showed the greatest total dry weight in both the crops. All the stresses, either alone or in combination, caused significant differences in total dry weight. The reductions in total dry weight under +UV-B, +T+UV-B+CO_2, and +UV-B+CO_2+T treatments were 64% and 60%; 25% and 9%; 28% and 55%; and 27% and 20%, in kale and mustard, respectively, when compared to the control. The total dry weight in both crops increased under the +CO_2 treatment and the +T+CO_2 treatment. Among the two crops, mustard had the higher total dry weight under all the treatments.

3.2. Root Growth and Developmental Attributes

3.2.1. Root Growth Traits

All the stress treatments alone or together led to significant differences in the root length of both crops. Both longest root length (LRL; 24% in kale and 17% in mustard) and total root length (TRL; 43% in kale and 39% in mustard) showed the highest reduction under the elevated UV-B treatment. The longest root length in kale decreased under all treatments except for +CO_2 treatment (14.7%) and +UV-B+CO_2 treatment (1.73%). In contrast, in mustard, LRL increased under all the treatments except +UV-B and +UV-B+CO_2 (3.2%), compared to their respective control (Table 4). The minimum adverse effect on total root length was observed under +UV-B+CO_2 treatment in the case of kale, with an average decrease of 1%. In mustard, the most negligible reduction of 1% each was observed under +CO_2 treatment and the CO_2 treatment, together with UV-B and elevated temperature (+UV-B+CO_2+T). Compared to the controls, TRL values increased for both crops under CO_2 treatment combined with high temperature (+T+CO_2).

Root surface area (RSA), average root diameter (RAD), root length per volume (RLPV), and root volume (RV) decreased significantly under +UV-B treatment, +UV-B+CO_2 treatment, and all three stresses together in both the crops. The +CO_2, alone or in combination with elevated temperature (+CO_2+T), exhibited an increase in RSA and RV compared to their control treatments in both crops (Table 4; Figure 2C). RSA and RLPV values for kale decreased by 22.6% and 12.6%, respectively, under the elevated temperature, whereas increases of 9.4% and 22.6% were observed in mustard. RAD and RV values for both crops decreased under high-temperature conditions.

3.2.2. Root Developmental Traits

All the root developmental traits showed significant reductions, accounting for 25% and 28% of the decline in the number of root tips (RT), 52% and 44% of the decrease in the number of root forks (RF), and 41.6%, and 37.5% of the decline in the number of root crossings (RC), in kale and mustard respectively, compared to their control under the +UV-B treatment. The +CO_2 treatment exhibited an 82% (kale) and 19.3% (mustard) increase in the number of root forks concerning its control treatment. Although the CO_2 treatment alone positively impacted the number of root tips and crossings in kale, it decreased both parameters in mustard. Under the high-temperature conditions, all the developmental traits fell (28% in RT, 25.3% in RF, and 6% in RC) in kale but showed an increase in mustard (4.6% in RT, 20% in RF, and 42.5% in RC; Table 4). Under the combination of all three stresses (+UV-B+CO_2+T), a decrease in the number of root tips and root forks in both crops and root crossings in kale was recorded; however, a slight increase (0.5%) was observed in the case of mustard.

Table 4. Mean values and percent change for the longest root length (LRL), total root length (TRL), root surface area (RSA), average root diameter (RAD), root length per volume (RLPV), root volume (RV), number of root tips (RT), number of root forks (RF) and number of root crossings (RC) measured under CO_2 concentration (control, 400 µmol mol^{-1} and +CO_2, 720 µmol mol^{-1}), elevated temperatures (25/17 °C and 35/27 °C (day/night)), and UV-B radiation (control, 0 kJ m^{-2} d^{-1}, and +UV-B, 10 kJ m^{-2} d^{-1}), and their interactions for kale and mustard at 26 DAS.

Root Traits	Crop	Treatments						
		Control	+CO_2	+T	+UV-B	+T+CO_2	+UV-B+CO_2	+UV-B+CO_2+T
LRL (cm plant^{-1})	Kale	38.3	44 (+14.7%)	31.5 (−17.8%)	29 (−24.3%)	29.8 (−22%)	39 (+1.7%)	31.7 (−17.3%)
	Mustard	41.2	42.7 (+3.6%)	48.2 (+17%)	34.2 (−17%)	42 (+2%)	39.8 (−3.2%)	45.7 (+10.9%)
TRL (cm plant^{-1})	Kale	3389	5327.1 (+57%)	2961.7 (−12.6%)	1924.7 (−43%)	3443.1 (+1.6%)	3352.3 (−1%)	2953.5 (−13%)
	Mustard	5717.9	5659.8 (−1%)	7012 (+22.6%)	3477.3 (−39%)	6011.6 (+5%)	3626 (−36.5%)	5665.1 (−1%)
RSA (cm^2 plant^{-1})	Kale	469.1	775.1 (+65%)	362.9 (−22.6%)	244.9 (−47.7%)	473 (+1%)	419.5 (−10.5%)	394.7 (−15.8%)
	Mustard	944.8	1169.9 (+24%)	1033.7 (+9.4%)	507.3 (−46.3%)	1259.2 (+33%)	491 (−48%)	924.6 (−2%)
RAD (mm plant^{-1})	Kale	0.4	0.5 (+4.3%)	0.4 (−11%)	0.4 (−7%)	0.4 (−1%)	0.4 (−9.2%)	0.4 (−1%)
	Mustard	0.5	0.7 (+28.7%)	0.5 (−10%)	0.4 (−12%)	0.7 (+30%)	0.4 (−16.5%)	0.5 (−1%)
RLPV (cm m^{-3})	Kale	3389	5327.1 (+57%)	2961.7 (−12.6%)	1924.7 (−43%)	3443.1 (+1.5%)	3352.3 (−1%)	2953.5 (−13%)
	Mustard	5717.9	5659.8 (−1%)	7012 (+22.6%)	3477.3 (−39%)	6011.6 (+5%)	3626 (−36.5%)	5655.1 (−1%)
RV (cm^3 plant^{-1})	Kale	5.2	9.1 (+75%)	3.5 (−31.5%)	2.5 (−52%)	5.2 (+0.6%)	4.2 (−19%)	4.2 (−19%)
	Mustard	12.5	19.9 (+58%)	12.2 (−2.5%)	5.9 (−52.5%)	21.1 (+68%)	5.3 (−57.4%)	12.1 (−3.3%)
RT (no. plant^{-1})	Kale	10,272.3	12,891.5 (+25.4%)	7382 (−28%)	7738.5 (−24.6%)	6579.3 (−36%)	7855.2 (−23.5%)	6836.8(−33.4%)
	Mustard	14,876	12,958.2 (−13%)	15,573.3 (+4.6%)	10,666.7 (−28%)	11,595.7 (−22%)	9788.2 (−34%)	11,546.1(−22.3%)
RF (no. plant^{-1})	Kale	31,897.8	12,891.5 (+25.4%)	23,822.2 (−25.3%)	15,334.3 (−52%)	31,979 (−0.3%)	30,935.3 (−3%)	24,168 (−24%)
	Mustard	64,775.2	77,329.1 (+19.3%)	77,665.8 (+20%)	36,247.7 (−44%)	83,101.7 (+28.3%)	37,047.7 (−43%)	62,078.5 (−4%)
RC (no. plant^{-1})	Kale	2791.2	4958.2 (+77.6%)	2619.8 (−6%)	1628.2 (−41.6%)	3037.5 (+9%)	3403.3 (+22%)	2380.6 (−14.7%)
	Mustard	4531	4387.2 (−3%)	6458.2 (+42.5%)	2830.2 (−37.5%)	4713.3 (+4%)	3436.7 (−24%)	4552. 83 (+0.5%)

3.3. Combined Stress Response Indices (CSRI)

The combined stress response index is the sum of relative individual stress responses at each treatment. CSRI values ranged from −7.4 to 8.3 in kale and −6.8 to 3.7 in mustard. The lowest CSRI values for both crops were observed under +UV-B treatment, suggesting higher deleterious effects of UV-B treatment on all the parameters (Figure 4). In comparison, the highest value for kale was observed under +CO_2 treatment and for mustard under +T+CO_2 treatment, pointing towards the positive impacts of elevated CO_2 concentrations. CSRI values under all the treatments except +CO_2 treatment and its combination with high temperature (+CO_2+T) were negative in kale. However, in mustard, CSRI values under +UV-B, +UV-B+CO_2 treatment, and under all the three stress together (+UV-B+CO_2+T) were negative (Figure 4).

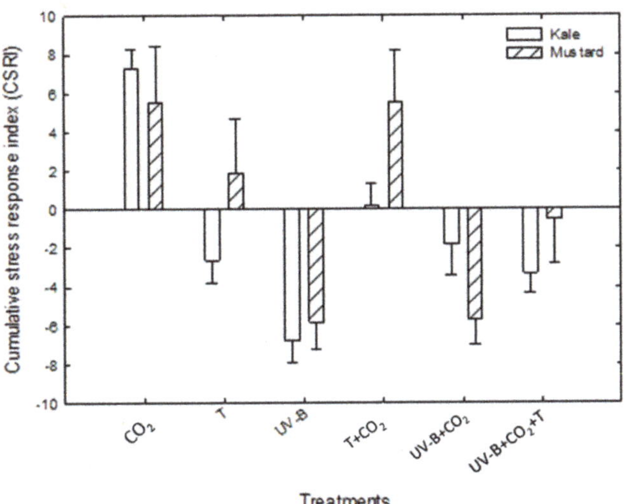

Figure 4. Cumulative stress response index (CSRI) calculated over all the treatments of kale and mustard in response to elevated carbon dioxide (720 ppm) (+CO_2), high temperature (35/27 °C, day/night) (+T), and increased UV-B radiation (10 kJ m^{-2} d^{-1}) (+UV-B) and their interactions.

4. Discussion

Brassica plants are often exposed to multiple stresses co-occurring during their growing season. Thus, it is imperative to conduct experiments in growth chambers in an environment that mimics natural conditions. In the current experiment, growing plants in SPAR chambers under fully controlled conditions permitted us to identify the functional relationships of growth and developmental responses of kale and mustard plants in response to the interaction of multiple abiotic stresses similar to those under natural conditions. Thus, this information may be helpful in management decisions and for crop model improvements to stimulate the vegetative growth of these crops in the field environment.

UV-B and mostly elevated temperatures drastically affect crop shoot, root growth, and developmental traits. However, CO_2 masked most of the other stresses' adverse effects (Tables 3 and 4). Plant height, leaf number, leaf area, and dry weight traits were affected mainly by UV-B stresses alone or combined with the other two stresses. Reduction in plant height under higher UV-B levels has been recently reported in *Capsicum annuum* [41,42] and *Brassica napus*. The shorter plants may be due to specific photomorphogenic responses of plants to elevated UV-B radiation via a UV-B photoreceptor [43]. Moreover, low photosynthetically active radiation (PAR, 400–700 nm) may have also affected the plant. It has been shown that increased PAR decreases the impacts of UV-B radiation on plant height [44]. Reduced plant height was also observed in two other brassica species

(*B. rapa* and *B. nigra*) exposed to UV-B radiation [45]. Studies on crops such as *Vigna mungo*, *V. radiata*, and *Glycine max* [46–48], *Triticum aestivum* and *Amaranthus tricolor* [49], and *Oryza sativa* [50] have shown retarded growth and reduced leaf area expansion in response to UV-B radiations [51]. The most likely reason for the reduction in growth is direct damage to DNA [52]. It has been observed that a plant under heat stress can delay cell division through reduced cell elongation. This affects the shoot net assimilation rates and the plant's total dry weight, ultimately reducing plant growth [53]. Heat stress also decreases stem growth, resulting in reduced plant height [54]. Following our results, similar reductions in plant height at higher temperatures have been reported in other crops, including recent reports in *Brassica juncea* [55] and *Oryza sativa* [56].

The results of a reduction in major growth and developmental parameters, such as total leaf area and fresh and dry weights, number of leaves, and height of plants (Table 3), obtained in this study under higher UV-B levels corroborate those found in *Arabidopsis thaliana* leaves [57], soybean [33], cotton [23], maize [27,58], *Phaseolus vulgaris* [59] and sweet potato [60]. One of the most common responses to UV-B is a leaf area decrease because of a reduction in cell division and expansion [61–63]. The decline in leaf area associated with the lower concentration of photosynthetic pigments seems to be the cause of the decrease in growth and the reduction in stem length and root dry mass as recorded under UV-B radiation treatment in *P. vulgaris*. These factors can result in lower absorption of sunlight and affect photosynthetic activity, leading to a decrease in photosynthesis, indirectly affecting plant growth [63]. In line with our findings, growth reduction under UV-B has also been reported in *Arabidopsis thaliana* [64] and *Capsicum annuum* [42]. In contrast, Nedunchezhian and Kulandaivelu [65] reported that slightly elevated UV-B radiation increases leaf area in cowpea. An increase in plant height, leaf area, and dry weight under elevated UV-B was also reported in *Ocimum basilicum* [66]. Leaf area reduction in rice has been recently reported under high temperatures alone and with elevated CO_2 [67]. High temperature and UV-B interaction decreased leaf area in *Brassica napus* [68].

The horticultural crops having a C_3 photosynthetic metabolism have shown beneficial effects indicating the increase in growth traits in onion [69,70] and tomato at 550 ppm CO_2 [71]. In perennial crops such as coconut (*Cocos nucifera* L.), studies indicate an increase in shoot height, leaf area, and shoot dry weight due to elevated CO_2 of up to 36% over chamber control [72,73]. We found all the shoot and root parameters increased under high CO_2 levels in both crops.

Approximately 20% of crops are sensitive to UV-B radiation regarding dry mass reduction [60,74]. We observed a decrease in marketable fresh weight and dry weight traits under all the treatments except CO_2 alone in the present study. A reaction to stress caused by UV-B radiation in plant development and metabolism could explain the reduction in dry and fresh leaf mass [75]. Fresh and dry weight reductions by 10–12% were reported in *Beta vulgaris* under UV-B [76]. Dai et al. [77] reported that, after a few weeks of UV-B exposure, the plant dry weight of rice was significantly reduced. Zuk-Golaszewska et al. [78] also reported a decrease in dry weight under high UV-B levels in *Avenafatua* and *Setariaviridis*.

On the contrary, a different response was found on broad bean and wheat, in which the plant's dry mass increased with the rising UV-B [79]. Like studies on brood bean and wheat, Zhang et al. [80] also reported an increase in whole plant dry weight in *Prunella vulgaris* plants when exposed to 15-day UV-B radiations in a growth chamber. This suggests that the UV-B effect is species/cultivar specific, and sometimes it benefits the growth and development of plants [81]. An increase in total dry weight was also observed in *Ocimum basilicum* and *Mentha piperita* under elevated temperatures [82]. Reduced plant weight under high temperature can also be related to decreased photosynthesis, increased transpiration [83], and, in turn, reduced water use efficiency [84]. Like our study under elevated temperatures, a decrease in dry weight has been recently reported in three *Brassica* sp. [85], *Brassica oleracea* [86], *Raphanus sativus* [87], and *Chenopodium quinoa* [88]. High temperature and UV-B interaction decreased leaf weight in *Brassica napus* [68]. Interaction of elevated temperatures and CO_2 increased plant height, the number of leaves, and the leaf area in *Fragaria* × *ananassa* [89], *Capsicum annuum* [90], and *Solanum lycopersicum* [91]. In contrast to our results, high temperature and elevated CO_2 reduced the whole-plant dry weight in rice [67].

The different effects of UV-B radiation and CO_2 on plant growth and development have been extensively studied for a wide range of horticultural and agronomic crops. Still, little work has considered UV-B and CO_2 interaction [41,92–97]. In an experiment conducted by Teramura et al. [92], soybean, wheat, and rice grown under two levels of CO_2 and UV-B showed increased total plant biomass in all three species under elevated CO_2. However, under the interaction of elevated CO_2 with enhanced UV-B radiation, these effects were eliminated in wheat and rice but remained in soybean. This indicates that the combined effects of UV-B and CO_2 are species-specific, and that UV-B can modify the positive effects of CO_2. Moreover, Ziska and Teramura [98] with rice, and Van de Staaij et al. [99] with wild ryegrass, have shown that the effects of CO_2 and UV-B radiation are independent. Our study revealed that elevated CO_2 can partially alleviate some of the adverse effects of UV-B radiation in *Brassica* sp. Similar results were observed by Brand et al. [100] in cotton. Qaderi et al. [101] experimented on the interactive effects of high temperature and UV-B levels in *Brassica napus*. They observed that the higher temperature with enhanced UVB negatively affected all the growth traits. These results confirm our findings.

Understanding root responses to changes in the aerial environment is essential in deciphering the crop responses to predicted climate changes [100]. Little is known about the effects of abiotics on the root system of kale and mustard compared to other major crops such as corn, rice, and cotton in the US Midsouth during seedling growth [102–104]. In the present study, elevated CO_2 concentration stimulated root growth, whereas high temperature and +UV-B either individually or in combination suppressed most root traits. Previous root studies on sorghum [105] and tomato [106] have reported significant root diameter, root volume, root length, and dry weight density under elevated CO_2 concentrations. Many studies have extensively documented the responses of plants to increasing atmospheric CO_2. However, the effects of elevated CO_2 on root dynamics have not been explored much, despite their importance for global carbon budgets and nutrient cycling in ecosystems. Therefore, the current data on the effects of elevated atmospheric CO_2 levels on root dynamics are insufficient to draw any conclusions. Different studies indicated a general increase in root growth under elevated CO_2 compared to ambient CO_2 levels [107,108].

In an experiment conducted by Sindhøj et al. [109], increased root growth was observed in nutrient-poor semi-natural grassland at an elevated CO_2 concentration. Moreover, elevated CO_2 affected root architecture through increased branching compared to ambient CO_2 in a shortgrass steppe. Root length, the number of roots, and the diameter of roots in the upper soil profile were also greater under such conditions [108]. In contrast to our findings, Ostonen et al. [110] reported decreased specific root lengths under elevated levels of CO_2.

Total root length, RSA, and RAD have been used to characterize root systems and evaluate their functional size [111]. These characteristics help predict nutrient uptake ability and performance under stress conditions. The root systems of kale and mustard at early developmental stages have not been adequately characterized. The present study investigated root structural parameters, including root development concerning RL, RCL, RSA, RV, RAD, root distribution pattern in the soil column, RS, and root branching. UV-B treatment alone caused significant reductions in all the root growth parameters. Reductions in root length under elevated temperatures have also been observed in *Solanum lycopersicum*, *Cucumis sativus*, and *Solanum melongena* [112].

Extreme temperatures have been observed to profoundly impact the growth and development of plant root systems (Table 4). The well-documented relationships between extreme temperatures and specific plant functions include nutrient uptake, photosynthesis, and carbon partitioning [29]. Elevated temperatures appear to impact mustard roots positively but led to a decrease in all root parameters in kale. Like our results in kale, Choi et al. [87] reported a reduction in root length and diameter under high temperatures in *Raphanus sativus* and *Brassica campestris*. A significant decrease in root parameters under high temperatures has also been reported in canola [113]. RLPV is directly related to the plants' water uptake ability because water is mainly absorbed passively, and generally reflects the development of lateral roots [114]. Plant roots optimize their root architecture to acquire water and essential nutrients. The number of root tips, forks, and crossings plays a vitally important role in the root architecture of a plant. They can enhance penetration through soil layers, ultimately leading to a positive effect on plant nutrient uptake. In the present study, root tips, forks, and crossings decreased significantly under UV-B treatment alone and together with elevated temperature and CO_2 treatment, indicating the harmful effects of multiple stresses on root architecture.

Although the adverse effects of CO_2 and the increase in other greenhouse gases in the environment, which appear to be a cause of global warming, are a global concern, elevated CO_2 may positively affect plants by mitigating the detrimental effects caused by UV-B radiation. Generally, high levels of atmospheric CO_2 have shown beneficial effects on plants, whereas enhanced levels of UV-B radiation are detrimental [115]. However, the relationship between these and other environmental factors, including temperature, light, drought, and salinity, is complex and has not been studied much. Therefore, multifactorial experiments must be undertaken to have a better understanding of plant growth and physiological responses to environmental stressors.

5. Conclusions

In the present study, the interactive effects of elevated CO_2, temperature, and UV-B radiation on the development of two *Brassica* sp. were quantified under a sunlit environment, similar to field conditions under optimum nutrient conditions. Plants grown at +UV-B alone or elevated temperatures produced shorter plants, with smaller leaf areas and shorter roots and reduced biomass. Elevated UV-B conditions had significant adverse effects on most shoot and root parameters, whereas +CO_2 resulted in an increase in all vegetative traits. The current study results indicate that high temperature and +UV-B may be major abiotic stressors that impact kale and mustard growth and development during the early season. Kale and mustard are some of the oldest green leafy vegetables globally, known for their health benefits; however, not much information is available on the interactions of various abiotic stresses in these two crops. Therefore, more research is required in this arena to better understand these two *Brassica* species. Moreover, improving early seedling growth and developmental response to abiotic stresses would benefit the current environment. Such improvements will be more apparent for vigorous plant growth in future projected climates.

Author Contributions: Conceptualization, K.R.R. and T.C.B.; Methodology, K.R.R., C.H.W. and T.C.B.; Software, A.S., C.H.W. and T.C.B.; Validation, K.R.R. and T.C.B.; Formal analysis, K.R.R. and T.C.B.; Investigation, A.S., K.R.R., C.H.W. and T.C.B.; Resources, K.R.R. and T.C.B.; Data curation, A.S., K.R.R., S.B. and T.C.B.; writing—original draft preparation, A.S.; writing—reviewing and editing, A.S., K.R.R., C.H.W., T.C.B., S.B., D.C. and W.G.; Visualization, A.S. and K.R.R.; Supervision, K.R.R.; Project administration, K.R.R.; Fund acquisition, K.R.R., T.C.B. and W.G. All authors have read and agreed to the published version of the manuscript.

Funding: This research is partially funded by the National Institute of Food and Agriculture, NIFA 2019-34263-30552 and MIS 043050, and the USDA-NIFA Hatch Project's work under accession number 149210.

Institutional Review Board Statement: Not applicable in this study.

Informed Consent Statement: Not applicable.

Data Availability Statement: Data is available with the corresponding author and will be shared upon request.

Acknowledgments: We thank David Brand and Thomas Horgan for their technical assistance and graduate students at the Environmental Plant Physiology Laboratory to help during data collection.

Conflicts of Interest: The authors declare no conflict of interest.

References

1. Lobell, D.B.; Field, C.B. Global scale climate–crop yield relationships and the impacts of recent warming. *Environ. Res. Lett.* **2007**, *2*, 014002. [CrossRef]
2. Battisti, D.S.; Naylor, R.L. Historical warnings of future food insecurity with unprecedented seasonal heat. *Science* **2009**, *323*, 240–244. [CrossRef] [PubMed]
3. Lobell, D.B.; Schlenker, W.; Costa-Roberts, J. Climate trends and global crop production since 1980. *Science* **2011**, *333*, 616–620. [CrossRef] [PubMed]
4. Deryng, D.; Conway, D.; Ramankutty, N.; Price, J.; Warren, R. Global crop yield response to extreme heat stress under multiple climate change futures. *Environ. Res. Lett.* **2014**, *9*, 034011. [CrossRef]
5. Lesk, C.; Rowhani, P.; Ramankutty, N. Influence of extreme weather disasters on global crop production. *Nature* **2016**, *529*, 84–87. [CrossRef]
6. Intergovernmental Panel on Climate Change, IPCC. Global warming of 1.5 °C. An IPCC special report on the impacts of global warming of 1.5 °C above pre-industrial levels and related global greenhouse gas emission pathways. In *The Context of Strengthening the Global Response to the Threat of Climate Change, Sustainable Development, and Efforts to Eradicate Poverty*; IPCC: Geneva, Switzerland, 2018.
7. Sewelam, N.; Brilhaus, D.; Bräutigam, A.; Alseekh, S.; Fernie, A.R.; Maurino, V.G. Molecular plant responses to combined abiotic stresses put a spotlight on unknown and abundant genes. *J. Exp. Bot.* **2020**, *71*, 5098–5112. [CrossRef]
8. Ahuja, I.; de Vos, R.C.; Bones, A.M.; Hall, R.D. Plant molecular stress responses face climate change. *Trend Plant Sci.* **2010**, *15*, 664–674. [CrossRef]
9. Ahammed, G.J.; Li, X.; Liu, A.; Chen, S. Physiological and defense responses of tea plants to elevated CO_2: A review. *Front. Plant Sci.* **2020**, *11*, 305. [CrossRef]
10. Satheesh, N.; Fanta, S. Kale: Review on nutritional composition, bioactive compounds, anti-nutritional factors, health beneficial properties and value-added products. *Cogent Food Agric.* **2020**, *6*, 1811048. [CrossRef]
11. Cartea, M.E.; Francisco, M.; Soengas, P.; Velasco, P. Phenolic compounds in *Brassica* vegetables. *Molecules* **2011**, *16*, 251–280. [CrossRef]
12. Lin, L.Z.; Harnly, J.M. Identification of the phenolic components of collard greens, kale, and Chinese broccoli. *J. Agric. Food Chem.* **2009**, *57*, 7401–7408. [CrossRef] [PubMed]
13. Fahey, J.W. *Brassica Encyclopedia of Food Sciences and Nutrition*; Caballero, B., Ed.; Academic Press: London, UK, 2003; pp. 606–615.
14. Cleary, B.V.M. Dietary fibre analysis. *Proc. Nutr. Soc.* **2003**, *62*, 3–9. [CrossRef]
15. Kim, Y.T.; Kim, B.K.; Park, K.Y. Antimutagenic and anti-cancer effects of leaf mustard and leaf mustard kimchi. *Prev. Nutr. Food Sci.* **2007**, *12*, 84–88. [CrossRef]
16. Dinkova-Kostova, A.T.; Kostov, R.V. Glucosinolates and isothiocyanates in health and disease. *Trends Mol. Med.* **2012**, *18*, 337–347. [CrossRef] [PubMed]
17. Tian, Y.; Deng, F. Phytochemistry and biological activity of Mustard (*Brassica juncea*): A review. *CyTA-J. Food.* **2020**, *18*, 704–718. [CrossRef]
18. Brandenberger, L.; Shrefler, J.; Rebek, E.; Damicone, J. *Cool Season Greens Production (Spinach, Collard, Kale, Mustard, Turnip, Leaf Lettuce)*; Decision of Agricultural Science and Natural Resources; Oklahoma State University: Stillwater, OK, USA, 2016; pp. 1–5.
19. Banadyga, A.A. Greens or "Potherbs"-Chard, Collards, Kale, Mustard, Spinach, New Zealand Spinach. *Agric. Inf. Bull.-US Dept. Agric.* **1977**, *409*, 163–170. Available online: https://naldc.nal.usda.gov/catalog/IND44315407 (accessed on 18 March 2022).
20. Allen, L.H., Jr.; Boote, K.J. Crop ecosystem responses to climatic change: Soybean. In *Climate Change and Global Crop Productivity*; Reddy, K.R., Hodges, H.F., Eds.; CABI Publishing: Wallingford, UK, 2000; pp. 133–160. [CrossRef]
21. Reddy, K.R.; Hodges, H.F. Crop ecosystem responses to climatic change: Cotton. In *Climate Change and Global Crop Productivity*; CABI Publishing: Wallingford, UK, 2000; pp. 161–187.
22. Ainsworth, E.A.; Davey, P.A.; Bernacchi, C.J.; Dermody, O.C.; Heaton, E.A.; Moore, D.J.; Long, S.P. A meta-analysis of elevated [CO_2] effects on soybean (*Glycine max*) physiology, growth and yield. *Glob. Change Biol.* **2002**, *8*, 695–709. [CrossRef]
23. Kakani, V.G.; Reddy, K.R.; Zhao, D.; Mohammed, A.R. Effects of ultraviolet-B radiation on cotton (*Gossypium hirsutum* L.) morphology and anatomy. *Ann. Bot.* **2003**, *91*, 817–826. [CrossRef]
24. Kakani, V.G.; Reddy, K.R.; Zhao, D.; Sailaja, K. Field crop responses to ultraviolet-B radiation: A review. *Agric. For. Meteorol.* **2003**, *120*, 191–218. [CrossRef]
25. Li, Y.; He, L.; Zu, Y. Intraspecific variation in sensitivity to ultraviolet-B radiation in endogenous hormones and photosynthetic characteristics of 10 wheat cultivars grown under field conditions. *S. Afr. J. Bot.* **2010**, *76*, 493–498. [CrossRef]

26. Reddy, K.R.; Singh, S.K.; Koti, S.; Kakani, V.G.; Zhao, D.; Gao, W.; Reddy, V.R. Quantifying corn growth and physiological responses to ultraviolet-B radiation for modeling. *Agron. J.* **2013**, *105*, 1367–1377. [CrossRef]
27. Singh, S.K.; Reddy, K.R.; Reddy, V.R.; Gao, W. Maize growth and developmental responses to temperature and ultraviolet-B radiation interaction. *Photosynthetica* **2014**, *52*, 262–271. [CrossRef]
28. Kataria, S.; Jajoo, A.; Guruprasad, K.N. Impact of increasing ultraviolet-B (UV-B) radiation on photosynthetic processes. *J. Photochem. Photobiol. B* **2014**, *137*, 55–66. [CrossRef] [PubMed]
29. Franco, J.A.; Bañón, S.; Vicente, M.J.; Miralles, J.; Martínez-Sánchez, J.J. Root development in horticultural plants grown under abiotic stress conditions—A review. *J. Hortic. Sci. Biotechnol.* **2011**, *86*, 543–556. [CrossRef]
30. Regent Instruments. *WinRHIZO Pro Software*; Version 2009c; Regent Instruments, Inc.: Québec, QC, Canada, 2009.
31. Reddy, K.R.; Kakani, V.G.; Zhao, D.; Koti, S.; Gao, W. Interactive effects of ultraviolet-B radiation and temperature on cotton physiology, growth, development and hyperspectral reflectance. *Photochem. Photobiol.* **2004**, *79*, 416–427. [CrossRef]
32. Koti, S.; Reddy, K.R.; Reddy, V.R.; Kakani, V.G.; Zhao, D. Interactive effects of carbon dioxide, temperature, and ultraviolet-B radiation on soybean (*Glycine max* L.) flower and pollen morphology, pollen production, germination, and tube lengths. *J. Exp. Bot.* **2005**, *56*, 725–736. [CrossRef]
33. Koti, S.; Reddy, K.R.; Kakani, V.G.; Zhao, D.; Gao, W. Effects of carbon dioxide, temperature and ultraviolet-B radiation and their interactions on soybean (*Glycine max* L.) growth and development. *Environ. Exp. Bot.* **2007**, *60*, 1–10. [CrossRef]
34. Reddy, K.R.; Kakani, V.G.; Hodges, H.F. Exploring the use of the environmental productivity index concept for crop production and modeling. In *Response of Crops to Limited Water: Understanding and Modeling Water Stress Effects on Plant Growth Processes*; Ahuja, L.R., Reddy, V.R., Saseendran, S.A., Yu, Q., Eds.; ASA-CSSA-SSSA: Madison, WI, USA, 2008; Volume 1, pp. 387–410. [CrossRef]
35. Reddy, K.R.; Hodges, H.F.; Read, J.J.; McKinion, J.M.; Baker, J.T.; Tarpley, L.; Reddy, V.T. Soil-Plant-Atmosphere-Research (SPAR) facility: A tool for plant research and modeling. *Biotronics* **2001**, *30*, 27–50. Available online: https://www.spar.msstate.edu/class/EPP-2008/Chapter%201/Reading%20material/Facilities/SPAR_Biotronics.pdf (accessed on 18 March 2022).
36. Murray, F.W. On the computation of saturation vapor pressure. *J. Appl. Meteorol.* **1967**, *6*, 203–204. [CrossRef]
37. Hewitt, E.J. Sand and water culture methods used in the study of plant nutrition. In *Technical Communications*; CABI Publishing: Wallingford, UK, 1952; p. 241.
38. McKinion, J.M.; Hodges, H.F. Automated system for measurement of evapotranspiration from closed environmental growth chambers. *Trans. ASAE* **1985**, *28*, 1825–1828. [CrossRef]
39. Timlin, D.; Fleisher, D.; Kim, S.H.; Reddy, V.; Baker, J. Evapotranspiration measurement in controlled environment chambers. *Agron. J.* **2007**, *99*, 166–173. [CrossRef]
40. SAS Institute. *SAS Guide to Macro Processing*; SAS Inst.: Cary, NC, USA, 2011; Volume 11.
41. Qaderi, M.M.; Reid, D.M. Growth and physiological responses of canola (*Brassica napus*) to UV-B and CO_2 under controlled environment conditions. *Physiol. Plant.* **2005**, *125*, 247–259. [CrossRef]
42. Rodríguez-Calzada, T.; Qian, M.; Strid, Å.; Neugart, S.; Schreiner, M.; Torres-Pacheco, I.; Guevara-González, R.G. Effect of UV-B radiation on morphology, phenolic compound production, gene expression, and subsequent drought stress responses in chili pepper (*Capsicum annuum* L.). *Plant Physiol. Biochem.* **2019**, *134*, 94–102. [CrossRef]
43. Lercari, B.; Sodi, F.; Di Paola, M.L. Photomorphogenic responses to UV radiation: Involvement of phytochrome and UV photoreceptors in the control of hypocotyl elongation in *Lycopersicon esculentum*. *Physiol. Plant* **1990**, *79*, 668–672. [CrossRef]
44. Corlett, J.E. Assessing the impact of UV-B radiation on the growth and yield of field crops. In *Plants and UV-B: Responses to Environmental Change*; Woodfin, R., Mepsted, R., Paul, N.D., Eds.; Cambridge University Press: Cambridge, UK, 1997; pp. 195–211.
45. Conner, J.K.; Zangori, L.A. A garden study of the effects of ultraviolet-B radiation on pollination success and lifetime female fitness in Brassica. *Oecologia* **1997**, *111*, 388–395. [CrossRef] [PubMed]
46. Mazza, C.A.; Boccalandro, H.E.; Giordano, C.V.; Battista, D.; Scopel, A.L.; Ballaré, C.L. Functional significance and induction by solar radiation of ultraviolet-absorbing sunscreens in field-grown soybean crops. *Plant Physiol.* **2000**, *122*, 117–126. [CrossRef] [PubMed]
47. Amudha, P.; Jayakumar, M.; Kulandaivelu, G. Impacts of ambient solar UV (280–400 nm) radiation on three tropical legumes. *J. Plant Biol.* **2005**, *48*, 284–291. [CrossRef]
48. Guruprasad, K.; Bhattacharjee, S.; Kataria, S.; Yadav, S.; Tiwari, A.; Baroniya, S.; Mohanty, P. Growth enhancement of soybean (*Glycine max*) upon exclusion of UV-B and UV-B/A components of solar radiation: Characterization of photosynthetic parameters in leaves. *Photosynth. Res.* **2008**, *96*, 115–115. [CrossRef]
49. Kataria, S.; Guruprasad, K.N. Exclusion of solar UV components improves growth and performance of *Amaranthus tricolor* varieties. *Sci. Hortic.* **2014**, *174*, 36–45. [CrossRef]
50. Teramura, A.H.; Ziska, L.H.; Sztein, A.E. Changes in growth and photosynthetic capacity of rice with increased UV-B radiation. *Physiol. Plant.* **1991**, *83*, 373–380. [CrossRef]
51. Sharma, S.; Chatterjee, S.; Kataria, S.; Joshi, J.; Datta, S.; Vairale, M.G.; Veer, V. A review on responses of plants to UV-B radiation related stress. In *UV-B Radiation: From Environmental Stressor to Regulator of Plant Growth*; Singh, V.P., Singh, S., Prasad, S.M., Parihar, P., Eds.; John Wiley & Sons: West Sussex, UK, 2017; pp. 75–97.

52. Giordano, C.V.; Galatro, A.; Puntarulo, S.; Ballaré, C.L. The inhibitory effects of UV-B radiation (280–315 nm) on Gunneramagellanica growth correlate with increased DNA damage but not with oxidative damage to lipids. *Plant Cell Environ.* **2004**, *27*, 1415–1423. [CrossRef]
53. Wahid, A.; Gelani, S.; Ashraf, M.; Foolad, M.R. Heat tolerance in plants: An overview. *Environ. Exp. Bot.* **2007**, *61*, 199–223. [CrossRef]
54. Prasad, P.V.V.; Staggenborg, S.A.; Ristic, Z. Impacts of drought and/or heat stress on physiological, developmental, growth, and yield processes of crop plants. In *Response of Crops to Limited Water: Understanding and Modeling Water Stress Effects on Plant Growth Processes*; Ahuja, L.R., Reddy, V.R., Saseendran, S.A., Yu, Q., Eds.; ASA-CSSA-SSSA: Madison, WI, USA, 2008; pp. 301–355. [CrossRef]
55. Chauhan, J.S.; Meena, M.L.; Saini, M.K.; Meena, D.R.; Singh, M.; Meena, S.S.; Singh, K.H. Heat stress effects on morpho-physiological characters of Indian Mustard (*Brassica juncea* L.). In Proceedings of the 16th Australian Research Assembly on Brassicas, Ballarat, Australia, 14–16 September 2009; pp. 91–97.
56. Schaarschmidt, S.; Lawas, L.M.F.; Glaubitz, U.; Li, X.; Erban, A.; Kopka, J.; Zuther, E. Season affects yield and metabolic profiles of rice (*Oryza sativa*) under high night temperature stress in the field. *Int. J. Mol. Sci.* **2020**, *21*, 3187. [CrossRef] [PubMed]
57. Poulson, M.E.; Boeger, M.R.T.; Donahue, R.A. Response of photosynthesis to high light and drought for Arabidopsis thaliana grown under a UV-B enhanced light regime. *Photosynth. Res.* **2006**, *90*, 79–90. [CrossRef]
58. Wijewardana, C.; Henry, W.B.; Gao, W.; Reddy, K.R. Interactive effects on CO_2, drought, and ultraviolet-B radiation on maize growth and development. *J. Photochem. Photobiol.* **2016**, *160*, 198–209. [CrossRef]
59. Wanzeler, R.B.; Zanetti, L.V.; Fantinato, D.E.; Gama, V.N.; Arrivabene, H.P.; de Almeida Leite, I.T.; Milanez, C.R.D. How does UV-B radiation affect the initial growth of common bean (*Phaseolus vulgaris* L.)? *Physiological and structural aspects. Braz. J. Dev.* **2019**, *5*, 26947–26958. [CrossRef]
60. Chen, Z.; Gao, W.; Reddy, K.R.; Chen, M.; Taduri, S.; Meyers, S.L.; Shankle, M.W. Ultraviolet (UV) B effects on growth and yield of three contrasting sweet potato cultivars. *Photosynthetica* **2020**, *58*, 37–44. [CrossRef]
61. Wargent, J.J.; Moore, J.P.; Roland Ennos, A.; Paul, N.D. Ultraviolet radiation as a limiting factor in leaf expansion and development. *Photochem. Photobiol.* **2009**, *85*, 279–286. [CrossRef]
62. Nogués, S.; Allen, D.J.; Morison, J.I.; Baker, N.R. Characterization of stomatal closure caused by ultraviolet-B radiation. *Plant Physiol.* **1999**, *121*, 489–496. [CrossRef]
63. Raghuvanshi, R.; Sharma, R.K. Response of two cultivars of *Phaseolus vulgaris* L. (French beans) plants exposed to enhanced UV-B radiation under mountain ecosystem. *Environ. Sci. Pollut. Res.* **2016**, *23*, 831–842. [CrossRef]
64. Morales, L.O.; Brosché, M.; Vainonen, J.; Jenkins, G.I.; Wargent, J.J.; Sipari, N.; Aphalo, P.J. Multiple roles for UV RESISTANCE LOCUS8 in regulating gene expression and metabolite accumulation in Arabidopsis under solar ultraviolet radiation. *Plant Physiol.* **2013**, *161*, 744–759. [CrossRef] [PubMed]
65. Nedunchezhian, N.; Kulandaivelu, G. Changes induced by ultraviolet-B (280–320 nm) to vegetative growth and photosynthetic characteristics in filed grown *Vigna unguiculate* L. *Plant Sci.* **1997**, *123*, 85–92. [CrossRef]
66. Sakalauskaite, J.; Viškelis, P.; Duchovskis, P.; Dambrauskiene, E.; Sakalauskiene, S.; Samuoliene, G.; Brazaityte, A. Supplementary UV-B irradiation effects on basil (*Ocimum basilicum* L.) growth and phytochemical properties. *J. Food Agric. Environ.* **2012**, *10*, 342–346.
67. Wang, W.; Cai, C.; He, J.; Gu, J.; Zhu, G.; Zhang, W.; Liu, G. Yield, dry matter distribution and photosynthetic characteristics of rice under elevated CO_2 and increased temperature conditions. *Field Crops Res.* **2020**, *248*, 107605. [CrossRef]
68. Son, K.H.; Ide, M.; Goto, E. Growth characteristics and phytochemicals of canola (*Brassica napus*) grown under UV radiation and low root zone temperature in a controlled environment. *Hortic. Environ. Biotechnol.* **2020**, *61*, 267–277. [CrossRef]
69. Daymond, A.J.; Wheeler, T.R.; Hadley, P.; Ellis, R.H.; Morison, J.I.L. The growth, development and yield of onion (*Allium cepa* L.) in response to temperature and CO_2. *J. Hortic. Sci.* **1997**, *72*, 135–145. [CrossRef]
70. Wurr, D.C.E.; Hand, D.W.; Edmondson, R.N.; Fellows, J.R.; Hannah, M.A.; Cribb, D.M. Climate change: A response surface study of the effects of CO_2 and temperature on the growth of beetroot, carrots and onions. *J. Agric. Sci.* **1998**, *131*, 125–133. [CrossRef]
71. Mamatha, H.; Rao, N.S.; Laxman, R.H.; Shivashankara, K.S.; Bhatt, R.M.; Pavithra, K.C. Impact of elevated CO_2 on growth, physiology, yield, and quality of tomato (*Lycopersicon esculentum* Mill) cv. Arka Ashish. *Photosynthetica* **2014**, *52*, 519–528. [CrossRef]
72. Kumar, S.N.; Bai, K.V.; Rajagopal, V.; Aggarwal, P.K. Simulating coconut growth, development and yield using infocrop-coconut model. *Tree Physiol.* **2008**, *28*, 1049–1058. [CrossRef]
73. Malhotra, S.K. Horticultural crops and climate change: A review. *Indian J. Agric. Sci.* **2017**, *87*, 12–22.
74. Teramura, A.H. Effects of ultraviolet-B radiation on the growth and yield of crop plants. *Physiol. Plant.* **1983**, *58*, 415–427. [CrossRef]
75. Boeger, M.R.T.; Poulson, M. Effects of ultraviolet-B radiation on leaf morphology of *Arabidopsis thaliana* (L.) Heynh. (Brassicaceae). *Acta Bot. Bras.* **2006**, *20*, 329–338. [CrossRef]
76. Karvansara, P.R.; Razavi, S.M. Physiological and biochemical responses of sugar beet (*Beta vulgaris* L.) to ultraviolet-B radiation. *PeerJ* **2019**, *7*, e6790. [CrossRef] [PubMed]
77. Dai, Q.; Peng, S.; Chavez, A.Q.; Vergara, B.S. Effects of UVB radiation on stomatal density and opening in rice (*Oryza sativa* L.). *Ann. Bot.* **1995**, *76*, 65–70. [CrossRef]

78. Zuk-Golaszewska, K.; Upadhyaya, M.K.; Golaszewski, J. The effect of UV-B radiation on plant growth and development. *Plant Soil Environ.* **2003**, *49*, 135–140. [CrossRef]
79. Al-Oudat, M.; Baydoun, S.A.; Mohammad, A. Effects of enhanced UV-B on growth and yield of two Syrian crops wheat (*Triticum durum* var. Horani) and broad beans (*Vicia faba*) under field conditions. *Environ. Exp. Bot.* **1998**, *40*, 11–16. [CrossRef]
80. Zhang, X.R.; Chen, Y.H.; Guo, Q.S.; Wang, W.M.; Liu, L.; Fan, J.; Li, C. Short-term UV-B radiation effects on morphology, physiological traits and accumulation of bioactive compounds in *Prunella vulgaris* L. *J. Plant Interact.* **2017**, *12*, 348–354. [CrossRef]
81. Hideg, É.; Jansen, M.A.; Strid, Å. UV-B exposure, ROS, and stress: Inseparable companions or loosely linked associates? *Trend Plant Sci.* **2013**, *18*, 107–115. [CrossRef]
82. Al Jaouni, S.; Saleh, A.M.; Wadaan, M.A.; Hozzein, W.N.; Selim, S.; Abd Elgawad, H. Elevated CO_2 induces a global metabolic change in basil (*Ocimum basilicum* L.) and peppermint (*Mentha piperita* L.) and improves their biological activity. *J. Plant Physiol.* **2018**, *224*, 121–131. [CrossRef]
83. Yang, X. Plants and Microclimate: A quantitative approach to environmental plant physiology. *Agric. For. Meteorol.* **1993**, *66*, 267–268. [CrossRef]
84. Shah, N.H.; Paulsen, G.M. Interaction of drought and high temperature on photosynthesis and grain-filling of wheat. *Plant Soil.* **2003**, *257*, 219–226. [CrossRef]
85. Angadi, S.V.; Cutforth, H.W.; Miller, P.R.; McConkey, B.G.; Entz, M.H.; Brandt, S.A.; Volkmar, K.M. Response of three Brassica species to high temperature stress during reproductive growth. *Can. J. Plant Sci.* **2000**, *80*, 693–701. [CrossRef]
86. Rodríguez, V.M.; Soengas, P.; Alonso-Villaverde, V.; Sotelo, T.; Cartea, M.E.; Velasco, P. Effect of temperature stress on the early vegetative development of *Brassica oleracea* L. *BMC Plant Biol.* **2015**, *15*, 145. [CrossRef] [PubMed]
87. Choi, E.Y.; Seo, T.C.; Lee, S.G.; Cho, I.H.; Stangoulis, J. Growth and physiological responses of Chinese cabbage and radish to long-term exposure to elevated carbon dioxide and temperature. *Hortic. Environ. Biotechnol.* **2011**, *52*, 376. [CrossRef]
88. Hinojosa, L.; Matanguihan, J.B.; Murphy, K.M. Effect of high temperature on pollen morphology, plant growth and seed yield in quinoa (*Chenopodium quinoa* Willd.). *J. Agron. Crops Sci.* **2019**, *205*, 33–45. [CrossRef]
89. Sun, P.; Mantri, N.; Lou, H.; Hu, Y.; Sun, D.; Zhu, Y.; Lu, H. Effects of elevated CO_2 and temperature on yield and fruit quality of strawberry (*Fragaria* × *ananassa* Duch.) at two levels of nitrogen application. *PLoS ONE* **2012**, *7*, e41000. [CrossRef]
90. Kumari, M.; Verma, S.C.; Bhardwaj, S.K. Effect of elevated CO_2 and temperature on crop growth and yield attributes of bell pepper (*Capsicum annuum* L.). *J. Agrometeorol.* **2019**, *21*, 1–6. [CrossRef]
91. Rangaswamy, T.C.; Sridhara, S.; Ramesh, N.; Gopakkali, P.; El-Ansary, D.O.; Mahmoud, E.A.; Abdel-Hamid, A.M. Assessing the impact of higher levels of CO_2 and temperature and their interactions on tomato (*Solanum lycopersicum* L.). *Plants* **2021**, *10*, 256. [CrossRef]
92. Teramura, A.H.; Sullivan, J.H.; Ziska, L.H. Interaction of elevated ultraviolet-B radiation and CO_2 on productivity and photosynthetic characteristics in wheat, rice, and soybean. *Plant Physiol.* **1990**, *94*, 470–475. [CrossRef]
93. Sullivan, J.H.; Teramura, A.H. The effects of UV-B radiation on loblolly pine. Interaction with CO_2 enhancement. *Plant Cell Environ.* **1994**, *17*, 311–317. [CrossRef]
94. Wand, S.J.; Midgley, G.F.; Musil, C.F. Physiological and growth responses of two African species, *Acacia karroo* and *Themedatriandra*, to combined increases in CO_2 and UV-B radiation. *Physiol. Plant* **1996**, *98*, 882–890. [CrossRef]
95. Gwynn-Jones, D.; Lee, J.A.; Callaghan, T.V. Effects of enhanced UV-B radiation and elevated carbon dioxide concentrations on a sub-arctic forest heath ecosystem. *Plant Ecol.* **1997**, *128*, 242–249. [CrossRef]
96. Visser, A.J.; Tosserams, M.; Groen, M.W.; Magendans, G.W.H.; Rozema, J. The combined effects of CO_2 concentration and solar UV-B radiation on faba bean grown in open-top chambers. *Plant Cell Environ.* **1997**, *20*, 189–199. [CrossRef] [PubMed]
97. Tosserams, M.; Visser, A.; Groen, M.; Kalis, G.; Magendans, E.; Rozema, J. Combined effects of CO_2concentration and enhanced UV-B radiation on faba bean. *Plant Ecol.* **2001**, *154*, 197–210. [CrossRef]
98. Ziska, L.H.; Teramura, A.H. CO_2 enhancement of growth and photosynthesis in rice (*Oryza sativa*): Modification by increased ultraviolet-B radiation. *Plant Physiol.* **1992**, *99*, 473–481. [CrossRef]
99. Van de Staaij, J.W.M.; Lenssen, G.M.; Stroetenga, M.; Rozema, J. The combined effects of elevated CO_2 levels and UV-B radiation on growth characteristics of *Elymus athericus* (=*E. pycnanathus*). *Vegetatio* **1993**, *104*, 433–439. [CrossRef]
100. Brand, D.; Wijewardana, C.; Gao, W.; Reddy, K.R. Interactive effects of carbon dioxide, low temperature, and ultraviolet-B radiation on cotton seedling root and shoot morphology and growth. *Front. Earth Sci.* **2016**, *10*, 607–620. [CrossRef]
101. Qaderi, M.M.; Basraon, N.K.; Chinnappa, C.C.; Reid, D.M. Combined effects of temperature, ultraviolet-B radiation, and watering regime on growth and physiological processes in canola (*Brassica napus*) seedlings. *Int. J. Plant Sci.* **2010**, *171*, 466–481. [CrossRef]
102. Wijewardana, C.; Hock, M.; Henry, B.; Reddy, K.R. Screening corn hybrids for cold tolerance using morphological traits for early-season seeding. *Crop Sci.* **2015**, *55*, 851–867. [CrossRef]
103. Reddy, K.R.; Brand, D.; Wijewardana, C.; Gao, W. Temperature effects on cotton seedling emergence, growth, and development. *Agron. J.* **2017**, *109*, 1379–1387. [CrossRef]
104. Singh, B.; Norvell, E.; Wijewardana, C.; Wallace, T.; Chastain, D.; Reddy, K.R. Assessing morphological characteristics of elite cotton lines from different breeding programmes for low temperature and drought tolerance. *J. Agron. Crop Sci.* **2018**, *204*, 467–476. [CrossRef]

105. Pritchard, S.G.; Prior, S.A.; Rogers, H.H.; Davis, M.A.; Runion, G.B.; Popham, T.W. Effects of elevated atmospheric CO_2 on root dynamics and productivity of sorghum grown under conventional and conservation agricultural management practices. *Agric. Ecosyst. Environ.* **2006**, *113*, 175–183. [CrossRef]
106. Cohen, I.; Rapaport, T.; Berger, R.T.; Rachmilevitch, S. The effects of elevated CO_2 and nitrogen nutrition on root dynamics. *Plant Sci.* **2018**, *272*, 294–300. [CrossRef] [PubMed]
107. Arnone, J.A.; Zaller, J.G.; Spehn, E.M.; Niklaus, P.A.; Wells, C.E.; Körner, C. Dynamics of root systems in native grasslands: Effects of elevated atmospheric CO_2. *New Phyol.* **2000**, *147*, 73–85. [CrossRef]
108. Milchunas, D.G.; Morgan, J.A.; Mosier, A.R.; LeCain, D.R. Root dynamics and demography in shortgrass steppe under elevated CO_2, and comments on minirhizotron methodology. *Glob. Change Biol.* **2005**, *11*, 1837–1855. [CrossRef]
109. Sindhøj, E.; Hansson, A.C.; Andrén, O.; Kätterer, T.; Marissink, M.; Pettersson, R. Root dynamics in a semi-natural grassland in relation to atmospheric carbon dioxide enrichment, soil water and shoot biomass. *Plant Soil.* **2000**, *223*, 255–265. [CrossRef]
110. Ostonen, I.; Püttsepp, Ü.; Biel, C.; Alberton, O.; Bakker, M.R.; Lõhmus, K.; Brunner, I. Specific root length as an indicator of environmental change. *Plant Biosyst.* **2007**, *141*, 426–442. [CrossRef]
111. Costa, C.; Dwyer, L.M.; Zhou, X.; Dutilleul, P.; Hamel, C.; Reid, L.M. Root morphology of contrasting maize genotypes. *Agron. J.* **2002**, *94*, 96–101. [CrossRef]
112. Canbay, S.; Polat, E. The effects of UV-B irradiation on development and quality of tomato, cucumber and eggplant seedlings. *Mediterr. Agric. Sci.* **2019**, *32*, 79–84. [CrossRef]
113. Wu, W.; Duncan, R.W.; Ma, B.L. Quantification of canola root morphological traits under heat and drought stresses with electrical measurements. *Plant Soil.* **2017**, *415*, 229–244. [CrossRef]
114. Gowda, V.R.; Henry, A.; Vadez, V.; Shashidhar, H.E.; Serraj, R. Water uptake dynamics under progressive drought stress in diverse accessions of the *Oryza* SNP panel of rice (*Oryza sativa*). *Funct. Plant Biol.* **2012**, *39*, 402–411. [CrossRef] [PubMed]
115. Rozema, J.; Van de Staaij, J.; Björn, L.O.; Caldwell, M. UV-B as an environmental factor in plant life: Stress and regulation. *Trend Ecol. Evol.* **1997**, *12*, 22–28. [CrossRef]

Article

Green-Synthesized Zinc Oxide Nanoparticles Mitigate Salt Stress in *Sorghum bicolor*

Tessia Rakgotho [1,2], Nzumbululo Ndou [1,2], Takalani Mulaudzi [1,*], Emmanuel Iwuoha [2], Noluthando Mayedwa [2] and Rachel Fanelwa Ajayi [2]

[1] Life Sciences Building, Department of Biotechnology, University of the Western Cape, Private Bag X17, Bellville 7535, South Africa; 3985506@myuwc.ac.za (T.R.); 3992677@myuwc.ac.za (N.N.)

[2] SensorLab, Department of Chemical Sciences, University of the Western Cape, Private Bag X17, Bellville 7535, South Africa; eiwuoha@uwc.ac.za (E.I.); 2548096@myuwc.ac.za (N.M.); fngece@uwc.ac.za (R.F.A.)

* Correspondence: tmulaudzi@uwc.ac.za; Tel.: +27-219-592-557

Abstract: Salinity is an abiotic stress that is responsible for more than 50% of crop losses worldwide. Current strategies to overcome salinity in agriculture are limited to the use of genetically modified crops and chemicals including fertilizers, pesticides and herbicides; however these are costly and can be hazardous to human health and the environment. Green synthesis of nanoparticles (NPs) is an eco-friendly and cost-effective method, and they might serve as novel biostimulants. This study investigated for the first time the efficiency of ZnO NPs, synthesized from *Agathosma betulina* to mitigate salt stress in *Sorghum bicolor*. Hexagonal wurtzite ZnO NPs of about 27.5 nm, were obtained. Sorghum seeds were primed with ZnO NPs (5 and 10 mg/L), prior to planting on potting soil and treatment with high salt (400 mM NaCl). Salt significantly impaired growth by decreasing shoot lengths and fresh weights, causing severe deformation on the anatomical (epidermis and vascular bundle tissue) structure. Element distribution was also affected by salt which increased the Na^+/K^+ ratio (2.9). Salt also increased oxidative stress markers (reactive oxygen species, malondialdehyde), enzyme activities (SOD, CAT and APX), proline, and soluble sugars. Priming with ZnO NPs stimulated the growth of salt-stressed sorghum plants, which was exhibited by improved shoot lengths, fresh weights, and a well-arranged anatomical structure, as well as a low Na^+/K^+ ratio (1.53 and 0.58) indicating an improved element distribution. FTIR spectra confirmed a reduction in the degradation of biomolecules correlated with reduced oxidative stress. This study strongly suggests the use of green-synthesized ZnO NPs from *A. betulina* as potential biostimulants to improve plant growth under abiotic stress.

Keywords: abiotic stress; green synthesis; priming; osmolytes; oxidative stress; salt; sorghum; buchu extract; ZnO NPs; antioxidant

Citation: Rakgotho, T.; Ndou, N.; Mulaudzi, T.; Iwuoha, E.; Mayedwa, N.; Ajayi, R.F. Green-Synthesized Zinc Oxide Nanoparticles Mitigate Salt Stress in *Sorghum bicolor*. *Agriculture* **2022**, *12*, 597. https://doi.org/10.3390/agriculture12050597

Academic Editors: Daniele Del Buono, Primo Proietti and Luca Regni

Received: 14 March 2022
Accepted: 21 April 2022
Published: 24 April 2022

1. Introduction

Nanotechnology is an emerging technology that has recently led to the synthesis of nano compounds (nanomaterials and nanoparticles) with special properties and the potential to be used as plant growth regulators and biostimulants in agriculture [1,2]. A few studies have demonstrated the effective use of different nanoparticles in the improvement of crop production under abiotic stresses [3–8]. Hence nanotechnology can be used in agriculture to meet targets in food production to feed the growing population [1].

Green synthesis of nanoparticles must be considered as the method of choice for plant application, due to its several advantages over conventional methods, since it is cost effective, eco-friendly, and can be easily scaled-up for increased production [9]. Additionally, there is no usage of high temperatures, energy, pressure, and toxic chemicals in green synthesis [10]. Plant extracts are the most common biological substrates used in the synthesis of nanoparticles since they are easily accessible and less toxic than microbes. These extracts contain secondary metabolites such as polysaccharides, polyphenolic compounds, amino acids, vitamins, and alkaloids among other compounds that act as reducing, stabilizing, and capping agents [11,12].

Metal oxide nanoparticles, specifically Zinc oxide nanoparticles (ZnO NPs) have recently gained a lot of attention in nanoscience due to their unique physicochemical properties and their various applications in biology, chemistry, medicine, and physics [11,13–15] and in recent years they have crossed over to agriculture [16,17]. Zinc oxide is nonorganic, cheap, and it has been regarded as a safe and non-toxic metal by the Food and Drug Administration (FDA). Several studies have revealed that a lower concentration of ZnO NPs can increase the growth and development of plants [18,19].

Salinity is among the major abiotic stresses that affect plant growth and development by altering several physiological and metabolic processes of plants [20–23]. It has affected approximately 6% of the total land surface globally and 20% of agricultural land, making salt stress the most serious environmental factor limiting the productivity of cultivated crops [24]. Salinity causes osmotic and ionic stress, which enhances the over-production of reactive oxygen species (ROS) causing lipid membrane damage, damage to biomolecules, and hence, cell death [23,25]. Plants overcome these effects by regulating several defense mechanisms including osmotic adjustments, induction of the antioxidant machinery, modulation of hormones, and some morphological and anatomical adaptations [23,26]

Sorghum bicolor (L.) Moench is an important grain crop ranked fifth in the world and second in Africa [24]. In African and Asian countries, sorghum is mainly used as a source of food for humans, whereas in other countries such as Australia, Brazil, and the United States of America, it is mostly used as a source of animal feed and energy production. It is an adaptable and a moderately drought and salt-tolerant cereal crop [27,28]; however, continuous exposure to abiotic stress can affect its growth and hence its productivity. Cultivating sorghum is very important because this will ensure its continuous use as a food source to combat food insecurity and in the production of bioenergy to solve the looming global energy crisis.

This study is the first to investigate the application of green synthesized ZnO NPs in *Sorghum bicolor* plant growth under salt stress. To better understand the effects of ZnO NPs as biostimulants, we evaluated their efficiency under normal and high salt (400 mM NaCl) stress conditions by analyzing growth attributes, anatomical structure, and the content of macronutrients. The extent of oxidative stress and the scavenging capacity on sorghum were also evaluated.

2. Materials and Methods

2.1. Preparation of the Plant Extract and the Green Synthesis of Zinc Oxide Nanoparticles

The synthesis of ZnO NPs completed in this study followed the original method described in [29] with modifications. *Agathosma betulina* (Buchu leaves) were purchased online at the Natural Essential Products Ltd. [https://essentiallynatural.co.za (accessed on 20 March 2020)] and used to prepare the green extract, which served as the capping agent [30]. About 10 g of grounded Buchu leaves were mixed with 250 mL distilled water (dH_2O) and boiled at 80 °C for 2 h. The mixture was filtered and centrifuged for 10 min at 6000 rpm and the supernatant was collected and used immediately or stored at 4 °C for further use.

Zinc nitrate hexahydrate (N_2O_6Zn) purchased from Sigma-Aldrich (Cat# 96482-500G, Lot # BCBJ7666V) was used as a precursor for the synthesis of ZnO nanoparticles (NPs). Synthesis was initiated by adding 3 g of Zinc nitrate into 250 mL of aqueous Buchu leaf extract and the mixture was boiled at 80 °C for 5 h. A color change of the mixture from pale yellow to dark brown appeared signifying the formation of ZnO NPs [31]. Synthesized ZnO NPs were freeze-dried to obtain ZnO NPs in powdered form, then calcined at 600 °C for 2 h to obtain a more crystallized structure of the nanoparticles. White powdered ZnO NPs were obtained and stored in airtight containers at room temperature when not in use or until further processing.

2.2. Characterization of Zinc Oxide Nanoparticles

The formation of the ZnO NPs was confirmed using Ultraviolet-Visible spectroscopy (Nicolett Evolution 100 from Thermo Electron Corporation, Johannesburg, South Africa) by observing peak formation within the 200–700 nm wavelength range. The phytochemical compositions that participated in the synthesis and the newly formed chemical compositions were determined using Fourier-transform infrared (FTIR) spectroscopy (PerkinElmer Spectrum 100-FTIR Spectrometer from PerkinElmer (Pty) Ltd., Midrand, South Africa) using a 400 to 4000 cm^{-1} spectral range. The morphology and size of the nanoparticles were determined using High-Resolution Scanning Electron Microscopy (Zeiss Auriga HR-SEM purchased from Carl Zeiss Microscopy GmbH, Jena, Germany), the quantity and presence of ZnO NPs were determined on SEM using Energy Dispersive X-ray Spectroscopy (EDX) performed on the Zeiss Auriga detector (Oxford Link-ISIS 300, Concord, MA, USA) [32], and High-Resolution Transmission Electron Microscopy (HRTEM) (Tecnai G2 F2O X-Twin HR-TEM purchased from FEI Company, Hillsboro, OR USA).

The phase purity and particle size of ZnO NPs were determined using the X-ray diffractometer, Brucker AXS (Germany) D8 advanced diffractometer unit. The Scherrer's equation (Equation (1)) was used to determine the crystalline size of the synthesized ZnO NPs [33]:

$$D = \frac{0.9\lambda}{\beta \cos \theta} \quad (1)$$

where D is the crystallite size, λ is the wavelength of x-ray used (1.5406 Å), β is the full width at half maximum (FWHM), and θ is the Bragg's angle.

2.3. Sorghum Seed Germination and Growth Conditions

Sorghum (*Sorghum bicolor* L. Moench) seeds purchased from Agricol, Brackenfell Cape Town, South Africa were disinfected as described previously [24].

For priming treatments, seeds were imbibed in distilled water (ddH_2O) only as the control and 5 and 10 mg/L ZnO NPs solutions for experiments. This was followed by overnight incubation in the dark at 25 °C with shaking at 600 rpm. Seeds were dried under the laminar flow and sown on filter paper placed on a plastic container (28 × 21 and 6 cm height) containing 50 mL of ddH_2O and germinated in the dark at 25 °C for 7 days.

Germinated seeds were sown on pots of sizes (18 × 14 and 6 cm height) and put inside a vessel container sized (21 × 16 and 5 cm height) containing a mixture of potting soil and vermiculite (2:1) and grown in the green house under controlled conditions (26 °C/22 °C day/night; 16 h/8 h light/dark regimes). After 14 days of growth, sorghum plants were treated with 100 mL of salt solution (400 mM NaCl-containing solution) every second day for 7 days. Sorghum plants were harvested on day seven after treatment, rinsed thoroughly with dH_2O and used immediately or stored at −80 °C for future use.

2.4. Growth Parameters

Shoot lengths were measured using a ruler in the mm range. Fresh weights (FW) of the shoots were weighed using a Mettle Toledo AE50 analytical balance (Marshall Scientific, Hampton, VA, USA). Dry Weights (DW) were determined after oven-drying the shoots at 70 °C for 72 h until a constant weight was obtained.

The anatomical structure (epidermis, xylem, and phloem) and element distribution were analyzed at the University of Cape Town, South Africa using High Resolution Scanning Electron Microscopy (HRSEM) and HRSEM-EDX as previously described in [34,35]. All spectra were analyzed using the built in Oxford Inca software suite. Samples were then imaged and collected using a Tescan MIRA field emission gun scanning electron microscope, operated at an accelerating voltage of 5 kV using an in-lens secondary electron detector.

2.5. *Physiological and Biochemical Analysis*

All spectrophotometric measurements in this study were performed using a Helios® Epsilon visible 8 nm bandwidth spectrophotometer (Thermo scientific Waltham, MA, USA) unless otherwise stated.

2.5.1. Histochemical Detection of Reactive Oxygen Species (ROS)

Histochemical detection of ROS was determined as described by [15]. For the detection of superoxide ($O_2^{\bullet-}$), leaves were immersed in 0.1% Nitroblue tetrazolium chloride (NBT) solution and incubated at 25 °C for 2 h in the dark.

The detection of H_2O_2 was performed by immersing sorghum leaves in 1 mg/mL 3′,3′-diaminobenzidine (DAB) solution and the mixture was incubated overnight in the dark. After incubation, all histochemical samples were boiled in 80% ethanol at 90 °C for 15 min to remove chlorophyll.

2.5.2. Malondialdehyde Content

Lipid peroxidation was determined by measuring malondialdehyde (MDA) content using the method described in [36]. A total of 100 mg of fresh sorghum plant materials were homogenized in 1 mL of 0.1% trichloroacetic acid (TCA (*w/v*)). Tubes were vortexed to mix well and centrifuged at 13,000 rpm (4°C) for 10 min. Small holes were created in the cap of the 2 mL Eppendorf tubes using a syringe needle to prevent the tubes from bursting due to pressure from the heat. About 400 uL of the supernatant was added into a 2 mL Eppendorf tube containing 1 mL of 0.5% TBA, followed by boiling at 80 °C in a water bath for 30 min. After incubation, the tubes were placed on ice for 5 min and centrifuged for another 5 min at 13,500 rpm (4 °C) to precipitate any remaining TBA. About 200 uL of the supernatant was transferred into a 96 well microtiter plate. The optical density was measured spectrophotometrically at 532 nm and 600 nm.

2.5.3. Fourier-Transform Infrared Spectroscopy (FTIR) Analysis of Biomolecules

The FTIR spectrum of the sorghum shoots was analyzed using a PerkinElmer Spectrum 100-FTIR Spectrometer [PerkinElmer (Pty) Ltd., Midrand, South Africa]. About 2 g of dry sorghum shoot tissues was analyzed where a wider window between 450 and 4000 cm^{-1} was considered.

2.5.4. Enzyme Activity Assays

Enzyme extraction was performed as described by the authors of [37], with a slight modification. About 100 mg of grounded sorghum plant material was homogenized in 100 mM sodium phosphate buffer (pH 8.0). The homogenized samples were centrifuged at 9000 rpm for 20 min at 4 °C. The supernatant was collected and was stored at 4 °C for future use. Activities of selected antioxidant enzymes including Superoxide dismutase (SOD, EC, 1.15.1.11), were estimated as described in [38]. Catalase (CAT, EC 1.11.16) and Ascorbate peroxidase (APX, EC 1.11.1.11) were estimated as described previously [38–40].

2.5.5. Proline Content

Proline content was measured as described in [34,41] with slight modifications. To be specific, pure proline was used as the standard to construct the standard curve instead of using the traditional toluene solution.

2.6. *Statistical Analysis*

All experiments were repeated at least four times and the data were statistically analyzed by the two-way ANOVA using GraphPad prism 9 (https://www.graphpad.com (accessed on 20 October 2021)). Data in the Figures and Tables represent the mean ± standard deviation. Statistical significance between the control and treated plants was determined by the Bonferroni's multiple comparison test and represented as *** = $p \leq 0.001$, ** = $p \leq 0.01$, and * = $p \leq 0.05$.

3. Results

3.1. Characterization of Zinc Oxide Nanoparticles

The Ultraviolet-Visible absorption spectra were used to determine the optical properties of the green-synthesized ZnO NPs (Figure S1A). Initially, the color of the reaction mixture was pale yellow, which changed to dark brown indicating the formation of nanoparticles. ZnO NPs and Buchu extracts have a wide absorption value and strong absorption peaks at 370 nm and 320 nm, respectively (Figure S1B).

Fourier-transform infrared spectroscopy (FTIR) investigated the functional groups of the phytoconstituents found in the Buchu extract that were responsible for capping and stabilizing the nanoparticles (Figure 1A). The spectrum revealed several bands from 500 to 4000 cm^{-1}. Both the Buchu extract (black line) and the ZnO NPs (red line) spectra exhibited strong peaks including those attributed to the O-H stretch, hydroxyl group, and H-bond (3427 cm^{-1}), to the C-H stretch (2926 cm^{-1}), and the C=C alkene group (1386 cm^{-1}). Some weak peaks were also observed at 1500 cm^{-1} and 1437 cm^{-1} that were attributed to the C=C and aromatic compound and at 1062 cm^{-1} attributed to the OH bending group. The spectrum evidently indicated ZnO NPs formation with an absorption peak at 460 cm^{-1}.

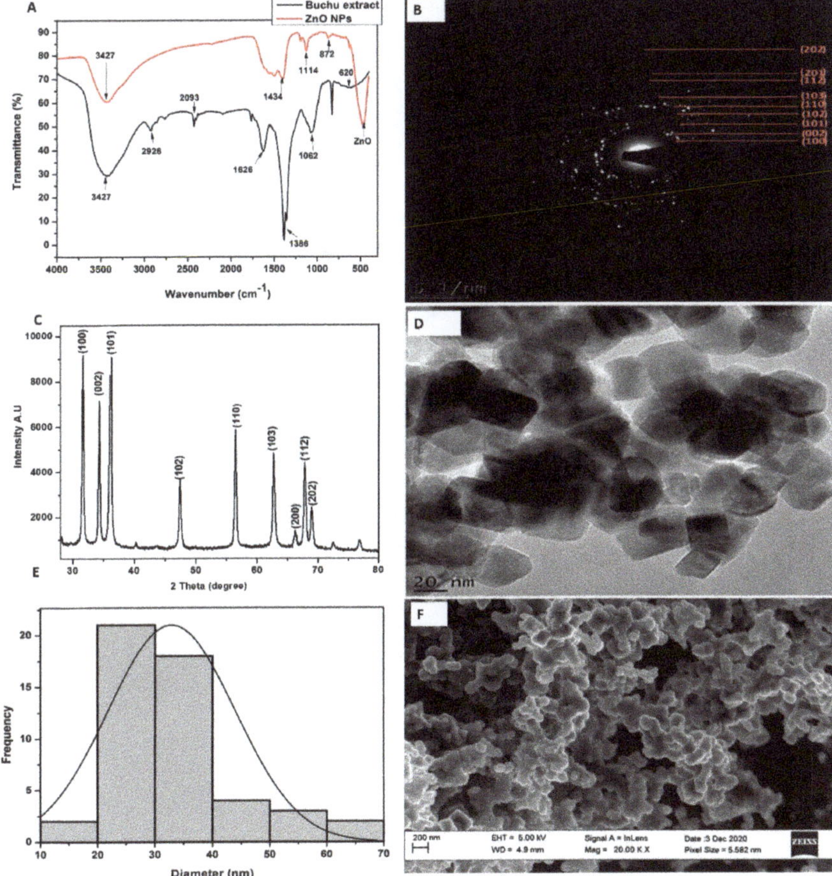

Figure 1. Optical and structural characterization of green-synthesized ZnO NPs using (**A**) FTIR, (**B**) SAED, (**C**) XRD, (**D**) HRTEM, (**E**) HRTEM micrograph, and (**F**) HRSEM analysis.

The selected area diffraction (SAED) pattern (Figure 1B) revealed diffraction peaks at (31.7°), (34.38°), (36.21°), (47.52°), (56.54°), (62.84°), (66.90°), (67.95°), and (69.05°), and were assigned to the (100), (002), (101), (102), (110), (103), (200), (112), and (201) planes on the *x*-ray Diffraction (XRD) spectrum (Figure 1C), respectively. The XDR spectrum of the synthesized ZnO NPs confirmed that the diffraction peaks were well matched with the hexagonal wurtzite structure of ZnO of the Joint Committee on Powder Diffraction Standards (JCPDS) Card Number 36-1451) as shown in Figure 1C and as described previously [42,43].

The average particle size of the prepared ZnO NPs as calculated based on the Debye-Scherrer's formula was 26 nm and matched with the TEM micrograph, which revealed the size of ZnO NPs to be between 20 nm and 30 nm (Figure 1E). HRTEM (Figure 1D) and HRSEM (Figure 1F) images of ZnO NPs revealed that the ZnO NPs were hexagonal, spherical, were of a granular nature and were agglomerated. The EDX spectrum indicated a high percentage of zinc and oxygen, which confirmed the presence of the elemental zinc and the oxygen signal from the ZnO NPs with a weight composition of 82.69% and 17.31%, respectively (Figure S1C).

3.2. The Effect of Salt and the Priming of the ZnO NPs on the Growth Attributes of *S. bicolor*

Plant growth was severely affected by salt stress (Figure 2A), as confirmed by a significant decrease in shoot length (46%) and fresh weight (73.75%) under salt stress (Figure 2B). However, a significant increase in shoot length was observed in plants primed with 5 mg/L (40.9%) and 10 mg/L (51.3%) ZnO NPs when treated with salt (Figure 2B). There was also a significant increase in the fresh weights of salt-treated plants when primed with 5 mg/L (38%) and 10 mg/L (60%) ZnO NPs (Figure 2C). There were no significant changes in the dry weights of all samples.

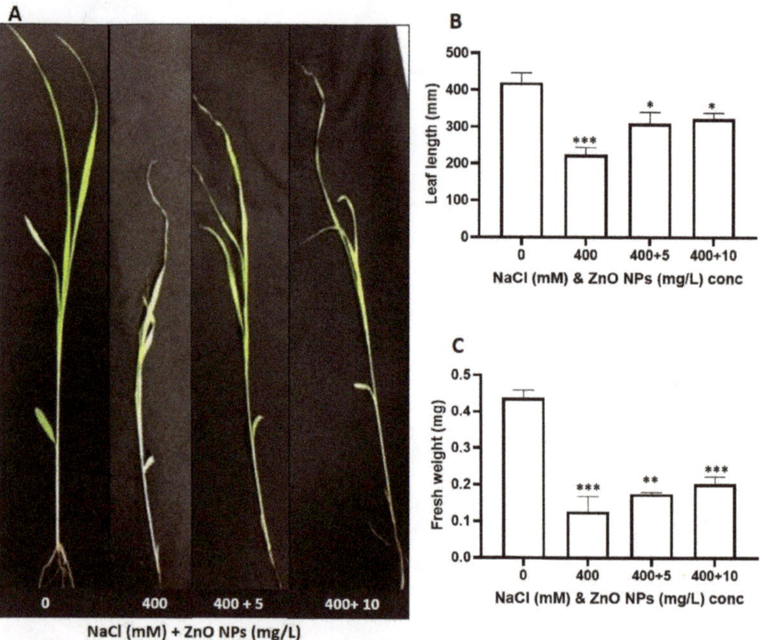

Figure 2. Effect of salt and the priming of the ZnO NPs on the growth parameters of sorghum. (**A**) Visual image, (**B**) shoot length, and (**C**) shoot fresh weight of sorghum plants in response to salt and the priming of the ZnO NPs. Error bars represent the SD calculated from three biological replicates. Statistical significance between the control and treated plants was determined by a two-way ANOVA performed on GraphPad 9.2.0, shown as *** = $p \leq 0.001$ ** = $p \leq 0.01$, and * = $p \leq 0.05$ according to the Bonferroni's multiple comparisons test.

To further understand the effect of salt stress and ZnO NPs on the growth of sorghum, the anatomical structure (epidermis, xylem, and phloem) and element distribution were analyzed using HRSEM and HRSEM/-EDX (Figure 3). HRSEM images revealed well-arranged and smooth epidermis layers in control (0 mM NaCl) plants (Figure 3A), whereas the epidermis from salt-treated plants revealed severe damage and showed signs of shrinkage (Figure 3B). Sorghum shoots from seeds that were primed with ZnO NPs before salt treatment showed an improvement as demonstrated by less deformation and shrinkage of the epidermis especially for the 5 mg/L ZnO NPs treatment (Figure 3C,D). The vascular bundle consisting of the xylem and phloem layers plays a significant role in the transport of water and nutrients [44]. The results revealed large, round, and wider openings of the xylem (Figure 3E, xylem presented by red arrows and phloem presented by white arrows), whereas the xylem of salt-treated sorghum plants was oval shaped, as if the walls had collapsed (Figure 3F). The xylem layers of the sorghum plants primed with ZnO NPs before salt treatment showed an improved surface structure and wider and round openings for both ZnO NPs concentrations (Figure 3G,H).

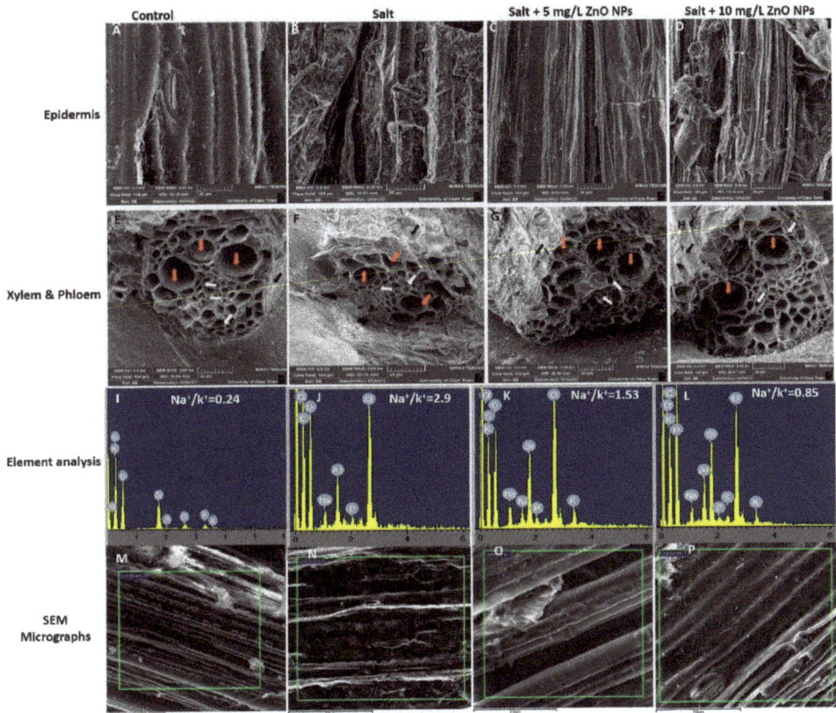

Figure 3. Effect of salt and the priming of the ZnO NPs on the anatomy and element distribution in sorghum plants. (**A–D**) Epidermis layers, (**E–H**) Xylem and phloem, (**I–L**) element content, (**M–P**) SEM micrographs for the EDX-investigated area.

The effect of salt and ZnO NPs on the absorption and transport of macronutrients was determined by analyzing the element distribution using HRSEM/-EDX, and more importantly, focusing on the Na^+ and K^+ and hence calculating the Na^+/K^+ ratio based on the Weight% (Figure 3I–L; Table S1). Sorghum plants treated with salt resulted in a 1.31-fold increase in Na^+, whereas the K^+ content decreased by 1.0-fold resulting in a Na^+/K^+ ratio of 2.9 (Figure 3J) as compared to the control (Na^+/K^+ = 2.13; Figure 3I). Priming with ZnO NPs prior to salt treatment resulted in a 0.56-fold decrease in Na^+, followed by a low Na^+/K^+ ratio of 1.53 (5 mg/L) (Figure 3K) and 0.85 (10 mg/L) (Figure 3L), when compared to plants treated with salt only (Figure 3I). SEM images for the EDX investigated area revealed significant morphological changes (Figure 3M–P). HRSEM images for control plants showed a smooth surface area (Figure 3M) as compared to the salt-treated plants (Figure 3N), which showed severe shrinkage. SEM images from the seedlings primed with ZnO NPs prior to salt treatment also revealed smooth epidermis layers with an improved surface area and less deformation (Figure 3O,P).

3.3. The Effect of Salt and the Priming of the ZnO NPs on Oxidative Damage in S. bicolor

Oxidative damage was determined by assaying the overproduction of Reactive Oxygen Species (ROS), which eventually led to the damage of lipid membranes, and other biomolecules (Figure 4). Histochemical detection of the superoxide anion ($O_2^{\bullet-}$) was performed using NBT whereby the production of $O_2^{\bullet-}$ was observed based on the appearance of blue spots on sorghum leaves (Figure 4A). Salt-treated plants had a high degree of $O_2^{\bullet-}$ production as seen by dark blue spots on the analyzed leaves as compared to the control, whereas the leaves of plants primed with ZnO NPs (5 mg/L and 10 mg/L) had reduced levels of $O_2^{\bullet-}$ under salt treatment (Figure 4A). Overproduction of H_2O_2 was detected using DAB staining, which was observed by the production of dark brownish spots (Figure 4B). Leaves of salt-treated plants exhibited more pronounced dark brownish color spots as compared to the control. However, priming with ZnO NPs (5 mg/L and 10 mg/L), prevented the over-accumulation of H_2O_2 since the dark brown spots were reduced to a higher degree as compared to the leaves of plants treated with salt only.

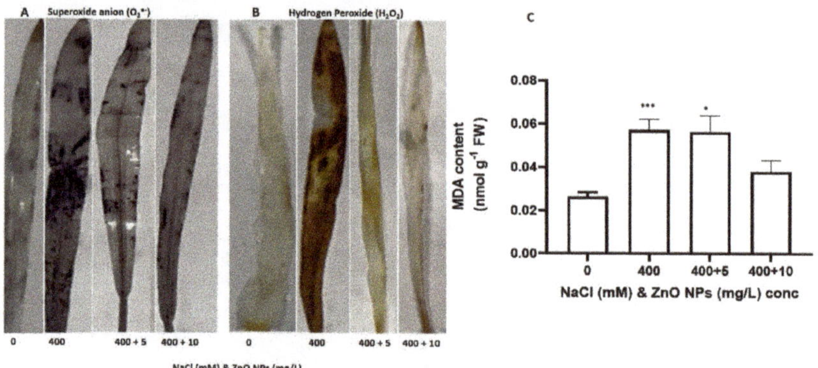

Figure 4. Effect of salt and the priming of ZnO NPs on ROS ($O_2^{\bullet-}$ and H_2O_2) accumulation and MDA content in sorghum. Histochemical detection of (**A**) $O_2^{\bullet-}$ and (**B**) H_2O_2. (**C**) Measurement of MDA content in salt-treated plants and salt-treated plants primed with ZnO NPs. Error bars represent the SD calculated from three biological replicates. Statistical significance between control and treated plants was determined by a two-way ANOVA performed on GraphPad 9.2.0, shown as *** = $p \leq 0.001$ and * = $p \leq 0.05$ according to the Bonferroni's multiple comparisons test.

Malondialdehyde content was measured to determine the extent of the oxidative damage on membrane lipids (Figure 4C). The results showed that a high salt concentration (400 mM NaCl) significantly increased MDA content by 115.38% as compared to the control. This indicates that salt stress caused significant and serious damage to sorghum membranes. Priming with 5 mg/L showed no difference, but 10 mg/L ZnO NPs decreased the MDA content by 32% in salt-treated plants, indicating the effectiveness of the ZnO NPs to protect plants from severe damage caused by salt.

To understand the effect of salt on the damage to the biomolecules, including carbohydrate, proteins, lipids, and phenolics, this study used FTIR spectroscopy and analyzed a wide spectral region from 450 to 4000 cm^{-1}, which was consistent in all samples (Figure 5). Peaks at 2916.39, 2103.69, and 1637.12 cm^{-1} showed the C-H and -C=C- stretching vibration, confirming the presence of alkanes and alkenes, respectively, thus confirming the presence of carbohydrates. The peak at 1370.74 cm^{-1} can be assigned a C-F stretching vibration for the alkyl halide group. Proteins were confirmed by the presence of amines seen by peaks at 1250.84 cm^{-1} and 1050.68 cm^{-1} (assigned C-N stretching, confirming the presence of aliphatic amines) and 899.62 cm^{-1} and 582.11 cm^{-1} (assigned N-H stretching vibration, confirming the presence of primary and secondary amines). These peaks were consistent in all samples, but big shifts were observed in the peaks that corresponded to carbohydrates and proteins between in all samples. Pronounced differences in the FTIR spectrum of control (black line) plants and those treated with salt only (red line) were clearly observed. Priming with 10 mg/L ZnO NPs (green) showed a better improvement in the spectral peak shift as compared to priming with 5 mg/L ZnO NPs (blue line) in salt-treated sorghum plants.

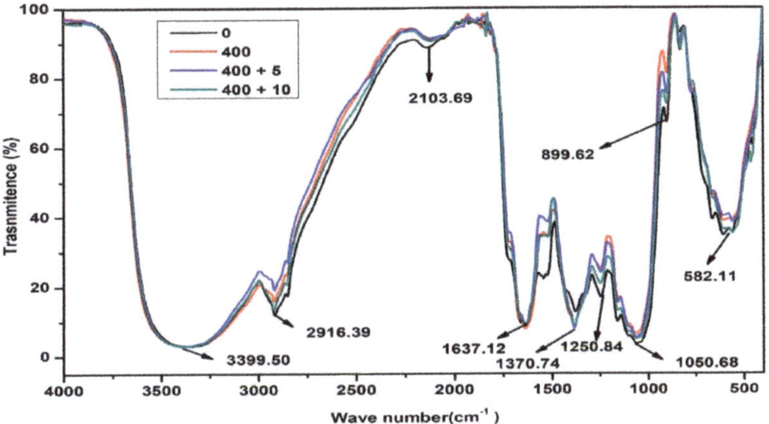

Figure 5. FTIR analysis of the effect of salt and the priming of the ZnO NPs on biomolecules in sorghum plants. Control (0 mM NaCl) plants (black), 400 mM NaCl/salt-treated (red), salt-treated plants primed with 5 mg/L ZnO NPs (blue) and 10 mg/L (green) ZnO NPs.

3.4. Effect of Salt and the Priming of the ZnONPs on the Antioxidative Capacity of S. bicolor

Salt markedly increased the antioxidant capacity of sorghum plants by inducing the activities of SOD (170%), CAT (131%), and APX (208%) as compared to their controls (Figure 6A–C). Priming with 5 mg/L ZnO NPs significantly reduced the antioxidant activities of SOD (58%), CAT (90%), and APX (61%), whereas priming with 10 mg/L ZnO NPs reduced the activities of SOD (68%) and APX (66%) (Figure 6B) to a higher degree, except for CAT (Figure 6C).

3.5. Effect of Salt and the Priming of ZnO NPs on the Osmoregulation in S. bicolor

Osmolytes such as proline and soluble sugars were analyzed to determine the level of osmotic balance in sorghum plants under salt (400 mM NaCl) stress and the effect of priming with ZnO NPs (Figure 7). Salt stress induced the accumulation of proline by 200% as compared to the control (Figure 7A). Compared with plants treated with salt only, plants primed with ZnO NPs prior to salt treatment resulted in very low proline content of 59.45% for 5 mg/L and 60.29% for 10 mg/L ZnO NPs priming (Figure 7A). Salt stress also induced the accumulation of soluble sugars by 160.60%, as compared to the control (Figure 7B). The priming of ZnO NPs had no significant effect on the content of soluble sugars (Figure 7B).

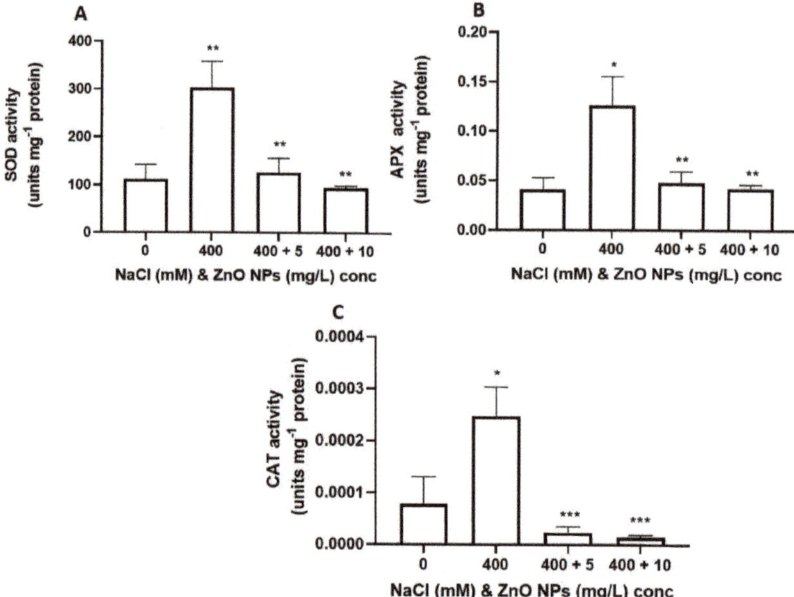

Figure 6. Effect of salt and the priming of ZnO NPs on the antioxidant enzyme activities of sorghum. (**A**) Superoxide Dismutase, (**B**) Ascorbate peroxidase and (**C**) Catalase activity in salt-treated plants, and those primed with ZnO NPs. Error bars represent the SD calculated from three biological replicates. Statistical significance between control and treated plants was determined by a two-way ANOVA performed on GraphPad 9.0, shown as *** = $p \leq 0.001$, ** = $p \leq 0.01$, and * = $p \leq 0.05$ according to the Bonferroni's multiple comparisons test.

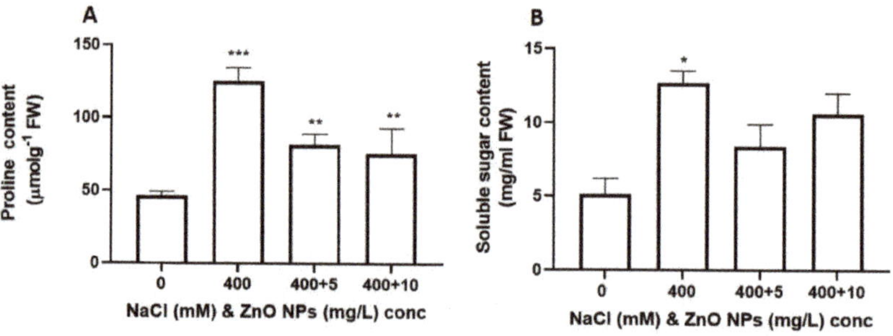

Figure 7. Effect of salt and ZnO NPs on osmolyte accumulation in sorghum. (**A**) Proline content and (**B**) soluble sugars in salt-treated plants and those primed with ZnO NPs. Error bars represent the SD calculated from three biological replicates. Statistical significance between control and treated plants was determined by a two-way ANOVA performed on GraphPad 9.2.0, shown as *** = $p \leq 0.001$ ** = $p \leq 0.01$, and * = $p \leq 0.05$ according to the Bonferroni's multiple comparisons test.

4. Discussion

In this study, ZnO NPs were synthesized following a green synthesis method using *Agathosma betulina* (Buchu) extract, which acted as a capping or reducing agent and a precursor for zinc nitrate [30,45–47]. As zinc nitrate was added to the extract, the color changed from pale yellow to dark brown, which indicated the formation of ZnO NPs [48].

The formation of ZnO NPs was further confirmed by observing the absorption peak at ~370 nm (Figure S1A). This is consistent with previous studies, which reported the absorption of ZnO NPs between 320 nm and 370 nm when synthesized using *Atalantia monophylla* [29], *Cinnamomum Tamala* leaf [49], and *Averrhoa bilimbi (L)* [50].

The FTIR spectra revealed the chemical bands that coded for physiochemical properties found in the plant extract that were responsible for the reduction process to form ZnO NPs as described previously [51]. The disappearance of 2926 and 2093 cm^{-1} spectral bands and the reduction in the intensity of other bands (1060 and 872 cm^{-1}) in ZnO NPs confirmed that phytochemicals were responsible for the reduction of irons to form NPs [31,52]. This was further confirmed by the appearance of the peak at 460 cm^{-1} that can be attributed to the presence of hexagonal wurtzite ZnO NPs, as reported in previous studies, which indicated that ZnO NPs exhibited peaks with similar shapes within the range of 400 to 680 cm^{-1}.

The size and structure of the synthesized ZnO NPs was estimated to be 26.03 nm and the hexagonal, spherical, granular nature, and agglomerated shape is due to the polarity and electrostatic attraction of the ZnO NPs [3]. These nanoparticles were proven to be pure and polycrystalline in nature as analyzed by the SAED and XRD. Similar trends were also observed in other studies where ZnO NPs were synthesized from *Phoenix roebelenii* [5] and *Cissus quadrangularis* [28].

Salinity is one of the major abiotic stresses affecting crop production causing over 50% of agricultural losses [37]. *Sorghum bicolor* is an important staple food crop worldwide widely grown in arid and semi-arid regions [53,54]; thus, it is important to prevent yield losses and maintain its growth and production. The present study demonstrated for the first time the positive effects of priming with green-synthesized ZnO NPs to mitigate the effects of salt stress on the growth of *Sorghum bicolor*, as observed by enhanced growth and an overall tolerance to salt stress. Most studies have demonstrated the exogenous application of nanoparticles (NPs) to mitigate abiotic stress including chilling on *Oryza Sativa L* [29], drought on tomatoes [4,18], salinity on *Glycine max* [7], *Lycopersicon esculentum* [23], *Eleusine coracana* L [55], *Brassica napus* [6] and *Abelmoschus esculentus* L. Moench [8].

Salt stress significantly reduced the growth attributes of sorghum plants including plant height, shoot length, and fresh weight (Figure 2). However, plants primed with ZnO NPs (5 mg/L and 10 mg/L) showed substantial growth improvement compared to non-primed plants under salt stress. The reduced growth might be due to osmotic stress, which affects the absorption and transport of nutrients and water resulting in declined turgidity and cell expansion; hence, reducing growth [56]. The reduced growth might also be due to the diversion of energy meant for growth to homeostasis and other metabolic processes [57], and this is supported by the correlation observed between reduced growth and the high content of osmolytes (Figure 7) in salt-treated sorghum plants.

The anatomical structure of salt-treated sorghum plants was severely affected (Figure 3A,B). The epidermis is an important tissue on the leaf that prevents water loss and invasion by pathogens, while the vascular bundle (xylem and phloem) participates in the transport of water and nutrients [58,59]. Both these structures showed shrinkage and deformation in salt-treated sorghum plants, while these effects were reversed in ZnO NPs-primed plants. Taken together these observations clearly suggest the role of ZnO NPs in promoting plant growth in harsh environments by protecting tissues that are important for transport of nutrients. This is true since Zn is a key element required for plant growth and development by mediating the biosynthesis of growth hormones and eventually activating cell division and enlargement [60,61].

The study further investigated the distribution of macronutrients (Figure 3C,D; Table S1) to understand the growth reduction induced by salt stress and its link with the affected anatomical structure. A correlation between the increase in toxic ions (Na^+ and Cl^-) and a decrease in the absorption of essential elements that are required for growth was observed in this study and this is evident by the high Na^+/K^+ ratio of 2.9 for salt-treated sorghum plants. Surprisingly, element distribution was improved in ZnO NPs-primed sorghum plants under salt stress, and this was supported by a decrease in the Na^+/K^+ ratio of 1.53 (5 mg/L ZnO NPs) and 0.85 (10 mg/L ZnO NPs). SEM micrographs of the investigated areas corroborated the element distribution, as shown by substantial changes in the morphology of the epidermis associated with shrinkage in salt-treated sorghum plants. This is true since salt stress causes membrane damage due to oxidative stress and these effects were reversed by priming with ZnO NPs. This result is supported by the role of Zn in maintaining membrane integrity, reducing the entry of toxic ions, mediating the translocation of nutrients, and thus maintaining cellular homeostasis [62–66].

To gain more insights into the effect of salt stress and the effectiveness of ZnO NPs on the growth of sorghum, this study also aimed to determine the extent of oxidative damage by assaying ROS accumulation, lipid peroxidation (Figure 4), and damage to biomolecules (Figure 5). When plants are exposed to abiotic stresses, they are generally associated with a high accumulation of free radicals, which induces the activity of antioxidants to regulate homeostasis and reduce lipid peroxidation [20]. ROS (e.g., H_2O_2) are signaling molecules at physiological levels, but their overproduction leads to the oxidative damage of membranes by increasing lipid peroxidation and causes damage to biomolecules [26,49]. Following the morphological attributes, sorghum plants treated with salt accumulated high levels of ROS and MDA, whereas in ZnO NPs-primed sorghum plants, low levels of these markers were observed. Any damage to the biomolecules was assessed using FTIR to inspect any shift in the spectral peaks of sorghum shoots (Figure 5). The FTIR spectra indicated that biomolecules in salt-treated sorghum plants were degraded, since shifts in the spectral peaks corresponding to carbohydrates (2916.39, 2103.69, and 1637.12 cm^{-1}) and proteins (1250.84 cm^{-1}, 1050.68 cm^{-1}, 899.62 cm^{-1}, and 582.11 cm^{-1}) were observed [65]. Only the peaks from the 10 mg/L ZnO NPs-primed sorghum plants were closer to those of the control, suggesting that the priming of ZnO NPs reduced salt-induced oxidative stress improving sorghum's response to salt stress and preventing the degradation of biomolecules.

ROS production induces the expression of antioxidant genes, leading to an increase in antioxidants, which enhance the scavenging capacity of ROS at the cellular level; hence, conferring tolerance against stress [67,68]. The results showed that the activities of superoxide dismutase (SOD), catalase (CAT), and ascorbate peroxidase (APX) were greatly increased in the sorghum shoots treated with salt (Figure 6). Similarly, other studies have shown the high activities of antioxidant enzymes in plants under salt stress [67,69]. SOD activity was greater than CAT and APX, suggesting that in addition to its role as the first line of defense in ROS scavenging, by converting superoxide anion ($O_2^{\bullet-}$) into H_2O_2 [6], SOD is a major contributor to the mediation of salt tolerance in sorghum. H_2O_2 is scavenged by peroxidases and catalase; however, APX has a higher affinity for H_2O_2 than CAT [70,71]. In this study, high APX activity was observed than CAT, suggesting that APX played a principal role in scavenging H_2O_2 as observed previously in sorghum [28,34]. In sorghum plants primed with ZnO NPs under salt stress, the activities of antioxidant enzymes significantly decreased, suggesting that ZnO NPs were effective in reducing the production of ROS; thus, preventing oxidative damage [20]. This is consistent with other studies as the application of ZnO NPs induced a tolerance to salt stress in soybeans [7], *Brassica nupus* [6], and *Abelmoschus esculentus* L. Moench [8].

Plants also survive stress through osmoregulation controlled by the accumulation of osmolytes including proline, soluble sugars, and glycine, to promote an osmotic balance at cellular level [72,73]. A significant increase in proline and soluble sugars was observed in salt-treated sorghum plants as compared to control plants; however, ZnO NPs-primed sorghum plants showed a reduced proline content with no significant changes observed for soluble sugars (Figure 7). Proline is an important signaling molecule that functions as a molecular chaperone by stabilizing and protecting membranes and proteins under abiotic stresses [70]. While soluble sugars play a similar role to proline as osmoprotectants [20], their level remained the same in ZnO NPs-primed sorghum plants under salt stress (Figure 7D). These results indicated that priming with ZnO NPs was efficient in improving osmoregulation in sorghum under salt stress, and hence there was no need for the plant to produce a high concentration of osmolytes when primed with ZnO NPs. High levels of proline under salt stress have been reported previously for sorghum [34,74–76]; however, the role of ZnO NPs in reducing proline content and hence the effects of salt stress in sorghum is reported for the first time in this study. In soybean, priming with ZnO NPs-was reported to be efficient in decreasing the proline content under salt stress [7].

A positive correlation was observed between the induction of proline content and the over-accumulation of ROS and the induction of antioxidant enzymes activities under salt stress. Similarly, their reduction in ZnO NPs-primed sorghum plants, might suggest that the accumulation of proline under salt stress was partly stimulated by the accumulation of toxic ions (Na^+ and Cl^-) and ROS. Thus, proline played a dual role to scavenge ROS and promote the activities of the antioxidant enzymes [71,72]. This is true since the ZnO NPs-primed sorghum plants presented exceptionally low levels of these traits to almost the same magnitude as that of the control, suggesting that there was no need for high proline production.

5. Conclusions

In summary, the results of this study indicate that sorghum growth is affected by high salt but priming with ZnO NPs stimulated a tolerance to salt and hence improved growth. Under salt stress, Na^+ over-accumulates in cells and causes osmotic and ionic stress, causing damage to membrane layers and affects the absorption of essential elements such as K^+ into cells. As K^+ is a component of most enzymes, its unavailability disrupts the normal functioning and regulation of cells [77–79]. Thus, these results partly propose that the mechanism of ZnO NPs-induced tolerance in sorghum is that ZnO NPs prevents damage to the epidermal layers and vascular bundle tissue, which leads to minimized water loss, and improved nutrient and water transport. This maintains ion homeostasis, thereby restricting the transport of Na^+ to the shoots and ensuring a low Na^+/K^+ ratio in sorghum shoots. This leads to the proper functioning of cells, prevents ROS accumulation and damage to biomolecules and hence improves sorghum growth under salt stress. However, these observations will require further experimental analysis by assaying the transcripts of genes encoding the Na^+ and K^+ pumps. Most importantly, these results provide a novel insight into the mechanism of the salinity response of sorghum as mediated by priming with ZnO NPs and the use of 10 mg/L is recommended.

Supplementary Materials: The following supporting information can be downloaded at: https://www.mdpi.com/article/10.3390/agriculture12050597/s1, Figure S1: Characterization of green-synthesized ZnO NPs using, Ultraviolet-Visible absorption spectra for ZnO NPs and Buchu extract at (A) 400–4000 nm wavelength, (B) 300–500 nm wavelength range, (C) histogram analysis of the size of ZnO NPs, (C), element composition of the ZnO NPs sample; Table S1: Overall element distribution analyzed by SEM-EDX spectroscopy in sorghum shoots.

Author Contributions: Conceptualization, T.M. and R.F.A.; methodology, T.R., N.N., N.M.; software, E.I., T.M. and R.F.A.; validation, T.M., E.I., R.F.A., formal analysis and investigation, T.R., N.N., N.M.; resources, T.M., R.F.A., E.I., data curation, T.M.; writing—original draft preparation, T.R.; writing—review and editing, T.M., N.N., N.M., E.I. and R.F.A., visualization, and supervision, T.M., E.I. and R.F.A.; project administration, T.M., R.F.A., E.I.; funding acquisition, T.M., E.I., R.F.A. All authors have read and agreed to the published version of the manuscript.

Funding: This research was funded by the National Research Foundation of South Africa under the Black Academics Advancement (UID: 112201) and Thuthuka Institutional Funding (UID: 121939, 116258 and 118469), Grants.

Institutional Review Board Statement: Not applicable.

Informed Consent Statement: Not applicable.

Data Availability Statement: Not applicable.

Acknowledgments: We would like to acknowledge the colleagues at the department of Biotechnology, UWC and the SensorLab at Chemical Sciences department, University of the Western Cape, for providing the space and most of the equipment that are situated in the laboratories available for use.

Conflicts of Interest: The authors declare no conflict of interest. The funders had no role in the design of the study; in the collection, analyses, or interpretation of data; in the writing of the manuscript, or in the decision to publish the results.

References

1. Chhipa, H. Applications of nanotechnology in agriculture. In *Methods in Microbiology*; Elsevier: Amsterdam, The Netherlands, 2019; pp. 115–142.
2. Kumar, V.; Guleria, P.; Kumar, V.; Yadav, S.K. Gold nanoparticle exposure induces growth and yield enhancement in Arabidopsis thaliana. *Sci. Total Environ.* **2013**, *461*, 462–468. [CrossRef] [PubMed]
3. Song, Y.; Jiang, M.; Zhang, H.; Li, R. Zinc Oxide Nanoparticles Alleviate Chilling Stress in Rice (*Oryza sativa* L.) by Regulating Antioxidative System and Chilling Response Transcription Factors. *Molecules* **2021**, *26*, 2196. [CrossRef] [PubMed]
4. El-Zohri, M.; Al-Wadaani, N.A.; Bafeel, S.O. Foliar Sprayed Green Zinc Oxide Nanoparticles Mitigate Drought-Induced Oxidative Stress in Tomato. *Plants* **2021**, *10*, 2400. [CrossRef] [PubMed]
5. Haripriya, P.; Stella, P.M.; Anusuya, S. Foliar spray of zinc oxide nanoparticles improves salt tolerance in finger millet crops under glass-house condition. *SCIOL Biotechnol.* **2018**, *1*, 20–29.
6. El-Badri, A.M.; Batool, M.; Wang, C.; Hashem, A.M.; Tabl, K.M.; Nishawy, E.; Kuai, J.; Zhou, G.; Wang, B. Selenium and zinc oxide nanoparticles modulate the molecular and morpho-physiological processes during seed germination of *Brassica napus* under salt stress. *Ecotoxicol. Environ. Saf.* **2021**, *225*, 112695. [CrossRef]
7. Gaafar, Y.Z.A.; Ziebell, H. Comparative study on three viral enrichment approaches based on RNA extraction for plant virus/viroid detection using high-throughput sequencing. *PLoS ONE* **2020**, *15*, e0237951. [CrossRef]
8. Alabdallah, N.M.; Alzahrani, H.S. The potential mitigation effect of ZnO nanoparticles on [*Abelmoschus esculentus* L. Moench] metabolism under salt stress conditions. *Saudi J. Biol. Sci.* **2020**, *27*, 3132–3137. [CrossRef]
9. Tsegaye, M.M.; Chouhan, G.; Fentie, M.; Tyagi, P.; Nand, P. Therapeutic Potential of Green Synthesized Metallic Nanoparticles against *Staphylococcus aureus*. *Curr. Drug Res. Rev.* **2021**, *13*, 172–183. [CrossRef]
10. Jamkhande, P.G.; Ghule, N.W.; Bamer, A.H.; Kalaskar, M.G. Metal nanoparticles synthesis: An overview on methods of preparation, advantages and disadvantages, and applications. *J. Drug Deliv. Sci. Technol.* **2019**, *53*, 101174. [CrossRef]
11. Bandeira, M.; Giovanela, M.; Roesch-Ely, M.; Devine, D.M.; Da Silva Crespo, J. Green synthesis of zinc oxide nanoparticles: A review of the synthesis methodology and mechanism of formation. *Sustain. Chem. Pharm.* **2020**, *15*, 100223. [CrossRef]
12. Islam, F.; Shohag, S.; Uddin, M.J.; Islam, M.R.; Nafady, M.H.; Akter, A. Exploring the Journey of Zinc Oxide Nanoparticles (ZnO-NPs) toward Biomedical Applications. *Materials* **2022**, *15*, 2160. [CrossRef] [PubMed]
13. Agarwal, H.; Kumar, S.V.; Rajeshkumar, S. A review on green synthesis of zinc oxide nanoparticles—An eco-friendly approach. *Resour.-Effic. Technol.* **2017**, *3*, 406–413. [CrossRef]
14. Mirzaei, H.; Darroudi, M. Zinc oxide nanoparticles: Biological synthesis and biomedical applications. *Ceram. Int.* **2017**, *43*, 907–914. [CrossRef]
15. Jiang, J.; Pi, J.; Cai, J. The Advancing of Zinc Oxide Nanoparticles for Biomedical Applications. *Bioinorg. Chem. Appl.* **2018**, *2018*, 1062562. [CrossRef] [PubMed]
16. Faizan, M.; Bhat, J.A.; Chen, C.; Alyemeni, M.N.; Wijaya, L.; Ahmad, P.; Yu, F. Zinc oxide nanoparticles (ZnO-NPs) induce salt tolerance by improving the antioxidant system and photosynthetic machinery in tomato. *Plant Physiol. Biochem.* **2021**, *161*, 122–130. [CrossRef] [PubMed]
17. Zulfiqar, F.; Ashraf, M. Nanoparticles potentially mediate salt stress tolerance in plants. *Plant Physiol. Biochem.* **2021**, *160*, 257–268. [CrossRef]
18. Singh, A.; Singh, N.B.; Hussain, I.; Singh, H.; Singh, S.C. Plant-nanoparticle interaction: An approach to improve agricultural practices and plant productivity. *Int. J. Pharm. Sci. Invent.* **2015**, *4*, 25–40.
19. Raskar, S.V.; Laware, S.L. Effect of zinc oxide nanoparticles on cytology and seed germination in onion. *Int. J. Curr. Microbiol. Appl. Sci.* **2014**, *3*, 467–473.
20. Gupta, S.; Schillaci, M.; Walker, R.; Smith, P.; Watt, M.; Roessner, U. Alleviation of salinity stress in plants by endophytic plant-fungal symbiosis: Current knowledge, perspectives and future directions. *Plant Soil* **2020**, *461*, 219–244. [CrossRef]
21. Thakur, M.; Bhattacharya, S.; Khosla, P.K.; Puri, S. Improving production of plant secondary metabolites through biotic and abiotic elicitation. *J. Appl. Res. Med. Aromat. Plants* **2018**, *12*, 1–12. [CrossRef]
22. Arif, Y.; Singh, P.; Siddiqui, H.; Bajguz, A.; Hayat, S. Salinity induced physiological and biochemical changes in plants: An omic approach towards salt stress tolerance. *Plant Physiol. Biochem.* **2020**, *156*, 64–77. [CrossRef] [PubMed]
23. Yang, Q.; Chen, Z.-Z.; Zhou, X.-F.; Yin, H.-B.; Li, X.; Xin, X.-F.; Hong, X.-H.; Zhu, J.-K.; Gong, Z. Overexpression of SOS (Salt Overly Sensitive) Genes Increases Salt Tolerance in Transgenic Arabidopsis. *Mol. Plant* **2009**, *2*, 22–31. [CrossRef] [PubMed]
24. Satish, L.; Shilpha, J.; Pandian, S.K.; Rency, A.S.; Rathinapriya, P.; Ceasar, S.A.; Largia, M.J.V.; Kumar, A.A.; Ramesh, M. Analysis of genetic variation in sorghum (*Sorghum bicolor* (L.) Moench) genotypes with various agronomical traits using SPAR methods. *Gene* **2015**, *576*, 581–585. [CrossRef] [PubMed]
25. Rahman, A.; Nahar, K.; Hasanuzzaman, M.; Fujita, M. Calcium Supplementation Improves Na+/K+ Ratio, Antioxidant Defense and Glyoxalase Systems in Salt-Stressed Rice Seedlings. *Front. Plant Sci.* **2016**, *7*, 609. [CrossRef]

26. Foyer, C.H.; Noctor, G. Redox homeostasis and antioxidant signaling: A metabolic interface between stress perception and physiological responses. *Plant Cell* **2005**, *17*, 1866–1875. [CrossRef]
27. Krishnamurthy, V.; Shukla, J. Intraseasonal and Seasonally Persisting Patterns of Indian Monsoon Rainfall. *J. Clim.* **2007**, *20*, 3–20. [CrossRef]
28. Costa, P.H.A.D.; Neto, A.D.D.A.; Bezerra, M.A.; Prisco, J.T.; Gomes-Filho, E. Antioxidant-enzymatic system of two sorghum genotypes differing in salt tolerance. *Braz. J. Plant Physiol.* **2005**, *17*, 353–362. [CrossRef]
29. Thema, F.; Manikandan, E.; Dhlamini, M.; Maaza, M. Green synthesis of ZnO nanoparticles via *Agathosma betulina* natural extract. *Mater. Lett.* **2015**, *161*, 124–127. [CrossRef]
30. Khan, Z.U.H.; Sadiq, H.M.; Shah, N.S.; Khan, A.U.; Muhammad, N.; Hassan, S.U.; Tahir, K.; Safi, S.Z.; Khan, F.U.; Imran, M.; et al. Greener synthesis of zinc oxide nanoparticles using *Trianthema portulacastrum* extract and evaluation of its photocatalytic and biological applications. *J. Photochem. Photobiol. B Biol.* **2019**, *192*, 147–157. [CrossRef]
31. Chau, T.P.; Brindhadevi, K.; Krishnan, R.; Alyousef, M.A.; Almoallim, H.S.; Whangchai, N.; Pikulkaew, S. A novel synthesis, analysis and evaluation of *Musa coccinea* based zero valent iron nanoparticles for antimicrobial and antioxidant. *Environ. Res.* **2022**, *206*, 112770. [CrossRef]
32. Saad, R.; Gamal, A.; Zayed, M.; Ahmed, A.M.; Shaban, M.; BinSabt, M.; Rabia, M.; Hamdy, H. Fabrication of ZnO/CNTs for Application in CO_2 Sensor at Room Temperature. *Nanomaterials* **2021**, *11*, 3087. [CrossRef] [PubMed]
33. Holzwarth, U.; Gibson, N. The Scherrer equation versus the 'Debye-Scherrer equation'. *Nat. Nanotechnol.* **2011**, *6*, 534. [CrossRef] [PubMed]
34. Mulaudzi, T.; Hendricks, K.; Mabiya, T.; Muthevhuli, M.; Ajayi, R.F.; Mayedwa, N.; Gehring, C.; Iwuoha, E. Calcium Improves Germination and Growth of *Sorghum bicolor* Seedlings under Salt Stress. *Plants* **2020**, *9*, 730. [CrossRef] [PubMed]
35. Rui, M.; Ma, C.; Hao, Y.; Guo, J.; Rui, Y.; Tang, X.; Zhao, Q.; Fan, X.; Zhang, Z.; Hou, T.; et al. Iron Oxide Nanoparticles as a Potential Iron Fertilizer for Peanut (*Arachis hypogaea*). *Front. Plant Sci.* **2016**, *7*, 815. [CrossRef]
36. Nakano, Y.; Asada, K. Hydrogen Peroxide is Scavenged by Ascorbate-specific Peroxidase in Spinach Chloroplasts. *Plant Cell Physiol.* **1981**, *22*, 867–880.
37. Vijayakumar, S.; Mahadevan, S.; Arulmozhi, P.; Sriram, S.; Praseetha, P.K. Green synthesis of zinc oxide nanoparticles using Atalantia monophylla leaf extracts: Characterization and antimicrobial analysis. *Mater. Sci. Semicond. Process.* **2018**, *82*, 39–45. [CrossRef]
38. Ramanarayanan, R.; Bhabhina, N.M.; Dharsana, M.V.; Nivedita, C.V.; Sindhu, S. Green synthesis of zinc oxide nanoparticles using extract of *Averrhoa bilimbi* (L) and their photoelectrode applications. *Mater. Today Proc.* **2018**, *5*, 16472–16477. [CrossRef]
39. Agarwal, H.; Nakara, A.; Menon, S.; Shanmugam, V. Eco-friendly synthesis of zinc oxide nanoparticles using *Cinnamomum Tamala* leaf extract and its promising effect towards the antibacterial activity. *J. Drug Deliv. Sci. Technol.* **2019**, *53*, 101212. [CrossRef]
40. Aebi, H. [13] Catalase in vitro. In *Methods in Enzymology*; Elsevier: Amsterdam, The Netherlands, 1984; pp. 121–126.
41. Dhindsa, R.S.; Plumb-Dhindsa, P.; Thorpe, T.A. Leaf Senescence: Correlated with Increased Levels of Membrane Permeability and Lipid Peroxidation, and Decreased Levels of Superoxide Dismutase and Catalase. *J. Exp. Bot.* **1981**, *32*, 93–101. [CrossRef]
42. Shashanka, R.; Esgin, H.; Yilmaz, V.M.; Caglar, Y. Fabrication and characterization of green synthesized ZnO nanoparticle based dye-sensitized solar cells. *J. Sci. Adv. Mater. Devices* **2020**, *5*, 185–191. [CrossRef]
43. Al-Kordy, H.M.H.; Sabry, S.A.; Mabrouk, M.E.M. Statistical optimization of experimental parameters for extracellular synthesis of zinc oxide nanoparticles by a novel haloalaliphilic *Alkalibacillus* sp. W7. *Sci. Rep.* **2021**, *11*, 1–14. [CrossRef] [PubMed]
44. Słupianek, A.; Dolzblasz, A.; Sokołowska, K. Xylem Parenchyma—Role and Relevance in Wood Functioning in Trees. *Plants* **2021**, *10*, 1247. [CrossRef] [PubMed]
45. Gupta, M.; Tomar, R.S.; Kaushik, S.; Mishra, R.K.; Sharma, D. Effective Antimicrobial Activity of Green ZnO Nano Particles of *Catharanthus roseus*. *Front. Microbiol.* **2018**, *9*, 2030. [CrossRef] [PubMed]
46. Sharmila, G.; Thirumarimurugan, M.; Muthukumaran, C. Green synthesis of ZnO nanoparticles using *Tecoma castanifolia* leaf extract: Characterization and evaluation of its antioxidant, bactericidal and anticancer activities. *Microchem. J.* **2018**, *145*, 578–587. [CrossRef]
47. Ogunyemi, S.O.; Abdallah, Y.; Zhang, M.; Fouad, H.; Hong, X.; Ibrahim, E.; Masum, M.I.; Hossain, A.; Mo, J.; Li, B. Green synthesis of zinc oxide nanoparticles using different plant extracts and their antibacterial activity against *Xanthomonas oryzae* pv. oryzae. *Artif. Cells Nanomed. Biotechnol.* **2019**, *47*, 341–352. [CrossRef]
48. Lyimo, G.V. Green Synthesised Zinc Oxide Nanoparticles and Their Antifungal Effect on Candida albicans Biofilms. Master's Thesis, University of the Western Cape, Western Cape, South Africa, 2020.
49. Khan, M.T.; Ahmed, S.; Shah, A.A.; Shah, A.N.; Tanveer, M.; El-Sheikh, M.A.; Siddiqui, M.H. Influence of Zinc Oxide Nanoparticles to Regulate the Antioxidants Enzymes, Some Osmolytes and Agronomic Attributes in *Coriandrum sativum* L. Grown under Water Stress. *Agronomy* **2021**, *11*, 2004. [CrossRef]
50. Aldeen, T.S.; Mohamed HE, A.; Maaza, M. ZnO nanoparticles prepared via a green synthesis approach: Physical properties, photocatalytic and antibacterial activity. *J. Phys. Chem. Solids* **2022**, *160*, 110313. [CrossRef]
51. Sathappan, S.; Kirubakaran, N.; Gunasekaran, D.; Gupta, P.K.; Verma, R.S.; Sundaram, J. Green Synthesis of Zinc Oxide Nanoparticles (ZnO NPs) Using *Cissus quadrangularis*: Characterization, Antimicrobial and Anticancer Studies. *Proc. Natl. Acad. Sci. India Sect. B Boil. Sci.* **2021**, *91*, 289–296. [CrossRef]

52. Afzal, S.; Sharma, D.; Singh, N.K. Eco-friendly synthesis of phytochemical-capped iron oxide nanoparticles as nano-priming agent for boosting seed germination in rice (*Oryza sativa* L.). *Environ. Sci. Pollut. Res.* **2021**, *28*, 40275–40287. [CrossRef]
53. Rosenow, D.; Quisenberry, J.; Wendt, C.; Clark, L. Drought tolerant sorghum and cotton germplasm. *Agric. Water Manag.* **1983**, *7*, 207–222. [CrossRef]
54. Punia, S. Barley starch modifications: Physical, chemical and enzymatic—A review. *Int. J. Biol. Macromol.* **2020**, *144*, 578–585. [CrossRef] [PubMed]
55. Parihar, P.; Singh, S.; Singh, R.; Singh, V.P.; Prasad, S.M. Effect of salinity stress on plants and its tolerance strategies: A review. *Environ. Sci. Pollut. Res.* **2014**, *22*, 4056–4075. [CrossRef] [PubMed]
56. Yadav, A.N. Plant microbiomes for sustainable agriculture: Current research and future challenges. *Plant Microbiomes Sustain. Agric.* **2020**, *25*, 475–482. [CrossRef]
57. Atkin, O.K.; Macherel, D. The crucial role of plant mitochondria in orchestrating drought tolerance. *Ann. Bot.* **2008**, *103*, 581–597. [CrossRef]
58. Pacheco-Silva, N.V.; Donato, A.M. Morpho-anatomy of the leaf of *Myrciaria glomerata*. *Rev. Bras. Farm.* **2016**, *26*, 275–280. [CrossRef]
59. Hwang, B.G.; Ryu, J.; Lee, S.J. Vulnerability of Protoxylem and Metaxylem Vessels to Embolisms and Radial Refilling in a Vascular Bundle of Maize Leaves. *Front. Plant Sci.* **2016**, *7*, 941. [CrossRef]
60. Ashraf, U.; Mahmood, M.H.-U.; Hussain, S.; Abbas, F.; Anjum, S.A.; Tang, X. Lead (Pb) distribution and accumulation in different plant parts and its associations with grain Pb contents in fragrant rice. *Chemosphere* **2020**, *248*, 126003. [CrossRef]
61. Elsheery, N.I.; Helaly, M.N.; El-Hoseiny, H.M.; Alam-Eldein, S.M. Zinc Oxide and Silicone Nanoparticles to Improve the Resistance Mechanism and Annual Productivity of Salt-Stressed Mango Trees. *Agronomy* **2020**, *10*, 558. [CrossRef]
62. Jiang, G.; Keller, J.; Bond, P.L. Bond, Determining the long-term effects of H2S concentration, relative humidity and air temperature on concrete sewer corrosion. *Water Res.* **2014**, *65*, 157–169. [CrossRef]
63. Weisany, W.; Sohrabi, Y.; Heidari, G.; Siosemardeh, A.; Ghassemi-Golezani, K. Changes in antioxidant enzymes activity and plant performance by salinity stress and zinc application in soybean ('*Glycine max*' L.). *Plant Omics* **2012**, *5*, 60–67.
64. He, G.; Wu, Y.; Zhang, Y.; Zhu, Y.; Liu, Y.; Li, N.; Li, M.; Zheng, G.; He, B.; Yin, Q.; et al. Addition of Zn to the ternary Mg–Ca–Sr alloys significantly improves their antibacterial properties. *J. Mater. Chem. B* **2015**, *3*, 6676–6689. [CrossRef] [PubMed]
65. Aktaş, H.; Abak, K.; Öztürk, L.; Çakmak, İ. The effect of zinc on growth and shoot concentrations of sodium and potassium in pepper plants under salinity stress. *Turk. J. Agric. For.* **2007**, *30*, 407–412.
66. Tolay, I. The impact of different Zinc (Zn) levels on growth and nutrient uptake of Basil (*Ocimum basilicum* L.) grown under salinity stress. *PLoS ONE* **2021**, *16*, e0246493. [CrossRef] [PubMed]
67. Gill, S.S.; Tuteja, N. Reactive oxygen species and antioxidant machinery in abiotic stress tolerance in crop plants. *Plant Physiol. Biochem.* **2010**, *48*, 909–930. [CrossRef]
68. Venkatachalam, P.; Priyanka, N.; Manikandan, K.; Ganeshbabu, I.; Indiraarulselvi, P.; Geetha, N.; Muralikrishna, K.; Bhattacharya, R.; Tiwari, M.; Sharma, N.; et al. Enhanced plant growth promoting role of phycomolecules coated zinc oxide nanoparticles with P supplementation in cotton (*Gossypium hirsutum* L.). *Plant Physiol. Biochem.* **2017**, *110*, 118–127. [CrossRef] [PubMed]
69. Ali, S.; Rizwan, M.; Qayyum, M.F.; Ok, Y.S.; Ibrahim, M.; Riaz, M.; Arif, M.S.; Hafeez, F.; Al-Wabel, M.I.; Shahzad, A.N. Biochar soil amendment on alleviation of drought and salt stress in plants: A critical review. *Environ. Sci. Pollut. Res.* **2017**, *24*, 12700–12712. [CrossRef]
70. Hayat, S.; Hayat, Q.; Alyemeni, M.N.; Wani, A.S.; Pichtel, J.; Ahmad, A. Role of proline under changing environments: A review. *Plant Signal. Behav.* **2012**, *7*, 1456–1466. [CrossRef]
71. Mittler, R. Oxidative stress, antioxidants and stress tolerance. *Trends Plant Sci.* **2002**, *7*, 405–410. [CrossRef]
72. Slama, I.; Abdelly, C.; Bouchereau, A.; Flowers, T.; Savouré, A. Diversity, distribution and roles of osmoprotective compounds accumulated in halophytes under abiotic stress. *Ann. Bot.* **2015**, *115*, 433–444. [CrossRef]
73. Upadhyaya, H.; Dutta, B.K.; Panda, S.K. Zinc Modulates Drought-Induced Biochemical Damages in Tea [*Camellia sinensis* (L) O Kuntze]. *J. Agric. Food Chem.* **2013**, *61*, 6660–6670. [CrossRef]
74. El Omari, R.; Ben Mrid, R.; Chibi, F.; Nhiri, M. Involvement of phosphoenolpyruvate carboxylase and antioxydants enzymes in nitrogen nutrition tolerance in *Sorghum bicolor* plants. *Russ. J. Plant Physiol.* **2016**, *63*, 719–726. [CrossRef]
75. El-Haddad ES, H.; O'Leary, J.W. Effect of salinity and K/Na ratio of irrigation water on growth and solute content of *Atriplex amnicola* and *Sorghum bicolor*. *Irrig. Sci.* **1994**, *14*, 127–133. [CrossRef]
76. Heidari, A.; Younesi, H.; Mehraban, Z. Removal of Ni(II), Cd(II), and Pb(II) from a ternary aqueous solution by amino functionalized mesoporous and nano mesoporous silica. *Chem. Eng. J.* **2009**, *153*, 70–79. [CrossRef]
77. Horie, T.; Schroeder, J.I. Sodium Transporters in Plants. Diverse Genes and Physiological Functions. *Plant Physiol.* **2004**, *136*, 2457–2462. [CrossRef]
78. Ward, P.S.; Patel, J.; Wise, D.R.; Abdel-Wahab, O.; Bennett, B.D.; Coller, H.A.; Cross, J.R.; Fantin, V.R.; Hedvat, C.V.; Perl, A.E.; et al. The Common Feature of Leukemia-Associated IDH1 and IDH2 Mutations Is a Neomorphic Enzyme Activity Converting α-Ketoglutarate to 2-Hydroxyglutarate. *Cancer Cell* **2010**, *17*, 225–234. [CrossRef]
79. Zhu, J.-K. Salt and drought stress signal transduction in plants. *Annu. Rev. Plant Biol.* **2002**, *53*, 247–273. [CrossRef]

Article

Effects of Selenium-Methionine against Heat Stress in Ca^{2+}-Cytosolic and Germination of Olive Pollen Performance

Alberto Marco Del Pino [1], Luca Regni [1,*], Alessandro Di Michele [2], Alessandra Gentile [3], Daniele Del Buono [1], Primo Proietti [1] and Carlo Alberto Palmerini [1]

1 Department of Agricultural, Food and Environmental Sciences (DSA3), University of Perugia, Borgo XX Giugno 74, 06121 Perugia, Italy; alberto.delpino@unipg.it (A.M.D.P.); daniele.delbuono@unipg.it (D.D.B.); primo.proietti@unipg.it (P.P.); carlo.palmerini@unipg.it (C.A.P.)

2 Department of Physics and Geology, University of Perugia, Via Pascoli, 06123 Perugia, Italy; alessandro.dimichele@unipg.it

3 Department of Agriculture, Food and Environment, University of Catania, Piazza Università, 95131 Catania, Italy; gentilea@unict.it

* Correspondence: luca.regni@unipg.it

Abstract: Climate change (CC), which causes temperatures to rise steadily, is causing global warming. Rising temperatures can reduce plant yield and affect pollen characteristics. In particular, heat stress strongly influences pollen viability for its sensitivity to this extreme environmental condition. This work evaluated the effect of heat stress on olive pollen after in vitro incubation at different temperatures (20, 30, and 40 °C). Furthermore, the potential of selenium-methionine (Se-met) in mitigating the detrimental effects of heat stress on olive pollen was investigated. In particular, how thermal stress can affect pollen was evaluated by testing the effect of temperature on pollen germinability and morphology and cytosolic Ca^{2+} content. The results suggest that the heat stress at 40 °C caused a marked reduction in the germination rate, changes in the morphology of the external pollen wall, and a decreased response to Ca^{2+}-agonist agents. On the contrary, in vitro treatment of pollen with Se-met improved the germination rate and Ca^{2+}-cytosolic homeostasis under heat stress conditions and confirmed the protective role of this compound in containing the hydrogen peroxide (H_2O_2) toxicity. Therefore, this study revealed that organic selenium could play a crucial role in promoting heat tolerance in olive tree pollen.

Keywords: *Olea europaea* L.; selenium; heat stress; Ca^{2+}-cytosolic; pollen germination

Citation: Del Pino, A.M.; Regni, L.; Di Michele, A.; Gentile, A.; Del Buono, D.; Proietti, P.; Palmerini, C.A. Effects of Selenium-Methionine against Heat Stress in Ca^{2+}-Cytosolic and Germination of Olive Pollen Performance. *Agriculture* **2022**, *12*, 826. https://doi.org/10.3390/agriculture12060826

Academic Editor: Urs Feller

Received: 2 May 2022
Accepted: 2 June 2022
Published: 8 June 2022

1. Introduction

Environmental stress is a significant issue and is already considered one of the main factors limiting crop growth, production, and yield [1]. In addition, ongoing climate change (CC) must be considered in this context, as it can further exacerbate the adverse environmental stress on crop systems [2]. Among the effects of CC, rising temperatures, which significantly impact agricultural systems, will play an even more crucial role over time [3–5]. Indeed, global warming is expected to negatively affect agriculture, with a temperature increase of 1–3 °C expected by the 21st century [6]. This will result in a significant reduction in crop yields and quality [6].

High temperatures can cause significant crop changes, altering their morphology, physiology, and biochemistry [7]. In particular, high temperature can reduce plant growth (roots and shoots) and biomass production, cause premature leaves senescence, hinder the ability of seeds to germinate, and decrease pollen viability [8]. In addition, exposure to high temperatures can induce severe physiological and biochemical changes in crops. The most frequently observed events are an increase in respiration and membrane permeability, a decrease in photosynthesis, and the production and accumulation of reactive oxygen species (ROS) [9]. In fact, high temperatures can cause ROS overproduction and, consequently, their accumulation in cells to very high concentrations. In addition, ROS can be very toxic to cells due to their reactivity toward many cellular components [7].

ROS can also control specific molecular signals, including those related to cytosolic Ca^{2+}. Ca is essential for plant nutrition and plays a dual role as a structural component of cell walls and membranes and as an intracellular second messenger [7]. In particular, as a messenger, this element is involved in numerous processes concerning pollen tube growth and fertilisation and the response to abiotic stresses [10]. Therefore, Ca homeostasis must be finely controlled and maintained [7].

Higher plants have a specialised sexual reproduction system and can produce abundant pollen that is transported long distances by wind or insects during habitat colonisation [11]. After landing on the stigma in angiosperms, the dehydrated pollen rapidly hydrates and begins to germinate. Germination of the pollen grain and proper pollen tube elongation are essential processes in plant sexual reproduction [12,13]. Nevertheless, high temperatures can damage the reproductive tissues of plants, causing asynchrony between the development of male and female floral structures and the formation of defective gametes and fertility problems [14]. Likewise, floral receptivity has a critical role in pollination dynamics and reproductive success, with consequences for fruit production [5,14,15]. In this regard, ROS accumulation under stress conditions can lead to pollen infertility, with detrimental effects and repercussions on agricultural production [16,17]. The correlation between Ca^{2+} dynamics and ROS during pollination and pollen tube formation has been widely described [11,18,19]. ROS act as agonists, stimulating the Ca^{2+} mobilisation from internal stores and triggering its entry into the cell from the extracellular spaces [16,17,20–23].

About the olive tree, this crop is adaptable to severe summer conditions, i.e., excessive heat load, low rainfall, and high daily irradiation [24,25]. However, due to CC, the gradual increase in temperatures can compromise this plant, hampering some stages of reproductive growth and development and the quality of the olive oil [25]. In addition, high temperatures may anticipate full flowering and shorten the duration of the flowering period. Despite this, the effects on pollen production and yield have not been sufficiently studied and understood to date [25,26]. However, recent scientific evidence has revealed the involvement and positive action of selenium (Se) on the cytosolic Ca^{2+} homeostasis and olive pollen germination [10,17,27–29].

Se is a micronutrient that, although not required by higher plants, can positively affect olive trees by promoting plant growth, alleviating UV-induced oxidative stress, stimulating chlorophyll biosynthesis, increasing the antioxidative defences of senescent plants and regulating the water status of drought-exposed plants [30–32].

Concerning olive trees, some positive effects of Se were documented. In particular, in this crop, this element was found to improve drought and salt stress tolerance [33,34] and phenol content [35,36] and stimulate pollen germination [27,37]. However, in this context, and to the best of our knowledge, no studies have been performed on the possible beneficial effects of Se in reducing or mitigating the detrimental effect of high temperature on olive pollen. Therefore, in this work, the effects of high temperatures on Ca^{2+}-cytosolic germination and morphology of olive pollen and the possible beneficial effects of selenium in heat stress tolerance have been deeply investigated. Furthermore, concerning the study of the effects of oxidative stress in different temperatures, hydrogen peroxide (H_2O_2) was used, as it is considered one of the most critical ROS that accumulates when oxidative perturbations occur.

2. Materials and Methods

2.1. Reagents

FURA-2AM (FURA-2-pentakis (acetoxymethyl) ester), PBS (Phosphate Buffered Saline), Triton X-100, EGTA (ethylene glycol-bis (β-aminoethyl ether), selenium methionine, hydrogen peroxide (H_2O_2), sodium chloride (NaCl), potassium chloride (KCl), magnesium chloride ($MgCl_2$), glucose, Hepes, and dimethyl sulfoxide (DMSO) were obtained from

Sigma-Aldrich corporation (St. Louis, MO, USA). All other chemicals and reagents (reagent grade) were high-quality.

2.2. Plant Material, Growing Conditions, and Pollen Collection

The study was carried out in 2020 in a thirty-year-old orchard near Perugia (Central Italy, 42°57′39.2″ N, 12°25′02.5″ E) on Leccino cultivar. The planting distance was 5 × 6 m, and the training system adopted was the "vase" system (single trunk 1 m high and 3–4 main branches). The soil texture was clay loam. The climate of the area was semi-continental, and the average temperature difference between the coldest (January) and warmest (July) months was 19–20 °C. The average annual air temperature was 13–14 °C, while the average diurnal temperature range of 10–11 °C. The maximum and minimum temperatures were 36 °C and −7 °C, respectively. The precipitations were distributed mainly in autumn, winter, and spring, and the annual average precipitation was about 800 mm.

The starting of the olive flowering was assessed when the pollen was freely released by shaking the anthers of different branches at different tree canopy heights and exposures [28]. When the 1st flowering stage was reached (end of May), three branches (70–80 inflorescences each) for each tree were bagged with white double-layered paper bags (0.65 × 0.35 m) to collect pollen. The bags were removed at the end of the flowering phase, and then the pollen was filtered through a cell strainer (40 µm).

2.3. In Vitro Thermal Stress of Olive Pollen

Aliquots of olive pollen (100 mg) were incubated at 20, 30, and 40 °C. The incubation carried out for 20, 48, and 62 h allowed the appearance of the maximum effect on cytosolic Ca^{2+} and on germination to be highlighted.

2.4. Measurement of Cytosolic Ca^{2+}

FURA-2AM probe enabled the measurement of intracellular calcium levels [29]. In particular, 100 mg of control and thermal stressed olive pollen were placed in 10 mL PBS and left to hydrate for 2 days. Hydrated pollens were collected by centrifugation at 1000× *g* 4 min and then resuspended in 2 mL Ca^{2+}-free HBSS buffer (120 mM NaCl, 5.0 mM KCl, $MgCl_2$ 1 mM, 5 mM glucose, 25 mM Hepes, pH 7.4). The pollen suspensions were incubated in the absence of light with FURA-2AM (2 µL of a 2 mM solution in DMSO) for 120 min. Then the samples were centrifuged at 1000× *g* 4 min, and the pollens were collected and suspended in 10 mL of Ca^{2+}-free HBSS containing 0.1 mM EGTA. The latter was used to exclude or minimise the potential background due to contaminant ions (to obtain a suspension of 1×10^6 hydrated pollen granules per mL).

A Perkin-Elmer LS 50 B spectrofluorometer (Markham, ON, Canada) was used to determine fluorescence (excitation 340 and 380 nm, emission 510 nm), set with a slit width of 10 nm and a 7.5 nm in the excitation and emission windows, respectively. Fluorometric measures were taken after 300–350 s. $CaCl_2$, H_2O_2, and Se-met were added to pollen samples for specific purposes, as described in the Results section. Cytosolic calcium concentrations ($[Ca^{2+}]_c$) were calculated according to Grynkiewicz [38], while the concentration of Se-met and H_2O_2 were established based on previous studies [27,28] and allowed to obtain beneficial effects without toxicity risks that can occur at higher concentrations.

2.5. Pollen Germination

The olive pollen samples (control and heat-stressed) were rehydrated by incubation for 30 min at room temperature in a humid chamber [39]. Then, pollen samples were placed on culture plates (6-well culture plates with 1.0 mg of pollen per plate) containing 3 mL of an agar-solidified culture medium composed as follows: agar 1%, sucrose 10%, boric acid (H_3BO_3) 100 ppm, and calcium chloride ($CaCl_2$) 1 mM, at pH 5.5 [40]. Subsequently, a uniform distribution was obtained on the surface of the substrate using a brush. Then the pollen grains were incubated for 24–48 h in a growth chamber at 25 °C. The number of germinated and non-germinated pollen grains was counted using a microscope with a

10× objective lens. Germination rates were estimated using two replicates of 100 grains. In particular, the grains were considered germinated if the pollen tube size was larger than the diameter of the grain [40]. The experiments were carried out according to a completely randomised design with four replicates.

2.6. Pollen Morphology

The morphology of the pollen was investigated using Field Emission Gun Electron Scanning Microscopy LEO 1525 ZEISS (Zeiss, Jena, Germany) after the pollen deposition on conductive carbon tape and metallisation with chromium (8 nm).

2.7. Statistical Analysis

Graph Pad Prism 6.03 software for Windows (La Jolla, CA, USA) was used for statistical tests. For the variance assumptions, different tests were conducted. In particular, homogeneity of variance was assessed by Levene's test and normal distribution by D'Agostino-Pearson omnibus normality test. The obtained results are expressed as mean values ± standard error of the mean (SEM). The significance of differences was analysed with Fisher's least significant differences test after analysis of the variance according to the 2-way split-plot design with complete randomisation with temperatures as the main plot and the treatments as sub-plot. Differences with $p < 0.05$ were considered statistically significant.

3. Results

3.1. Scanning Electron Microscopy Analysis of Olive Pollen

Olive pollen grains incubated in vitro at 20 (Figure 1A) and 40 °C (Figure 1B) for 62 h were analysed by Field Emission Scanning Electron Microscopy (FE-SEM).

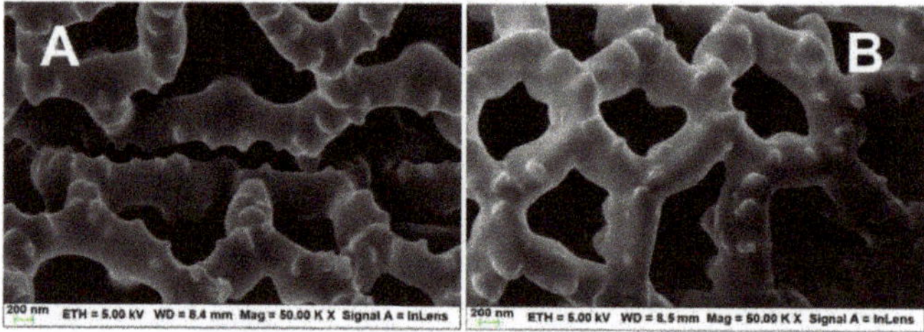

Figure 1. FE-SEM images of olive pollen incubated at 20 (**A**) and 40 °C (**B**) for 62 h.

The two populations showed no differences in size and shape, while differences appeared at high magnification (50 K×) in the sculpture of the outer pollen wall. In particular, the network of the reticulum showed a lower number of external elements (granules) in the pollen incubated at 40 °C than in that incubated at 20 °C (Figure 1). Images of pollen incubated at 30 °C were similar to those incubated at 20 °C (data not reported).

3.2. Ca^{2+}-Cytosolic ($[Ca^{2+}]_c$) Changes in Olive Pollen in Heat Stress

The effects of Se-met, H_2O_2, and Se-met + H_2O_2 were studied on pollen incubated in vitro at 20, 30, and 40 °C, investigating Ca^{2+}-cytosolic ($\Delta[Ca^{2+}]_c$). The $\Delta[Ca^{2+}]_c$ increased with Se-met, H_2O_2, Se-met + H_2O_2, mostly at 20 °C, less at 30 °C, and even less at 40 °C. Se-met and H_2O_2 individually increased Ca^{2+}-cytosolic, but they did not show an additive effect when both were present in the incubation medium (Figure 2).

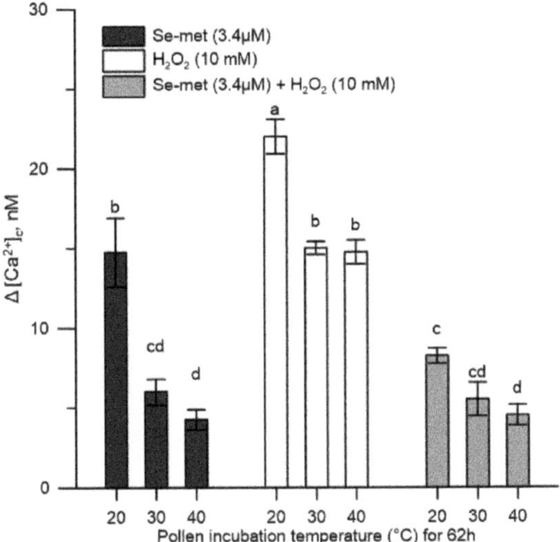

Figure 2. Changes in Ca^{2+}-cytosolic ($\Delta[Ca^{2+}]_c$) in olive pollen incubated at 20, 30, and 40 °C, in the presence of Se-met (3.4 μM), H_2O_2 (10 mM), and Se-met (3.4 μM) + H_2O_2 (10 mM). Values are expressed as means ± SEM. Different letters indicate statistically significant differences ($p < 0.05$).

3.3. Ca^{2+}-Entry in Olive Pollen in Heat Stress

The effects of Se-met, H_2O_2, and Se-met + H_2O_2 on Ca^{2+}-cytosolic were studied on pollen incubated at 20, 30, and 40 °C, with the addition of 1 mM $CaCl_2$ in the incubation medium. The entry of extracellular Ca^{2+} (Ca^{2+}-entry) was tested by monitoring the increase of $\Delta[Ca^{2+}]_c$. Under basal (control) conditions, Ca^{2+}-entry was similar in pollen incubated at all three temperatures. In contrast, Se-met promoted the extracellular Ca^{2+}-entry, while the effect of H_2O_2 was to reduce the Ca^{2+}-entry. Finally, Ca^{2+}-entry returned to values similar to the basal conditions when the H_2O_2 was added to the pollen pre-treated with Se-met (Figure 3).

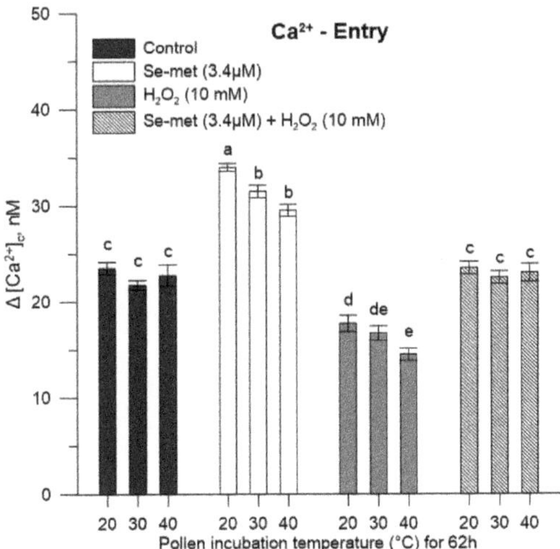

Figure 3. Ca^{2+}-entry in olive pollen incubated at 20, 30, and 40 °C, in basal conditions (control), in the presence of Se-met (3.4 μM), H_2O_2 (10 mM), and Se-met + H_2O_2 and $CaCl_2$ 1 mM in the incubation medium. Values are expressed as means ± SEM. Different letters indicate statistically significant differences ($p < 0.05$).

3.4. *Germination of Olive Pollen Subjected to Heat Stress*

Pollen collected from olive trees was subjected to heat stress. As a result, marked reductions in the germination rate were recorded compared to the control. In particular, samples incubated at 40 and 30 °C showed significant reductions in the germination of about 80% and 20%, respectively, compared to the control pollen. In addition, hydrogen peroxide strongly affected pollen germination, reducing it by about 90% at all three temperatures investigated. On the contrary, Se-met positively influenced pollen germination, increasing it in samples subjected to heat stress and oxidative stress (H_2O_2—Figure 4).

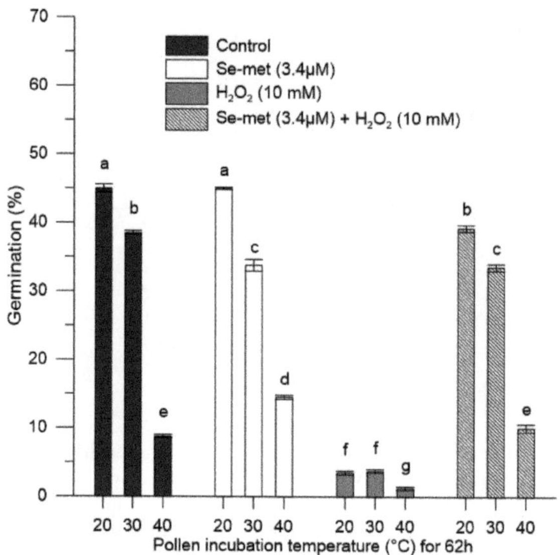

Figure 4. Germination of olive pollen incubated at 20, 30, and 40 °C. Control pollen (control), treated with H_2O_2 (10 mM), Se-met (3.4 μM), and Se-met (3.4 μM) + H_2O_2 (10 mM). Values are expressed as means ± SEM. Different letters indicate statistically significant differences ($p < 0.05$).

3.5. *Time-Course of High Temperature on Pollen Germination*

Prolonged heat stress at 40 °C significantly influenced the germination rate. In particular, after 24–48 h and 62 h of incubation at 40 °C, reductions of 75% and 80% were observed in control samples, respectively. Furthermore, the treatment with H_2O_2 severely reduced the capacity of pollen to germinate, which, in addition, showed no measurable fluctuations in the exposure time to high temperatures. In contrast, the treatment with Se-met improved the germination rate when the samples were subjected to thermal (40 °C) and oxidative stress (H_2O_2), regardless of the incubation time (Figure 5).

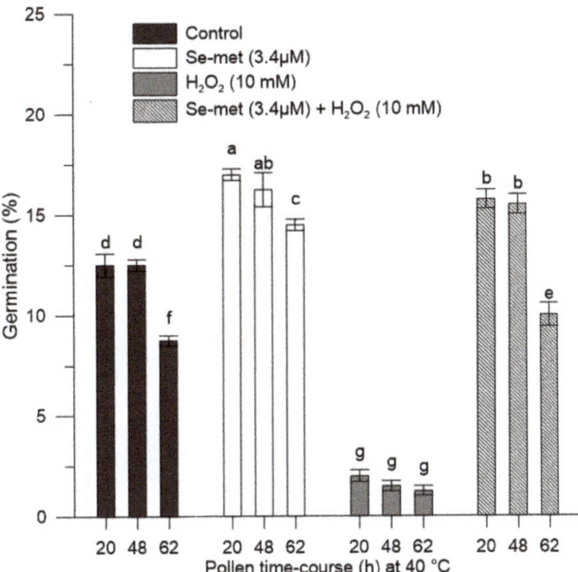

Figure 5. Germination of olive pollen incubated at 40 °C for 20, 48, and 62 h. Pollen was untreated (control) or treated with H_2O_2 (10 mM), Se-met (3.4 μM), and Se-met (3.4 μM) + H_2O_2 (10 mM). Values are expressed as means ± SEM. Different letters indicate statistically significant differences ($p < 0.05$).

4. Discussion

Adverse environmental conditions and abiotic stresses caused by global warming can lead to a progressive decrease in crop production [1,6]. Among the most impactful environmental stresses on crops, heat can play a detrimental role for plants. The magnitude of the effect of this stress on crops depends on the duration, fluctuations, and intensity of temperatures exceeding the optimal values for plant growth conditions [9].

4.1. Morphological Investigations in Olive Pollen Grains

In this work, to simulate the effects of heat stress on olive pollen, samples were incubated in vitro at high temperatures (30 and 40 °C), and the results were compared to those obtained for control samples (20 °C). The effect and consequences of heating were assessed by analysing pollen morphology, Ca^{2+}-cytosolic, and germination.

SEM analyses were carried out on pollen subjected or not to heat stress, as morphological alterations could influence pollen germination, olive tree fertilisation, and fruit set process [41]. The morphological investigations revealed changes in the sculpture of the outer wall of pollen incubated at 40 °C, but not in its size and shape. In quinoa, although no morphological differences were found in the pollen surface between heat-stressed and controls, the pollen wall thickness (intine and extine) increased due to thermal stress [42].

4.2. Fluctuations of Ca^{2+}-Cytosolic in Olive Pollen under Heat Stress Conditions

Ca^{2+}-cytosolic and germination were the parameters examined under thermal stress conditions, as the temperature can strongly influence them. However, numerous studies have shown that maintaining proper Ca^{2+}-cytosolic levels can promote pollen germination and tubules formation [11,18,19]. For these reasons, in this work, Ca^{2+}-cytosolic in pollen was measured using the "Ca^{2+} add-back" protocol [43]. In this respect, our experiments allowed us to discriminate increases in cytosolic Ca^{2+} due to the release of Ca^{2+} from intracellular stores from those resulting from the extracellular ion entry. In addition, as high temperatures can cause numerous changes in plant physiology and lead to increased ROS production [9], it seemed rational to evaluate the individual and combined effects of thermal and oxidative stress, the latter simulated by the treatments with hydrogen peroxide.

4.3. *Effects of Se-Met in Ca^{2+}-Cytosolic during Heat and Oxidative Stress*

Preceding studies have suggested the use of selenium in its organic form as Se-met due to its protective role against oxidative stress and its efficacy in maintaining Ca^{2+} homeostasis and olive pollen germination [27,28]. Furthermore, it should be noted that the other reason selenium has been used in this organic form is that it is less toxic than the inorganic forms. (Na-selenate and selenite) [27,28].

Our experiments showed that high temperatures, namely at 30 and 40 °C, attenuated the effects of H_2O_2 on the changes of Ca^{2+}-cytosolic, limiting the release of the stored element. Therefore, these results highlight that high temperatures improved the tolerance threshold for Ca^{2+} agonists, represented in this work by hydrogen peroxide. This was presumably due to the activation of antioxidant defences in response to high temperatures. In line with this, it has been documented that some antioxidant activities can be activated in pollen during environmental stress and can maintain cellular redox homeostasis, resulting in improved germination [44,45]. Moreover, this study showed that the treatment with Se-met restored Ca^{2+} homeostasis by counteracting the adverse effects of H_2O_2 at all the temperatures investigated. This action is considered beneficial as increases in Ca^{2+}-cytosolic are generally correlated with immediate increases in ROS content, particularly superoxide anion, the first reactive oxygen species produced under stress conditions [46]. Finally, it should be mentioned that Se-met, administered alone during heat stress, prevented alterations in Ca^{2+}-cytosolic, thus indicating that the compound mentioned above did not lose its antioxidant properties. In particular, these results align with those of Del Pino et al. [28], who highlighted the beneficial effects promoted by Na-selenate in preventing the onset of oxidative stress in internal pollen stores.

4.4. *Effects of Se-Met on Olive Pollen Germination Subjected to Heat Stress*

Numerous studies have reported that damage to reproductive tissues exposed to high temperatures leads to reduced productivity, yield, and crop quality [5,14,15]. Our experiments showed that high temperatures strongly affected the germination of olive pollen, which drastically lost performance. In fact, the pollen germination rate was reduced by 80% at 40 °C and 20% at 30 °C. Heat stress can reduce pollen viability and cause poor fertilisation; in particular, pollen viability during development is severely compromised if the temperature exceeds 25/35 °C [47]. In addition, our experiments showed that H_2O_2 strongly reduced the pollen germination at all the temperatures studied, whereas Se-met, when administered in combination with the oxidant, reversed its negative impact on pollen germination. Finally, when administered alone to pollen, Se-met counteracted the detrimental effect of heat stress at 40 °C. The stimulating effect of Se-met on pollen tolerance to abiotic stresses has already been documented. This compound essentially acts as a ROS scavenger, thus preventing oxidation-related alterations of Ca^{2+} channels [27]. This beneficial effect is significant for its potential consequences in agriculture, as several abiotic factors that can lead to ROS accumulation can influence pollen germination [27].

4.5. *Effect of Se-Met on Pollen Germination in Time-Course Experiment of Heat Stress*

Time-course experiments, in which the temperature was maintained at 40 °C for all the treatments, showed that thermal stress strongly impacted germination. In addition, pollen germination decreased further, regardless of the treatment applied, when the exposure time was extended to 62 h. However, Se-met was very effective in counteracting the negative impact of both high temperature and H_2O_2, and this beneficial effect may be related to the ability of this active compound to improve the oxidative status of pollen [27].

5. Conclusions

The results reported in this study demonstrate the protective role of Se-met in pollen to cope with heat stress, as evidenced by increased germination and improved Ca^{2+} homeostasis. Indeed, both high temperature and oxidative stress affected pollen Ca^{2+} signal but in different ways. Heat stress reduced the response to Ca^{2+} agonist stimuli, whereas oxidative stress increased Ca^{2+}-cytosolic by prompting the release of the ion from internal stores and depressing its entry. In contrast, Se contributed to the restoration of Ca^{2+} homeostasis by enhancing the Ca^{2+}-entry mechanism in both the abiotic stresses. The latter condition is necessary for the activation of the germination process. In light of the above, we have shown that Se-met is a possible candidate for improving heat tolerance in olive pollen.

Author Contributions: A.M.D.P.: Conceptualisation, Methodology, Formal analysis, Investigation, Data curation, Writing—Original Draft, Writing—Review and Editing. L.R.: Conceptualisation, Methodology, Formal analysis, Investigation, Data curation, Writing—Original Draft, Writing—Review and Editing. A.D.M.: Methodology, Formal analysis, Investigation, Data curation, Writing—Original Draft. A.G.: Conceptualisation, Writing—Original Draft, Writing—Review and Editing, Supervision. D.D.B.: Conceptualisation, Investigation, Writing—Original Draft, Writing—Review and Editing, Supervision. P.P.: Conceptualisation, Methodology, Formal analysis, Investigation, Data curation, Writing—Original Draft, Writing—Review and Editing, Supervision. C.A.P.: Conceptualisation, Methodology, Formal analysis, Investigation, Data curation, Writing—Original Draft, Writing—Review and Editing, Supervision. All authors have read and agreed to the published version of the manuscript.

Funding: This study was partially funded by the project "Ricerca di Base 2020" of the Department of Agricultural, Food and Environmental Sciences of the University of Perugia (Coordinator: Primo Proietti).

Data Availability Statement: Data will be available on request to the corresponding author.

Conflicts of Interest: The authors declare no conflict of interest.

References

1. Del Buono, D. Can Biostimulants Be Used to Mitigate the Effect of Anthropogenic Climate Change on Agriculture? It Is Time to Respond. *Sci. Total Environ.* **2021**, *751*, 141763. [CrossRef] [PubMed]
2. Mittler, R. Abiotic Stress, the Field Environment and Stress Combination. *Trends Plant Sci.* **2006**, *11*, 15–19. [CrossRef] [PubMed]
3. Ainsworth, E.A.; Ort, D.R. How Do We Improve Crop Production in a Warming World? *Plant Physiol.* **2010**, *154*, 526–530. [CrossRef] [PubMed]
4. Parrotta, L.; Faleri, C.; Cresti, M.; Cai, G. Heat Stress Affects the Cytoskeleton and the Delivery of Sucrose Synthase in Tobacco Pollen Tubes. *Planta* **2016**, *243*, 43–63. [CrossRef] [PubMed]
5. Carpenedo, S.; Raseira, M.D.C.B.; Franzon, R.C.; Byrne, D.H.; Da Silva, J.B. Stigmatic receptivity of peach flowers submitted to heat stress. *Acta Sci. Agron.* **2019**, *42*, e42450. [CrossRef]
6. Field, C.B.; Barros, V.; Stocker, T.F.; Dahe, Q.; Dokken, J.D.; Ebi, K.L.; Mastrandrea, M.D.; Mach, K.J.; Plattner, G.K.; Allen, S.K.; et al. *Managing the Risks of Extreme Events and Disasters to Advance Climate Change Adaptation: Special Report of the Intergovernmental Panel on Climate Change*; Cambridge University Press: Cambridge, UK, 2012; p. 582. ISBN 9781107025066.
7. Del Buono, D.; Regni, L.; Del Pino, A.M.; Bartucca, M.L.; Palmerini, C.A.; Proietti, P. Effects of Megafol on the Olive Cultivar 'Arbequina' Grown Under Severe Saline Stress in Terms of Physiological Traits, Oxidative Stress, Antioxidant Defenses, and Cytosolic Ca^{2+}. *Front. Plant Sci.* **2021**, *11*, 603576. [CrossRef]
8. Hasanuzzaman, M.; Nahar, K.; Alam, M.M.; Roychowdhury, R.; Fujita, M. Physiological, Biochemical, and Molecular Mechanisms of Heat Stress Tolerance in Plants. *Int. J. Mol. Sci.* **2013**, *14*, 9643–9684. [CrossRef]
9. Paupière, M.J.; van Heusden, A.W.; Bovy, A.G. The Metabolic Basis of Pollen Thermo-Tolerance: Perspectives for Breeding. *Metabolites* **2014**, *4*, 889–920. [CrossRef]
10. Steinhorst, L.; Kudla, J. Calcium—A Central Regulator of Pollen Germination and Tube Growth. *Biochim. Biophys. Acta Mol. Cell Res.* **2013**, *1833*, 1573–1581. [CrossRef]
11. Michard, E.; Alves, F.; Feijó, J.A. The Role of Ion Fluxes in Polarized Cell Growth and Morphogenesis: The Pollen Tube as an Experimental Paradigm. *Int. J. Dev. Biol.* **2009**, *53*, 1609–1622. [CrossRef]
12. Lazzaro, M.D.; Cardenas, L.; Bhatt, A.P.; Justus, C.D.; Phillips, M.S.; Holdaway-Clarke, T.L.; Hepler, P.K. Calcium Gradients in Conifer Pollen Tubes; Dynamic Properties Differ from Those Seen in Angiosperms. *J. Exp. Bot.* **2005**, *56*, 2619–2628. [CrossRef] [PubMed]

13. Wu, J.; Wang, S.; Gu, Y.; Zhang, S.; Publicove, S.J.; Franklin-Tong, V.E. Self-Incompatibility in *Papaver rhoeas* Activates Nonspecific Cation Conductance Permeable to Ca^{2+} and K^+. *Plant Physiol.* **2011**, *155*, 963–973. [CrossRef] [PubMed]
14. Zinn, K.E.; Tunc-Ozdemir, M.; Harper, J.F. Temperature Stress and Plant Sexual Reproduction: Uncovering the Weakest Links. *J. Exp. Bot.* **2010**, *61*, 1959–1968. [CrossRef] [PubMed]
15. Snider, J.L.; Oosterhuis, D.M.; Skulman, B.W.; Kawakami, E.M. Heat Stress-Induced Limitations to Reproductive Success in *Gossypium hirsutum*. *Physiol. Plant.* **2009**, *137*, 125–138. [CrossRef]
16. Carafoli, E. Intracellular Calcium Homeostasis. *Annu. Rev. Biochem.* **1987**, *56*, 395–433. [CrossRef]
17. Görlach, A.; Bertram, K.; Hudecova, S.; Krizanova, O. Calcium and ROS: A Mutual Interplay. *Redox Biol.* **2015**, *6*, 260–271. [CrossRef]
18. Campanoni, P.; Blatt, M.R. Membrane Trafficking and Polar Growth in Root Hairs and Pollen Tubes. *J. Exp. Bot.* **2007**, *58*, 65–74. [CrossRef]
19. Cheung, A.Y.; Wu, H.M. Structural and Signaling Networks for the Polar Cell Growth Machinery in Pollen Tubes. *Annu. Rev. Plant Biol.* **2008**, *59*, 547–572. [CrossRef]
20. Yan, Y.; Wei, C.-L.; Zhang, W.-R.; Cheng, H.-P.; Liu, J. Cross-Talk between Calcium and Reactive Oxygen Species Signaling. *Acta Pharmacol. Sin.* **2006**, *27*, 821–826. [CrossRef]
21. Clapham, D.E. Calcium Signaling. *Cell* **2007**, *131*, 1047–1058. [CrossRef]
22. Brini, M.; Calì, T.; Ottolini, D.; Carafoli, E. Intracellular Calcium Homeostasis and Signaling. *Met. Ions Life Sci.* **2013**, *12*, 119–168. [CrossRef] [PubMed]
23. Orrenius, S.; Gogvadze, V.; Zhivotovsky, B. Calcium and Mitochondria in the Regulation of Cell Death. *Biochem. Biophys. Res. Commun.* **2015**, *460*, 72–81. [CrossRef] [PubMed]
24. Brito, C.; Dinis, L.-T.; Moutinho-Pereira, J.; Correia, C.M. Drought Stress Effects and Olive Tree Acclimation under a Changing Climate. *Plants* **2019**, *8*, 232. [CrossRef] [PubMed]
25. Ben-Ari, G.; Biton, I.; Many, Y.; Namdar, D.; Samach, A. Elevated Temperatures Negatively Affect Olive Productive Cycle and Oil Quality. *Agronomy* **2021**, *11*, 1492. [CrossRef]
26. Selak, G.V.; Perica, S.; Goreta Ban, S.; Poljak, M. The Effect of Temperature and Genotype on Pollen Performance in Olive (*Olea europaea* L.). *Sci. Hortic.* **2013**, *156*, 38–46. [CrossRef]
27. Del Pino, A.M.; Regni, L.; D'Amato, R.; Tedeschini, E.; Businelli, D.; Proietti, P.; Palmerini, C.A. Selenium-Enriched Pollen Grains of *Olea europaea* L.: Ca^{2+} Signaling and Germination Under Oxidative Stress. *Front. Plant Sci.* **2019**, *10*, 1611. [CrossRef]
28. Del Pino, A.M.; Regni, L.; D'amato, R.; Di Michele, A.; Proietti, P.; Palmerini, C.A. Persistence of the Effects of Se-Fertilization in Olive Trees over Time, Monitored with the Cytosolic Ca^{2+} and with the Germination of Pollen. *Plants* **2021**, *10*, 2290. [CrossRef]
29. Regni, L.; Micheli, M.; Del Pino, A.M.; Palmerini, C.A.; D'Amato, R.; Facchin, S.L.; Famiani, F.; Peruzzi, A.; Mairech, H.; Proietti, P. The First Evidence of the Beneficial Effects of Se-Supplementation on In Vitro Cultivated Olive Tree Explants. *Plants* **2021**, *10*, 1630. [CrossRef]
30. Hartikainen, H.; Xue, T. The Promotive Effect of Selenium on Plant Growth as Triggered by Ultraviolet Irradiation. *J. Environ. Qual.* **1999**, *28*, 1372–1375. [CrossRef]
31. Terry, N.; Zayed, A.M.; De Souza, M.P.; Tarun, A.S. Selenium in Higher Plants. *Annu. Rev. Plant Biol.* **2000**, *51*, 401–432. [CrossRef]
32. Kuznetsov, V.V.; Kholodova, V.P.; Kuznetsov, V.V.; Yagodin, B.A. Selenium Regulates Water Relations of Plants under Drought. *Dokl. Akad. Nauk* **2003**, *390*, 713–716.
33. Proietti, P.; Nasini, L.; Del Buono, D.; D'Amato, R.; Tedeschini, E.; Businelli, D. Selenium Protects Olive (*Olea europaea* L.) from Drought Stress. *Sci. Hortic.* **2013**, *164*, 165–171. [CrossRef]
34. Regni, L.; Palmerini, C.A.; Del Pino, A.M.; Businelli, D.; D'Amato, R.; Mairech, H.; Marmottini, F.; Micheli, M.; Pacheco, P.H.; Proietti, P. Effects of Selenium Supplementation on Olive under Salt Stress Conditions. *Sci. Hortic.* **2021**, *278*, 109866. [CrossRef]
35. D'Amato, R.; Proietti, P.; Nasini, L.; Del Buono, D.; Tedeschini, E.; Businelli, D. Increase in the Selenium Content of Extra Virgin Olive Oil: Quantitative and Qualitative Implications. *Grasas Aceites* **2014**, *65*, e025. [CrossRef]
36. D'Amato, R.; Proietti, P.; Onofri, A.; Regni, L.; Esposto, S.; Servili, M.; Businelli, D.; Selvaggini, R. Biofortification (Se): Does It Increase the Content of Phenolic Compounds in Virgin Olive Oil (VOO)? *PLoS ONE* **2017**, *12*, e0176580. [CrossRef] [PubMed]
37. Tedeschini, E.; Proietti, P.; Timorato, V.; D'Amato, R.; Nasini, L.; Dei Buono, D.; Businelli, D.; Frenguelli, G. Selenium as Stressor and Antioxidant Affects Pollen Performance in *Olea europaea*. *Flora Morphol. Distrib. Funct. Ecol. Plants* **2015**, *215*, 16–22. [CrossRef]
38. Grynkiewicz, G.; Poenie, M.; Tsien, R.Y. A New Generation of $Ca_2{}^+$ Indicators with Greatly Improved Fluorescence Properties. *J. Biol. Chem.* **1985**, *260*, 3440–3450. [CrossRef]
39. Rejón, J.D.; Zienkiewicz, A.; Rodríguez-García, M.I.; Castro, A.J. Profiling and Functional Classification of Esterases in Olive (*Olea europaea*) Pollen during Germination. *Ann. Bot.* **2012**, *110*, 1035–1045. [CrossRef]
40. Martins, E.S.; Davide, L.M.C.; Miranda, G.J.; Barizon, J.O.; Souza Junior, F.A.; de Carvalho, R.P.; Gonçalves, M.C. In Vitro Pollen Viability of Maize Cultivars at Different Times of Collection. *Cienc. Rural* **2017**, *47*, e20151077. [CrossRef]
41. Khaleghi, E.; Karamnezhad, F.; Moallemi, N. Study of Pollen Morphology and Salinity Effect on the Pollen Grains of Four Olive (*Olea europaea*) Cultivars. *S. Afr. J. Bot.* **2019**, *127*, 51–57. [CrossRef]
42. Hinojosa, L.; Matanguihan, J.B.; Murphy, K.M. Effect of High Temperature on Pollen Morphology, Plant Growth and Seed Yield in Quinoa (*Chenopodium quinoa* Willd.). *J. Agron. Crop Sci.* **2019**, *205*, 33–45. [CrossRef]

43. Hecquet, C.M.; Ahmmed, G.U.; Vogel, S.M.; Malik, A.B. Role of TRPM2 Channel in Mediating H_2O_2-Induced Ca_2^+ Entry and Endothelial Hyperpermeability. *Circ. Res.* **2008**, *102*, 347–355. [CrossRef] [PubMed]
44. Bokszczanin, K.L.; Fragkostefanakis, S. Perspectives on Deciphering Mechanisms Underlying Plant Heat Stress Response and Thermotolerance. *Front. Plant Sci.* **2013**, *4*, 315. [CrossRef] [PubMed]
45. Xie, D.-L.; Zheng, X.-L.; Zhou, C.-Y.; Kanwar, M.K.; Zhou, J. Functions of Redox Signaling in Pollen Development and Stress Response. *Antioxidants* **2022**, *11*, 287. [CrossRef] [PubMed]
46. Demidchik, V. Mechanisms of Oxidative Stress in Plants: From Classical Chemistry to Cell Biology. *Environ. Exp. Bot.* **2015**, *109*, 212–228. [CrossRef]
47. Chu, Y.-C.; Chang, J.-C. Heat Stress Leads to Poor Fruiting Mainly Due to Inferior Pollen Viability and Reduces Shoot Photosystem II Efficiency in "Da Hong" Pitaya. *Agronomy* **2022**, *12*, 225. [CrossRef]

MDPI

Article

Heat Shock Treatment Promoted Callus Formation on Postharvest Sweet Potato by Adjusting Active Oxygen and Phenylpropanoid Metabolism

Qi Xin [1,2,3], Bangdi Liu [1,2,*], Jing Sun [1,2], Xinguang Fan [2,4], Xiangxin Li [2,5], Lihua Jiang [2,3], Guangfei Hao [2,3], Haisheng Pei [1,2] and Xinqun Zhou [1,2]

1 Academy of Agricultural Planning and Engineering, Ministry of Agriculture and Rural Affairs, Beijing 100125, China
2 Key Laboratory of Agro-Products Primary Processing, Ministry of Agriculture and Rural Affairs, Beijing 100125, China
3 College of Life Science and Food Engineering, Hebei University of Engineering, Handan 056038, China
4 College of Food Engineering, Ludong University, Yantai 264025, China
5 Institute of Apicultural Research, Chinese Academy of Agricultural Sciences, Beijing 100093, China
* Correspondence: bangdi.liu@foxmail.com

Abstract: This study aimed to investigate that rapid high-temperature treatment (RHT) at an appropriate temperature could accelerate callus formation by effectively promoting the necessary metabolic pathways in sweet potato callus. In this study, the callus of sweet potato was treated with heat shock at 50, 65, and 80 °C for 15 min. The callus formation was observed within 1, 3, and 5 days, and the accumulation of intermediates in the metabolism of phenylpropane and reactive oxygen species and changes in enzyme activities were determined. The results showed that appropriate RHT treatment at 65 °C stimulated the metabolism of reactive oxygen species at the injury site of sweet potato on the first day, and maintained a high level of reactive oxygen species production and scavenging within 5 days. The higher level of reactive oxygen species stimulated the phenylalanine ammonia-lyase (PAL), 4-coumarate-CoA ligase and cinnamate-4-hydroxylase activities of the phenylpropane metabolic pathway, and promoted the rapid synthesis of chlorogenic acid, *p*-coumaric acid, rutin, and caffeic acid at the injury site, which stacked to form callus. By Pearson's correlation analysis, catalase (CAT), PAL, and chlorogenic acid content were found to be strongly positively correlated with changes in all metabolites and enzymatic activities. Our results indicated that appropriate high-temperature rapid treatment could promote sweet potato callus by inducing reactive oxygen species and phenylpropane metabolism; moreover, CAT, PAL, and chlorogenic acid were key factors in promoting two metabolic pathways in sweet potato callus.

Keywords: Rapid high temperature; Phenylalanine ammonia-lyase; Catalase; Chlorogenic acid; Key metabolic mechanisms

Citation: Xin, Q.; Liu, B.; Sun, J.; Fan, X.; Li, X.; Jiang, L.; Hao, G.; Pei, H.; Zhou, X. Heat Shock Treatment Promoted Callus Formation on Postharvest Sweet Potato by Adjusting Active Oxygen and Phenylpropanoid Metabolism. *Agriculture* **2022**, *12*, 1351. https://doi.org/10.3390/agriculture12091351

Academic Editors: Daniele Del Buono, Luca Regni and Primo Proietti

Received: 25 July 2022
Accepted: 18 August 2022
Published: 1 September 2022

1. Introduction

Sweet potato (*Ipomoea batatas* (L.) Lam.) is now an important vegetable crop, as its yield is high and it is easily cultivated [1]. China is the leading global sweet potato producer. According to 2017 statistics, the total sweet potato planting area in China was 8.9373 million hm^2 and the total yield was 34.189 million tons. However, this root tuber crop is characterized by a thin epicarp, high water content, and poor storage performance. Sweet potato epicarp is readily damaged during mechanized or artificial harvesting and postharvest processing [2]. Consequently, sweet potato storage and transport are associated with storage decay by microbial pathogens, depletion of nutrients with high respiration rates, and some other quality losses [3]. The Ministry of Agriculture and Rural Affairs of China reported that the comprehensive loss of postharvest sweet potato roots caused by improper storage exceeded 30% in 2019 [4]. Therefore, mitigation of postharvest loss during storage and transportation is of paramount importance.

Callus formation is an internal defense mechanism of injured plant tissues. It occurs in roots, stems, leaves, and fruit [5]. Calli help prevent water loss and microbial infection. Wounded plant tissues generate transduction signals that induce reactive oxygen, phenylpropanoid, and fatty acid metabolism. Oxidative cross-linking of related metabolites then occurs and calli gradually form around the wounds [6]. Numerous studies have been conducted on callus formation of root and tuber crops such as white potato. The mechanism of phellem formation in potato callus has been elucidated [7,8]. Other studies demonstrated that potato callus formation was influenced by crop variety [7], harvest maturity [9], temperature [10], relative humidity, ambient light conditions [11], gas composition [12], and chemical stimulation [6,13]. However, few studies have investigated the sweet potato callus healing process. Prior research on sweet potato callus formation focused mainly on improving sweet potato storability through callus healing. Low-temperature heating (30 °C–35 °C) under cold storage for 3–7 days promoted sweet potato callus formation [14]. Mwanga found that sweet potatoes subjected to continuous callus treatment at 32 °C and 90% relative humidity (RH) for 4 days were more durable than those without callus treatment [15]. Amand reported that 18 different sweet potato varieties subjected to callus treatment at 30 °C and 85% RH for 7 days more readily formed wound calli than untreated samples [16]. Nevertheless, sweet potato is associated with strong harvest seasonality, long callus formation time, high heating equipment costs, and environmentally unsustainable crop protection methods. For these reasons, postharvest callus healing cannot fully meet market and production demands. In addition, some preharvest studies have pointed out that multiple fungicide stroby spray treatment [17] and sodium nitroprusside treatment [18] during the preharvest process could promote the wound healing for postharvested potatoes. Long-term callus maintenance at low temperatures may result in uneven heating and callus formation, yield loss, and tuber decay [17]. Therefore, callus healing time should be shortened and the conditions of the ambient environment and heat treatment should be harmonized to improve the efficiency and stability of callus formation.

The present study endeavored to produce rapid, uniform, and efficient sweet potato calli through rapid heat callus treatment (RHT) at comparatively higher temperatures and shorter treatment times. We observed sweet potato callus formation; analyzed changes in active oxygen metabolism, phenylpropanoids, and enzyme activity; and identified the key factors affecting sweet potato callus responses to RHT. The findings of the present study provide theoretical and methodological bases for rapid callus induction in postharvest sweet potato root tubers.

2. Results

2.1. Effect of RHT Treatment on Callus Wound Healing in Sweet Potato

Sweet potato callus consists of suberin and lignin, and the former is its principal component [4]. Suberin comprises fatty acid polymers and phenolic monomers produced by the fatty acid and phenylpropanoid metabolism pathways, respectively. Phenolic monomers combined with the cations in toluidine blue and turned purplish blue, and this reaction was used to evaluate suberin stacking. Figure 1 shows that after callus staining, only scant blue fluorescence appeared in the sweet potato wound at the original stage. In contrast, different fluorescence intensities appeared in the other groups and at the other stages. Compared with CK, all RHT groups presented with higher fluorescence on day 5. Therefore, RHT promoted suberin accumulation in sweet potato. The sweet potato wound tissue treated with RHT at 65 °C showed relatively more significant suberin deposition on days 1, 3, and 5. Hence, RHT at 65 °C most strongly promoted sweet potato callus. After RHT at 65 °C, the callus on the wound surface was grayish white with no obvious browning. In contrast, the sweet potato calli formed after RHT at 50 °C and 80 °C presented with different degrees of browning.

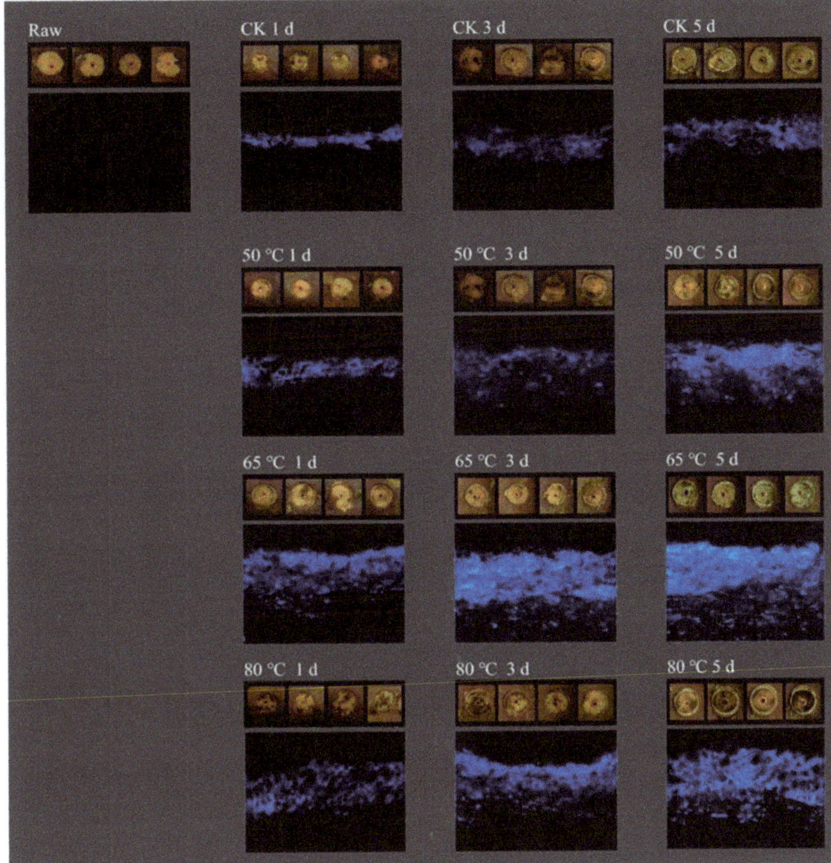

Figure 1. Effects of RHT treatments on suberin deposition in sweet potato tuber wound.

Lignin causes stacking and accumulates around the wound during callus healing in white potato and sweet potato [19]. Lignin is a glycerolipid polymer that forms during callus healing and is localized mainly to the cell walls and the plasma membranes. The composition and structure of lignin and wax are similar. Both prevent water and nutrient loss and resist bacterial infection in plants [20]. Therefore, the lignin content in healing tubers also affects callus formation. Figure 2A shows that from day 1, the lignin content in sweet potato had significantly increased compared with the raw samples under RHT and CK. However, on days 3 and 5, the lignin content was significantly higher under RHT at 65 °C and 80 °C than under CK and RHT at 50 °C. Figure 2B shows that callus deposition increased with lignin content and suberin stacking. The callus was composed of lignin and polyphenolsuberin. The higher the production of lignin and polyphenolsuberin, the thicker the callus. Thus, this finding and those illustrated in Figures 1 and 2A reveal that callus thickness was greatest under RHT at 65 °C. It reached 0.38 mm and 0.49 mm on days 3 and 5, respectively, and lignin–suberin stratification was higher than it was under the other treatments. The foregoing results indicate that the heat-shock treatment promoted lignin accumulation and suberin deposition in sweet potato wounds and that RHT at 65 °C might be the most effective stimulatory treatment for this purpose.

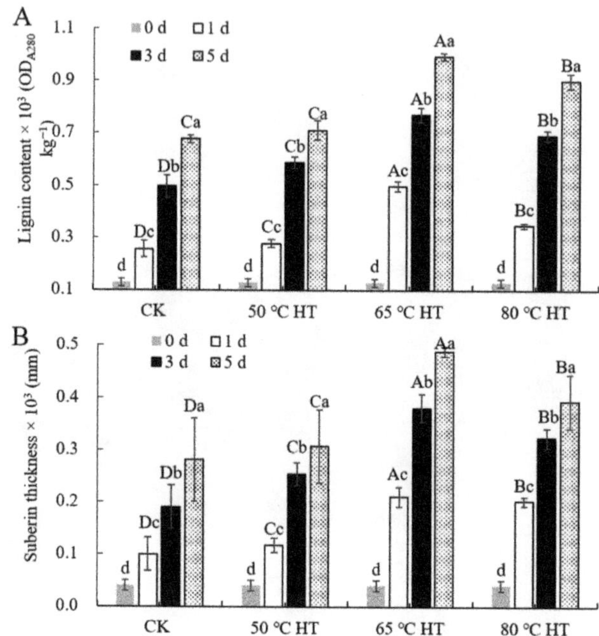

Figure 2. Effects of RHT treatment on suberin thickness and lignin content in sweet potato callus ((**A**): Liginn content; (**B**): Suberin thinckness. Note: Capital letters are comparisons at the same time at different temperatures, and lowercase letters are comparisons at different times at the same temperature).

2.2. Effects of RHT Treatment on Phenolic Compounds in Sweet Potato Callus

Phenolic compounds are precursors of suberin and lignin in sweet potato and the main components in sweet potato wounds and calli [3]. High-performance liquid chromatography (HPLC) was used to detect phenolic compounds in sweet potato callus. They were identified as *p*-coumaric acid, catechin, chlorogenic acid, and rutin. On day 5, the total phenol and flavonoid levels were higher in sweet potato subjected to RHT at 65 °C than in those subjected to other treatments. By contrast, the total phenol and flavonoid levels did not differ between sweet potatoes treated with RHT at 50 °C and those treated with RHT at 80 °C. For these reasons, the total phenols and flavonoids included substances implicated in suberin and lignin biosynthesis.

The chlorogenic acid content strongly responded to the temperature treatments and was higher in all RHT groups than in CK. From days 1–5, the chlorogenic acid content was the highest under the RHT at 80 °C (Figure 3C). The *p*-coumaric acid and rutin levels also substantially increased during callus healing. However, the changes in *p*-coumaric acid content were relatively more evident in response to RHT at 65 °C and 80 °C. The *p*-coumaric acid content in the sweet potato subjected to day 5 of RHT at 65 °C was twice that of the sweet potato under CK. Figure 3E shows no difference in rutin content under any treatment between days 1 and 3. By day 5, however, the rutin content was significantly higher under RHT at 65 °C than under the other treatments. Figure 3F shows no significant difference between the RHT at 50 °C and CK treatment at any sampling point in terms of catechin content. Therefore, RHT at 50 °C did not markedly induce catechin biosynthesis. There was no significant difference in the sweet potato subjected to RHT at 65 °C or RHT at 80 °C in terms of catechin content on day 3. By day 5, the catechin content was higher under RHT at 65 °C than under the other heat treatments. For these reasons, RHT at 65 °C and 80 °C promoted phenolic compound accumulation during sweet potato callus healing. Furthermore, the accumulation of phenolic compounds may help sweet potato wounds rapidly heal and form callus tissue. However, RHT at 80 °C might block total phenol, total flavonoid, *p*-coumaric acid, and catechin synthesis. Excessively high temperatures might inhibit the enzymes involved in phenylpropanoid metabolism. The present study showed that RHT at 65 °C was the optimal treatment for inducing phenolic compound synthesis in sweet potato wounds.

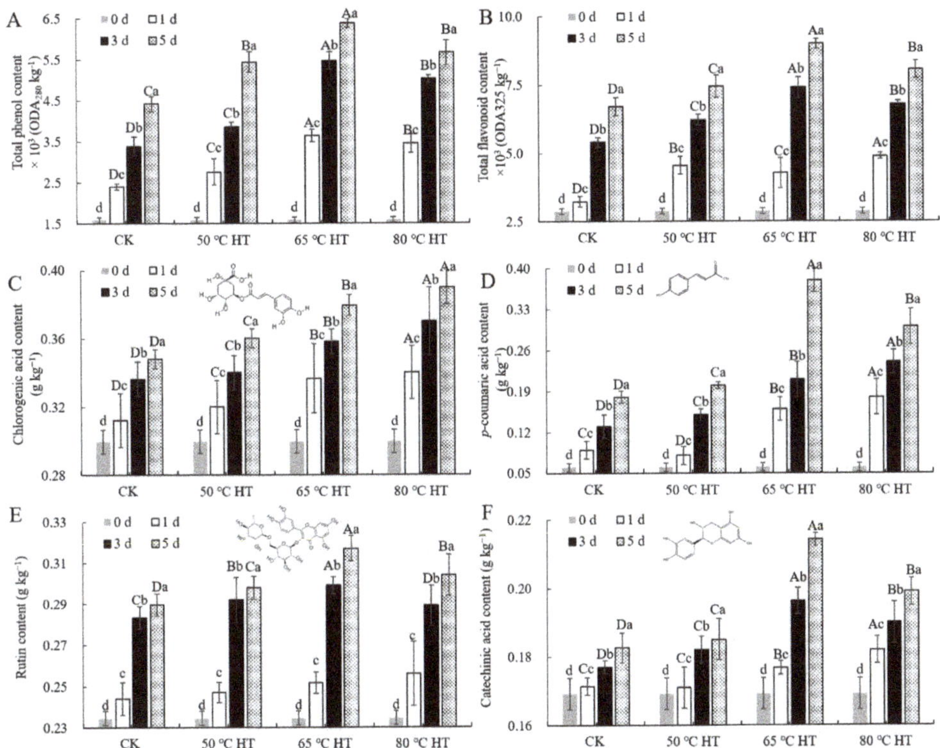

Figure 3. Effects of RHT treatment on total phenol, total flavonoid, chlorogenic acid, p-coumaric acid, rutin, and catechin content in sweet potato callus ((**A**): Total phenol content; (**B**): Total flavonoid content; (**C**): Chlorogenic acid content; (**D**): p-coumaric acid content; (**E**): Rutin content; (**F**): Catechinic acid content. Note: Capital letters are comparisons at the same time at different temperatures, and lowercase letters are comparisons at different times at the same temperature).

2.3. *Effects of RHT Treatment on Phenylpropanoid Metabolic Enzyme Activity during Sweet Potato Callus Healing*

Numerous studies demonstrated that plant callus formation is closely related to phenylpropanoid metabolism. Phenylalanine ammonia-lyase (PAL), *trans*-cinnamate 4-hydroxylase (C4H), 4-coumarate-CoA ligase (4CL), and cinnamyl alcohol dehydrogenase (CAD) are the major intermediate metabolic enzymes in phenylpropanoid metabolism [21]. Figure 3 shows that compared with the origin, the activity levels of the foregoing enzymes in all groups were increased during callus healing. PAL is a key rate-limiting enzyme at the beginning of phenylpropanoid metabolism. Figure 4C shows that PAL activity significantly increased under all treatments on day 1 and was highest under RHT at 80 °C. On days 3 and 5, however, PAL activity declined under RHT at 80 °C but increased under the other three treatments. On day 5, PAL activity under RHT at 65 °C was more than twice that under CK. The 4CL and C4H activity levels significantly increased on day 1. RHT at 80 °C inhibited 4CL and C4H activity on days 3 and 5. C4H activity peaked on day 1 and remained high under RHT at 65 °C on days 3 and 5. C4H activity significantly increased under RHT at 50 °C on day 5. Nevertheless, there was no significant difference between RHT at 65 °C and RHT at 50 °C in terms of C4H activity. The 4CL activity reached a peak on day 3 under RHT at 65 °C and was higher than that at other sampling points. CAD is implicated in lignin biosynthesis at the end of phenylpropanoid metabolism. Figure 3D shows that the change trend in CAD activity differed from those of the other three enzymes. CAD activity increased with time in all treatment groups. On days 3 and 5, there were no differences among RHT at 50 °C, 65 °C, and 80 °C in terms of CAD activity. In addition, CAD activity was high in the CK group at day 3. Hence, lignin content and phellem layer thickness also increased under CK in the absence of any high-temperature stimulus, and sweet potato injury may have promoted CAD activity. However, PAL, 4CL, and C4H were relatively more temperature sensitive. Appropriate RHT treatment may induce PAL, 4CL, and C4H in sweet potato. Several studies reported that wound injury and temperature stimulation promote ROS metabolism [4]. The influence of temperature on CAD, PAL, 4CL, and C4H may also be associated with ROS generation and clearance.

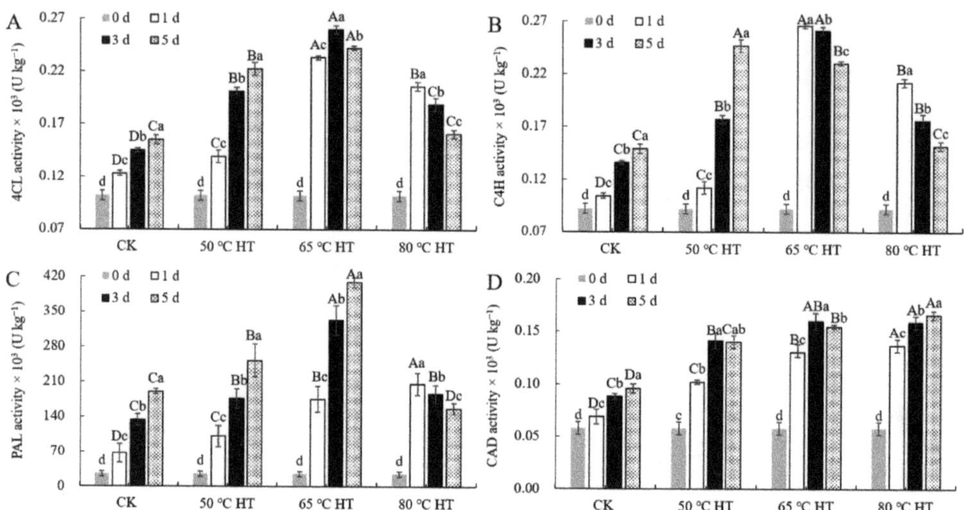

Figure 4. Effects of RHT treatment on PAL, C4H, 4CL, and CAD activity in sweet potato callus ((**A**): 4CL activity; (**B**): C4H activity; (**C**): PAL activity; (**D**): CAD activity. Note: Capital letters are comparisons at the same time at different temperatures, and lowercase letters are comparisons at different times at the same temperature).

2.4. *Effects of RHT Treatment on ROS Production and Scavenging in Sweet Potato Callus*

Neither low- nor high-temperature RHT treatment stimulated phenylpropanoid metabolism in sweet potato possibly because active oxygen metabolism occurred in the wounds [4]. Figure 5 shows the indices related to ROS generation and scavenging in sweet potato callus healing. Figure 5B,C shows ROS production capacity in sweet potato. ROS production was significantly higher under RHT at 80 °C than under other treatments. Therefore, temperature induces ROS generation in sweet potato. Figure 5A shows the ROS scavenging parameters during sweet potato callus healing. The 2,2-diphenyl-1-picrylhydrazyl (DPPH) scavenging assay displayed the same trends for CK, RHT at 50 °C, and RHT at 65 °C and exhibited a gradual increase from days 0 to 5. For the sweet potatoes treated with RHT at 80 °C, DPPH scavenging peaked on day 1 and gradually decreased thereafter. On day 1, DPPH scavenging under RHT at 80 °C was 1.25 times higher than it was under RHT at 65 °C. Nevertheless, DPPH scavenging under RHT at 65 °C was higher than that under CK and RHT at 50 °C. These results suggest that high temperature significantly induces free radical scavenging in sweet potato wounds. At day 3, DPPH scavenging significantly decreased in sweet potato under RHT at 80 °C possibly because of decreases in the activity of enzymes related to ROS scavenging, namely, catalase (CAT) and superoxide dismutase (SOD). Figure 5E,F shows that the changing trends were the same for CAT, SOD, and DPPH scavenging. RHT at 50 °C and RHT at 65 °C significantly induced CAT and SOD, whereas RHT at 80 °C significantly inhibited them and weakened the free radical-scavenging capacity of sweet potato. Maximum CAT and SOD activity levels were lower under RHT at 80 °C than under RHT at 65 °C on day 1. Peroxidase (POD) is another important ROS-scavenging enzyme (Figure 5D). POD activity was the same under both RHT at 80 °C and RHT at 65 °C. POD activity was essentially the same on days 1, 3, and 5. However, POD activity was significantly higher under RHT at 80 °C and RHT at 65 °C than under CK or RHT at 50 °C.

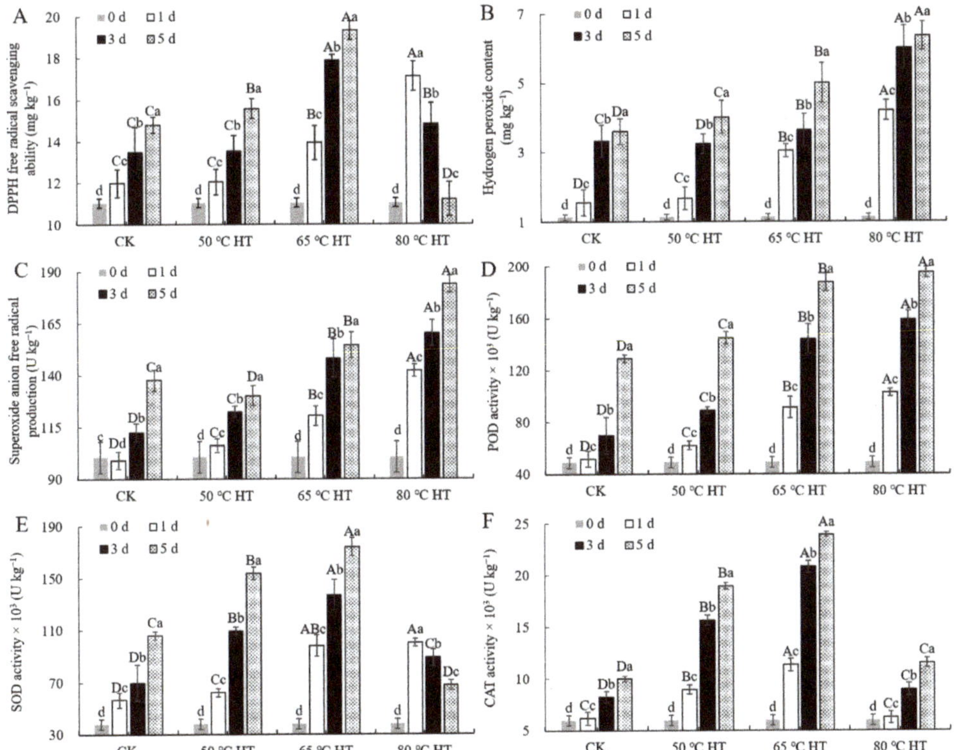

Figure 5. Effects of RHT treatment on DPPH, hydrogen peroxide, and superoxide anion scavenging and POD, SOD, and CAT activity in sweet potato callus ((**A**): DPPH free radical scavenging ability; (**B**): Hydrogen peroxide content; (**C**): Superoxide anion free radical production; (**D**): POD activity; (**E**): SOD activity; (**F**): CAT activity. Note: Capital letters are comparisons at the same time at different temperatures, and lowercase letters are comparisons at different times at the same temperature).

The foregoing results indicate that temperature had a significant impact on several active oxygen metabolism indices in sweet potato. Active oxygen metabolism, phenylpropane metabolism-related enzymes, phenolic compounds, and callus formation were also correlated. However, no valid conclusions could be drawn from direct data analysis. Hence, the key factors affecting sweet potato callus had to be identified through correlation data analysis.

2.5. Pearson's Correlation Analysis of Factors Affecting Sweet Potato Callus

Pearson's correlation analysis was conducted on all indices under all RHT treatments. The acceptance level was set to 0.01 to screen for key regulatory factors affecting the responses of sweet potato callus to RHT treatment. The results are shown in Figure 6. Compared with the lignin content, suberin stacking was more significantly correlated with other parameters in sweet potato callus healing. PAL and CAT were the most critical enzymes in phenylpropanoid and ROS metabolism, respectively, in terms of their effects on sweet potato callus formation. PAL and CAT activity levels were significantly correlated with suberin thickness. CAT activity was positively correlated with DPPH and superoxide anion scavenging as well as total phenol and total flavonoid production. PAL activity, lignin content, suberin thickness, and total phenol and total flavonoid production were also strongly positively correlated. The foregoing results suggest that CAT and PAL regulate ROS and phenylpropanoid metabolism, respectively; link these metabolic pathways; and improve sweet potato callus healing. In the study of sweet potato [22] and potato [23] on callus healing, it was pointed out that different physicochemical treatments can effectively stimulate the generation of reactive oxygen species in potato wounds, which can effectively promote potato root callus. In addition, Meng's [24] study on carrot pointed out that peroxidase in reactive oxygen species metabolism was the key enzyme in the carrot callus process and three peroxidases (DcPrx30, DcPrx32, and DcPrx62) were upregulated in the phloem of carrot.

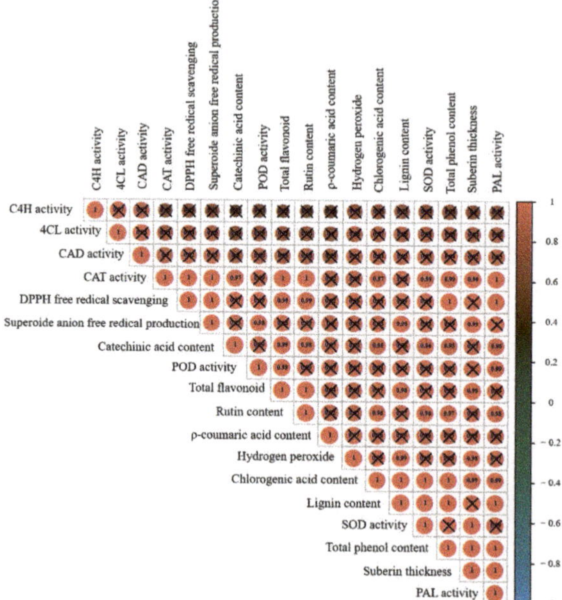

Figure 6. Correlation analysis of substances affecting sweet potato callus under different RHT treatments. Red: positive correlation. Blue: negative correlation. Numbers: significance of correlation. X: $p > 0.01$.

PAL and C4H are at the front end of phenylpropanoid metabolism and regulate phenolic compound biosynthesis [25]. Here, both enzymes were more strongly correlated with sweet potato callus than either 4CL or CAD. The latter enzymes were not significantly correlated with any index. In ROS metabolism, CAT and SOD scavenge free radicals [26]. Here they were significantly correlated with callus parameters. POD affects hydrogen peroxide production and was not significantly correlated with suberin stacking or lignin content. Thus, ROS production may not substantially contribute to sweet potato callus induction. ROS were generated under RHT at 80 °C. Nevertheless, the influence of this treatment on callus formation was weaker than that of RHT at 60 °C. Phenolic compounds are key substrates in suberin stacking. Total phenols and flavonoids were strongly correlated with suberin stacking thickness. However, only the chlorogenic acid content was significantly correlated with suberin stacking thickness. Thus, chlorogenic acid may be the key phenolic compound affecting sweet potato callus. There were significant correlations among chlorogenic acid content, PAL and CAT activity, and DPPH scavenging.

3. Discussion

Callus is vital to the successful storage and transport of root crops. Root crops enter into direct contact with the soil during their growth and can transmit soil-borne microorganisms at harvest. Root crops often undergo severe decay during storage and transportation [27]. Postharvest callus formation involves cell proliferation and differentiation, signal transduction, disease resistance, secondary metabolism, and energy generation [7]. However, relatively few reports have been published on postharvest calli to date. Several studies indicated that ROS, fatty acid, and phenylpropanoid metabolic pathways occur in the calli of root crops such as white potato and sweet potato [28]. Hence, the enzymes associated with these pathways are implicated in the entire callus formation process.

ROS may act both as signal molecules and oxidants in white potato and sweet potato callus formation [29]. In the present study, ROS production responded rapidly when the sweet potatoes were subjected to RHT at 65 °C and RHT at 80 °C. Superoxide anion production capacity significantly increased in sweet potato callus subjected to RHT at 65 °C. Accumulation of superoxide anion, singlet oxygen, and other ROS disrupted active oxygen metabolism, which, in turn, activated SOD, CAT, POD, and other antioxidant enzymes in sweet potato. SOD then disproportionated the superoxide anions into H_2O_2. Therefore, RHT accelerated the induction of the key metabolic pathways associated with callus formation in damaged sweet potato tissue. Phenylpropanoid and fatty acid metabolism are the key pathways affecting sweet potato and white potato callus formation [30]. Phenylpropanoid metabolism induces the generation of numerous monophenols in injured plant tissues. Other enzymes catalyze the polymerization of these monophenols into polyphenol suberin (SPP). Monophenols also participate in lignin metabolism, which synthesizes lignin stacks around the calli [31]. In fatty acid metabolism, the damaged tissues produce numerous short-chain fatty acids (SCFAs) that are then polymerized to form suberin polyaliphatics (SPA). The terminal SPA in fatty acid metabolism and the SPD generated by phenylpropanoid metabolism participate in suberin polymerization. Suberin then deposits the final sealing callus layer known as the polyphenolic domain (SPPD) [32]. We found that RHT at 50 °C did not induce the rapid production of ROS such as superoxide anion and hydrogen peroxide and could not, therefore, initiate phenylpropanoid metabolism. Though RHT at 80 °C induced ROS production in the early sweet potato stages, it significantly inhibited CAT and SOD. Hence, RHT at 80 °C disrupted ROS production and clearance by days 3–5 and could not continuously promote callus formation in the wound tissue. In addition, ROS imbalance may cause fruit and vegetable browning [33]. We discovered substantial melanin accumulation in the callus surface of the sweet potato subjected to RHT at 80 °C. The melanin might have been derived from polyphenol oxidation.

Abundant hydrogen peroxide was produced under all RHT treatments and induced sweet potato callus formation through phenylpropanoid metabolism. RHT at 65 °C induced PAL and 4CL, and phenylpropanoid metabolism generated phenolic monomers. Zhu showed relatively higher PAL and 4CL activity and phenolic compound production around white potato wounds [33]. Compared with RHT at 50 °C and RHT at 80 °C, RHT at 65 °C significantly increased total flavonoid, rutin, chlorogenic acid, *p*-coumaric acid, and other phenolic compounds at days 0–5. At the appropriate temperatures, heat shock treatments induce phenylpropanoid metabolism and promote SPDD stacking. Compared with CK and RHT at 50 °C, RHT at 65 °C RHT and RHT at 80 °C significantly increased the lignin content. However, there were no significant differences among CK, RHT at 50 °C, RHT at

65 °C, or RHT at 80 °C in terms of lignin content by day 5. Therefore, RHT temperature may only be weakly correlated with lignin synthesis and accumulation. The terminal steps in lignin synthesis are associated with phenylpropanoid metabolism [34]. Santos stated that lignin synthesis may be correlated with low temperature, whereas lignin metabolism is relatively less affected by temperature fluctuation or high temperature [35]. Here, we measured the parameters of ROS and phenylpropanoid metabolism and found that both pathways induced sweet potato callus formation. Hence, there may be a strong correlation between these pathways in sweet potato callus.

Pearson's correlation analysis confirmed the foregoing speculations. Suberin stacking, CAT, PAL, and chlorogenic acid may be key factors in RHT-induced sweet potato callus formation. PAL is a key rate-limiting enzyme in phenylpropanoid metabolism. It catalyzes the deamination of *L*-phenylalanine to *trans*-cinnamic acid [36]. C4H then hydroxylates *trans*-cinnamic acid to *p*-coumaric, chlorogenic, ferulic, erucic, and other phenolic acids that may participate in SPPD formation. Furthermore, 4CL catalyzes the conversion of these phenolic acids to various phenolic acid-CoA. CAD then transforms the latter into lignin-derived substrates such as cinnamyl alcohol, coniferol, and sinucinol. At this stage, callus formation in sweet potato is complete [37]. The key metabolic pathways and key factors that RHT treatment can use to promote sweet potato callus are shown in Figure 7. The present study demonstrated that in sweet potato callus formation, ROS metabolism rapidly responds to RHT treatment and CAT produces and removes ROS. Abundant ROS induce PAL, which synthesizes and causes the accumulation of chlorogenic acid. Suberin is then quickly stacked at the sweet potato wound and promotes callus formation there.

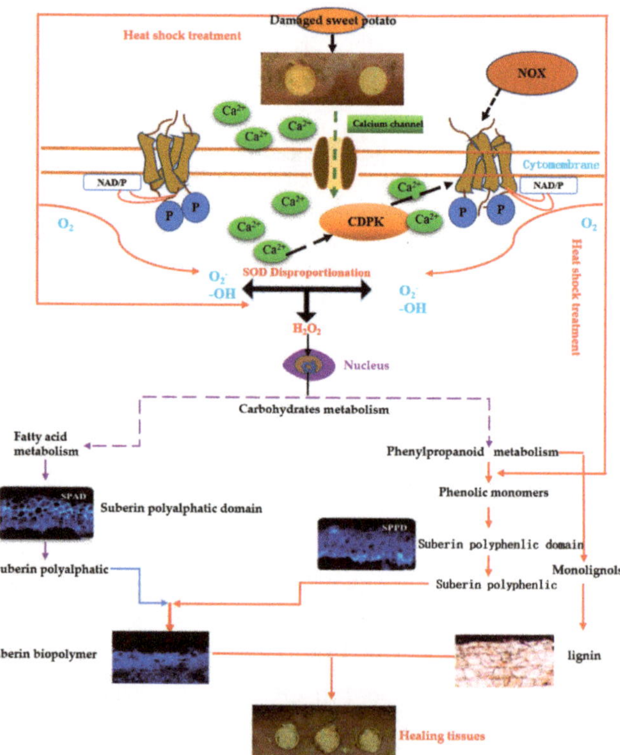

Figure 7. Pattern of metabolic pathways involved in callus of sweet potato tuber under heat shock treatment. CDPK: calcium-dependent protein kinase; NOX: NADPH oxidase; SOD: superoxide dismutase.

4. Materials and Methods

4.1. Sweet Potato Acquisition and Sample Preparation

Sweet potato (*Ipomoea batatas* (L.) Lam.). Xiguahong was purchased and transported from Fujian Province, China. Each tuber was packed in a bubble bag and transported within 1 day in a refrigerated truck (<13 ± 2 °C) from Fujian to Beijing. At the Beijing laboratory, tubers uniform in size; free of epicarp damage, disease, and insect pests; and weighing 250 ± 50 g were selected as the experimental materials.

4.2. Artificial Injury and Heat Shock Treatment of Sweet Potato Samples

The selected sweet potato roots were washed twice with tap water and distilled water and dried at room temperature (25 ± 1 °C). The knives (deli, Zhejiang, China) and hole (deli, Zhejiang, China) punches (15 mm diameter) used in the artificial wound experiment were disinfected with 95% (*v*/*v*) ethanol. Each tuber was perforated with the hole punches and 15 mm epicarp (deli, Zhejiang, China) disks were excised with the knives to a depth of 3 mm. Three wounds were artificially induced on two sides of each tuber.

The damaged sweet potatoes were arbitrarily divided into four groups. Unheated tubers served as the control (CK), whereas the other three groups were subjected to the RHT treatments.

The pretreatment method of sweet potato callus referred to the previous method in the laboratory [4]; the wounded tubers were placed in an independently designed sweet potato RHT machine and heated to 50 °C, 65 °C, or 80 °C for 15 min. After the RHT treatment, the tubers were cooled to 25 ± 1 °C and stored until the subsequent experiments.

All samples were transferred to cold storage (13 ± 1 °C; 55% RH) in the laboratory and wounds were allowed to heal for 5 days. The foregoing temperature and relative humidity were optimal for sweet potato storage [4]. The sample quantity of sweet potato in CK and the three RHT treatment groups were all more than 150. The sample healing preparation process was repeated thrice.

4.3. Lignin and Suberin Accumulation in Wounded Tissue

Sweet potato lignin staining was observed according to the method of Jiang with some modifications [6]. Tissue blocks with the injured surface were hand-sliced (0.4 mm–0.5 mm depth) vertically with a blade. The prepared slices were immediately rinsed with distilled water to remove starch granules and then immersed in 1% (*w*/*v*) phloroglucinol solution for 2 h staining on a glass slide with a few drops of concentrated hydrochloric acid. After 5 min, images of red-stained deposited lignin were captured with a microscope (DM500, Leica Shanghai limited company, Shanghai, China) under 10 × magnification. Suberin deposition was microscopically detected by its autofluorescence according to Fugate et al. [20]. The autofluorescence of the suberin was analyzed using a microscope (DM500, Leica Shanghai limited company, Shanghai, China) with a fluorescence excitation filter at 280 nm and an emission filter at 620 nm. The prepared sections (0.3 mm–0.4 mm depth) were rinsed with distilled water 2–3 times before capturing images under 10× magnification. Six potato tubers for each group were used to observe staining and autofluorescence.

4.4. Determination of Lignin Content in Wound Callus of Sweet Potato

The determination method of lignin content by Zhou was improved [20]. At the end of the callus, samples were taken from the roots of sweet potato in each treatment group with a stainless-steel knife. During the determination, 1 g of frozen callus was taken, and 4 mL 95% ethanol precooled at 4 °C was added and beaten evenly with a beater. The callus was transferred to a 10 mL centrifuge tube and centrifuged at 4 °C at 10,000× *g* for 20 min. After the supernatant was discarded, 2 mL 95% ethanol was added, mixed, centrifuged (4 °C 10,000× *g* 10 min), and repeated 3 times, and then ethanol was used. After the precipitation was collected and dried to a constant weight, the dry matter was moved to a small test tube. Before the measurement, the water bath was placed in a fume hood in advance and the temperature was adjusted to 70 °C for preheating. First, 1 mL 25% acetyl bromide solution was added to the test tube and mixed well. It was then immediately placed in a water bath for reaction for 30 min; 1 mL sodium hydroxide (2 mol·L^{-1}), 0.1 mL hydroxylamine hydrochloride (7.5 mol·L^{-1}), and 2 mL glacial acetic acid were added successively and centrifuged at 4 °C at 10,000× *g* for 20 min; and 0.5 mL supernatant was absorbed. The absorbance was measured at 280 nm with glacial acetic acid at a constant volume of 5 mL and repeated 3 times.

4.5. Determination of Phenylpropanoid Metabolism-Related Enzyme Activities in Sweet Potato Callus

The phenylalanine ammonia-lyase (PAL), cinnamate-4-hydroxylase (C4H), 4-coumarate-CoA ligase (4CL) activity was evaluated spectrophotometrically according to the methodology of Jiang et al. [6] by using a UV–2600 spectrophotometer (Shimadzu, Kyoto, Japan), with some modifications. Protein (pro) content in the enzyme extract solution was measured by the Coomassie brilliant blue G250 method. The PAL, C4H, and 4CL activity was reported in U·Kg^{-1}.

Cinnamyl alcohol dehydrogenase (CAD) was determined by Sarni with some modification [38]. The 1.0 g frozen powder was homogenized in 3 mL TRIS-HCl buffer (pH 8.8, containing 40 g·L^{-1} PVP), 15 mmol·L^{-1} β-mercaptoethanol, 10% methylene, and 2% (*w*/*v*) PEG in an ice bath. After standing for 30 min and centrifugation at 12 000× *g* for 30 min (4 °C), the supernatant was a crude enzyme solution. Reaction system: 0.2 mL crude enzyme solution, 0.8 mL reaction solution (containing 10 mmol·L^{-1} nicotinamide adenine dinucleotide phosphorus (NADP) and 5 mmol·L^{-1} *trans*-cinnamic acid), water bath at 37 °C for 30 min, 1 mol·L^{-1} HCl to terminate the reaction (if precipitate exists, after centrifugation), absorbance value measured at 400 nm, and 0.2 mL PBS and 0.8 mL reaction solution used as control. The enzyme activity unit (U) was defined as 0.001 change of light absorption value per minute, and the CAD activity was expressed as U·Kg^{-1}.

4.6. Determination of Antioxidant Capacity and ROS Metabolism-Related Enzymes of Sweet Potato Callus

The activity of superoxide dismutase (SOD) was determined by azoblue tetrazole photoreduction method with some modifications [39]. The activity of catalase (CAT) was determined by colorimetric method [39]. Peroxidase (POD) activity was measured by referring to the method of Dastmalchi et al. [7]. The above enzyme activities were expressed by U·Kg^{-1}.

The determination of 1,1-dipheny1-2-picryl-hydrazyl (DPPH) clearance capacity referred to Oirschot and was modified [31]. A total of 5 g of frozen callus was taken and placed in a centrifuge tube (50 mL). The 20 mL ethanol (70%) solution was added and mixed, ultrasonic treatment was carried out for 1 h, centrifugation was carried out (4 °C 10,000× *g* 20 min), and the supernatant was taken as the extract. The 0.05 mL extract was absorbed and placed in a test tube. A total of mL DPPH ethanol solution with a concentration of 0.3 mmol·L^{-1} was first added and mixed. After a water bath at 30 °C for 1 h, the reaction was immediately cooled and terminated. DPPH radical scavenging capacity was calculated using trolox as standard equivalent, expressed in mg·kg^{-1}.

4.7. Determination of Phenols, Hydrogen Peroxide, and Superoxide Anion in Sweet Potato Callus

Extraction of phenolics was carried out according to Liu with some modifications [40]. Ten grams of sweet potato tissue were homogenized with liquid nitrogen, then mixed with 20 mL of 70% (*v*/*v*) ethanol and placed under ultrasound assisted at 30 °C for 0.5 h; thereafter, the mixture was centrifuged and the supernatant was collected and concentrated with a rotary vacuum evaporator (RE-52; Yarong Biochemistry Instrument Factory, Shanghai, China) at 30 °C. Identification and quantification of the phenolic compounds in the extract were carried out using an HPLC method (Liu et al., 2020) with Shimadzu liquid chromatography (pumps, LC-20AT; diode array detection, SPD-M20A) and an RP C18 column (Venusil ASB, 5 µm, 4.6 mm 9 250 mm; Agela Technologies Inc.,Tianjing, China). The operating conditions were as follows: mobile phase, 1% (*v*/*v*) acetic acid (A), and methanol (B); gradient, 12–25% B from 0 min to 15 min, 25–35% B from 15 min to 25 min, 35–55% B from 25 min to 50 min, 55–65% B from 50 min to 60 min and 65–12% B from 60 min to 70 min; flow rate, 1.0 mL·min^{-1}; column temperature, 35 °C; injection volume, 20 L; and UV detection wavelength, 280 nm.

A total of 3.0 g of frozen sweet potato powder was extracted in 3 mL of cold acetone for 10 min before centrifugation at 4 °C and 10,000× g for 30 min; the supernatant was collected for H_2O_2 content determination. The reaction mixture contained 1 mL of supernatant, 100 μL of 20% $TiCL_4$, and 100 μL of concentrated ammonia. The mixture was reacted for 10 min and then centrifuged for 10 min. The resultant precipitate was washed three times with acetone and then dissolved with the addition of 2 mL of 1-mmol/L concentrated sulfuric acid. The OD value was measured at 410 nm, and H_2O_2 content was calculated from a standard curve. Thus, H_2O_2 content was expressed as mg·kg^{-1}.

Two grams of frozen callus were weighed and placed in a 10 mL centrifuge tube. The 5 mL 0.05 mol/L pH7.8 phosphoric acid buffer (containing 0.001 mol·L^{-1} ethylenediamine tetraacetic acid, 0.3% Triton X-100 and 2% polyvinylpyrrolidone) were added and mixed. Centrifugation (4 °C 10,000× g 20 min) was performed to collect supernatant. The content of superoxide anion (O^{2-}) was determined by using the Nanjing Jiancheng kit(Nanjing Jiancheng, Nanjing, China). A total of 0.8 mL supernatant was mixed with equal volume Tris-HCl buffer solution (50 mmol·L^{-1}, pH 8.2). The mixture was allowed to stand at 25 °C for 15 min, then 0.4 mL of 1.5 mmol·L^{-1} pyrogallic acid was added and thoroughly mixed. The absorbance value of the mixture was detected every 30 s at 550 nm, and the detection lasted for 5 min. Meanwhile, ascorbic acid was used as the positive control group, and 50 mmol·L^{-1} Tris-HCl buffer was used as the blank group. The content of superoxide anion in sweet potato was expressed as U·kg^{-1}.

4.8. Data Statistics and Analysis

The above experiments were repeated three times. Excel 2010 software (Microsoft Corporation, Washington, WA, USA) was used to make statistics on all data, calculate mean value and standard deviation, and plot. SPSS 26.0 software (IBM, New York, NY, USA) was used to conduct analysis of variance and multiple difference significance analysis of the experimental data; $p < 0.05$ indicated significant difference. Pearson's correlation heat map was analyzed and plotted using HIPLOT.

5. Conclusions

Damaged sweet potato tubers were subjected to RHT at 50 °C, 65 °C, and 80 °C. Changes in enzyme activity and intermediate accumulation in ROS and phenylpropanoid metabolism were evaluated. RHT at 65 °C significantly promoted callus formation in injured postharvest sweet potato root tubers. PAL in phenylpropanoid metabolism, CAT in ROS metabolism, and chlorogenic acid were the key factors inducing and developing callus in response to RHT treatment. The results of this study may provide theoretical and methodological bases for rapid callus induction in postharvest sweet potato. However, it remains to be established how PAL- and CAT-related genes respond as transduction signals to RHT treatment. Future research should determine how phenolic acids contribute to suberin stacking. Subsequent studies should aim to clarify the mechanisms by which heat shock promotes sweet potato callus and devise novel methods of reducing postharvest loss during sweet potato production, storage, and transport.

Author Contributions: B.L. and Q.X. designed the experimental trials. Q.X., J.S., X.F., X.L. and B.L. performed the experiments and collected the samples. B.L., Q.X., L.J. and G.H. wrote the article. B.L., H.P. and X.Z. revised and edited the article. All authors have read and agreed to the published version of the manuscript.

Funding: This work was supported by the National Key R & D Program of China (No. 2016YFD0401301 and 2017YFD0401305).

Institutional Review Board Statement: Not applicable.

Informed Consent Statement: Not applicable.

Data Availability Statement: Not applicable.

Acknowledgments: All experiments in this work were supported by the Key Laboratory of Agro-Products Primary Processing.

Conflicts of Interest: The authors declare no conflict of interest.

References

1. Pérez-Pazos, J.V.; Rosero, A.; Martínez, R.; Pérez, J.; Morelo, J.; Araujo, H.; Burbano-Erazo, E. Influence of morpho-physiological traits on root yield in sweet potato (*Ipomoea batatas* Lam.) genotypes and its adaptation in a sub-humid environment. *Sci. Hortic.* **2020**, *275*, 109703. [CrossRef]
2. Ji, C.Y.; Kim, H.S.; Lee, C.-J.; Kim, S.-E.; Lee, H.-U.; Nam, S.-S.; Li, Q.; Ma, D.-F.; Kwak, S.-S. Comparative transcriptome profiling of tuberous roots of two sweetpotato lines with contrasting low temperature tolerance during storage. *Gene* **2019**, *727*, 144244. [CrossRef] [PubMed]
3. Parmar, A.; Kirchner, S.M.; Sturm, B.; Hensel, O. Pre-harvest Curing: Effects on Skin Adhesion, Chemical Composition and Shelf-life of Sweetpotato Roots under Tropical Conditions. *East Afr. Agric. For. J.* **2017**, *82*, 130–143. [CrossRef]
4. Xin, Q.; Sun, J.; Feng, X.; Zhao, Z.; Liu, B.; Jiang, L.; Hao, G. Heat Shock Treatment Promotion Wound Curing and Metabolic Mechanism of Sweet Potato. *Food Sci.* **2022**. (In Chinese with Abstract) [CrossRef]
5. Xue, W.; Liu, N.; Zhang, T.; Li, J.; Chen, P.; Yang, Y.; Chen, S. Substance metabolism, IAA and CTK signaling pathways regulating the origin of embryogenic callus during dedifferentiation and redifferentiation of cucumber cotyledon nodes. *Sci. Hortic.* **2021**, *293*, 110680. [CrossRef]
6. Jiang, H.; Wang, B.; Ma, L.; Zheng, X.; Gong, D.; Xue, H.; Bi, Y.; Wang, Y.; Zhang, Z.; Prusky, D. Benzo-(1, 2, 3)-thiadiazole-7-carbothioic acid s-methyl ester (BTH) promotes tuber wound healing of potato by elevation of phenylpropanoid metabolism. *Postharvest Biol. Technol.* **2019**, *3*, 125–132. [CrossRef]
7. Dastmalchi, K.; Rodriguez, M.P.; Lin, J.; Yoo, B.; Stark, R.E. Temporal resistance of potato tubers: Antibacterial assays and metabolite profiling of wound-healing tissue extracts from contrasting cultivars. *Phytochemistry* **2018**, *159*, 75–89. [CrossRef]
8. Woolfson, K.N.; Bjelica, A.; Haggitt, M.L.; Kachura, A.; Zhang, Y.N. Differential induction of polar and non-polar metabolism during wound-induced suberization in potato (*Solanum tuberosum* L.) tubers. *Plant J.* **2018**, *93*, 931–942. [CrossRef]
9. Kumar, G.N.M.; Lulai, E.C.; Suttle, J.C.; Knowles, N.R. Age-induced loss of wound-healing ability in potato tubers is partly regulated by ABA. *Planta* **2010**, *232*, 1433–1445. [CrossRef]
10. Feng, X.; Hansen, J.D.; Biasi, B.; Tang, J.; Mitcham, E.J. Use of hot water treatment to control codling moths in harvested California 'Bing' sweet cherries. *Postharvest Biol. Technol.* **2004**, *31*, 41–49. [CrossRef]
11. Jakubowski, T. Use of UV-C radiation for reducing storage losses of potato tubers. *Bangladesh J. Bot.* **2018**, *47*, 533–537. [CrossRef]
12. Lulai, E.C.; Suttle, J.C.; Pederson, S.M. Regulatory involvement of abscisic acid in potato tuber wound healing. *J. Exp. Bot.* **2008**, *59*, 1175–1186. [CrossRef] [PubMed]
13. Lobato, M.C.; Daleo, G.R.; Andreu, A.B.; Olivieri, F.P. Cell Wall Reinforcement in the Potato Tuber Periderm After Crop Treatment with Potassium Phosphite. *Potato Res.* **2017**, *61*, 19–29. [CrossRef]
14. Goffner, D.; Joffroy, I.; Grima-Pettenati, J.; Halpin, C.; Knight, M.E.; Schuch, W.; Boudet, A.M. Purification and characterization of isoforms of cinnamyl alcohol dehydrogenase from Eucalyptus xylem. *Planta* **1992**, *188*, 48–53. [CrossRef] [PubMed]
15. Mwanga, R.; Andrade, M.; Carey, E.; Low, J.; Yencho, C.; Grüneberg, W. Sweetpotato (*Ipomoea batatas* L.). In *Genetic Improvement of Tropical Crops*; Springer: Cham, Switzerland, 2017; pp. 181–218. [CrossRef]
16. St. Amand, P.C.; Randle, W.M. Ethylene production as a possible indicator of wound healing in roots of several sweet potato. cutivars. *Euphytica* **1991**, *53*, 97–102. [CrossRef]
17. Ge, X.; Zhu, Y.; Li, Z.; Bi, Y.; Yang, J.; Zhang, J.; Prusky, D. Preharvest multiple fungicide stroby sprays promote wound healing of harvested potato tubers by activating phenylpropanoid metabolism. *Postharvest Biol. Technol.* **2021**, *171*, 111328. [CrossRef]
18. Wang, B.; Jiang, H.; Bi, Y.; He, X.; Wang, Y.; Li, Y.; Zheng, X.; Prusky, D. Preharvest multiple sprays with sodium nitroprusside promote wound healing of harvested muskmelons by activation of phenylpropanoid metabolism. *Postharvest Biol. Technol.* **2019**, *158*, 110988. [CrossRef]
19. Zhou, F.H.; Jiang, A.L.; Feng, K.; Gu, S.T.; Xu, D.Y.; Hu, W.Z. Effect of methyl jasmonate on wound healing and resistance in fresh-cut potato cubes. *Postharvest Biol. Technol.* **2019**, *157*, 110958. [CrossRef]
20. Fugate, K.K.; Ribeiro, W.S.; Lulai, E.C.; Deckard, E.L.; Finger, F.L. Cold Temperature Delays Wound Healing in Postharvest Sugarbeet Roots. *Front Plant Sci.* **2016**, *7*, 499. [CrossRef]
21. Zheng, X.Y.; Jiang, H.; Bi, Y.; Wang, B.; Wang, T.L.; Li, Y.C.; Gong, D.; Wei, Y.N.; Li, Z.C.; Prusky, D. Comparison of wound healing abilities of four major cultivars of potato tubers in China. *Postharvest Biol. Technol.* **2020**, *164*, 111167. [CrossRef]
22. Irungu, F.G.; Tanga, C.M.; Ndiritu, F.G.; Mathenge, S.G.; Kiruki, F.G.; Mahungu, S.M. Enhancement of potato (*Solanum tuberosum* L.) postharvest quality by use of magnetic fields—A case of shangi potato variety. *Appl. Food Res.* **2022**, *2*, 100191. [CrossRef]

23. Zheng, X.; Zhang, X.; Jiang, H.; Zhao, S.; Silvy, E.M.; Yang, R.; Han, Y.; Bi, Y.; Prusky, D. Kloeckera apiculata promotes the healing of potato tubers by rapidly colonizing and inducing phenylpropanoid metabolism and reactive oxygen species metabolism. *Postharvest Biol. Technol.* **2022**, *192*, 112033. [CrossRef]
24. Meng, G.; Fan, W.; Rasmussen, S.K. Characterisation of the class III peroxidase gene family in carrot taproots and its role in anthocyanin and lignin accumulation. *Plant Physiol. Biochem.* **2021**, *167*, 245–256. [CrossRef] [PubMed]
25. Sun, M.; Jiang, F.; Zhou, R.; Wen, J.; Cui, S.; Wang, W.; Wu, Z. Respiratory burst oxidase homologue-dependent H2O2 is essential during heat stress memory in heat sensitive tomato. *Sci. Hortic.* **2019**, *258*, 108777. [CrossRef]
26. Vreugdenhil, D.; Bradshaw, J.E.; Gebhardt, C.; Govers, F.; MacKerron, D.K.L.; Taylor, M.; Ross, H.A. *Potato Biology and Biotechnology: Advances and Perspectives*; Elsevier: Amsterdam, The Netherlands, 2007.
27. Liu, B.; Lyu, X.; Wang, C.; Sun, J.; Jiang, W. Process optimization of high temperature short-time hot air treatment to promote sweet potato callus. *Trans. Chin. Soc. Agric. Eng.* **2020**, *36*, 313–322. (In Chinese with Abstract) [CrossRef]
28. Liang, W.; Zhu, Y.; Chai, X.; Kong, R.; Li, B.; Li, Y.; Bi, Y.; Prusky, D. p-Coumaric Acid Promoted Wound Healing of Potato Tubers by Accelerating the Deposition of Suberin Poly Phenolic and Lignin at Wound Sites. *Sci. Agric. Sin.* **2021**, *54*, 4434–4445.
29. Razem, F.A.; Bernards, M.A. Reactive oxygen species production in association with suberization: Evidence for an NADPH-dependent oxidase. *J. Exp. Bot.* **2003**, *54*, 935–941. [CrossRef]
30. Chai, X.; Kong, R.; Zheng, X.; Zhu, Y.; Liang, W.; Zhao, S.; Li, B.; Bi, Y.; Prusky, D. Chlorine dioxide accelerates the wound healing process of potato tubers by eliciting phenylpropanoid metabolism. *Food Sci.* **2022**, 1–11. Available online: http://kns.cnki.net/kcms/detail/11.2206.TS.20220321.1634.034.html (accessed on 22 March 2022). (In Chinese with Abstract)
31. Van Oirschot, Q.E.A.; Rees, D.; Aked, J.; Kihurani, A. Sweetpotato cultivars differ in efficiency of wound healing. *Postharvest Biol. Technol.* **2006**, *42*, 65–74. [CrossRef]
32. Lulai, E.C. Skin-Set, Wound Healing, and Related Defects. In *Potato Biology and Biotechnology*; Elsevier: Amsterdam, The Netherlands, 2007; Chapter 22; pp. 471–500.
33. Zhou, F.; Xu, D.; Liu, C.; Chen, C.; Tian, M.; Jiang, A. Ascorbic acid treatment inhibits wound healing of fresh-cut potato strips by controlling phenylpropanoid metabolism. *Postharvest Biol. Technol.* **2021**, *181*, 111644. [CrossRef]
34. Vanholme, R.; De Meester, B.; Ralph, J.; Boerjan, W. Lignin biosynthesis and its integration into metabolism. *Curr. Opin. Biotechnol.* **2019**, *56*, 230–239. [CrossRef]
35. dos Santos, A.B.; Bottcher, A.; Vicentini, R.; Mayer, J.L.S.; Kiyota, E.; Landell, M.A.; Creste, S.; Mazzafera, P. Lignin biosynthesis in sugarcane is affected by low temperature. *Environ. Exp. Bot.* **2015**, *120*, 31–42. [CrossRef]
36. Deng, Y.; Lu, S. Biosynthesis and Regulation of Phenylpropanoids in Plants. *Crit. Rev. Plant Sci.* **2017**, *36*, 257–290. [CrossRef]
37. Soler, M.; Serra, O.; Fluch, S.; Molinas, M.; Figueras, M. A potato skin SSH library yields new candidate genes for suberin biosynthesis and periderm formation. *Planta* **2011**, *233*, 933–945. [CrossRef]
38. Sarni, F.; Grand, C.; Boudet, A.M. Purification and properties of cinnamoyl-CoA reductase and cinnamyl alcohol dehydrogenase from poplar stems (*Populus* × *euramericana*). *Eur. J. Biochem.* **1984**, *139*, 259–265. [CrossRef] [PubMed]
39. Yang, R.; Han, Y.; Han, Z.; Ackah, S.; Li, Z.; Bi, Y.; Yang, Q.; Prusky, D. Hot water dipping stimulated wound healing of potato tubers. *Postharvest Biol. Technol.* **2020**, *167*, 111245. [CrossRef]
40. Liu, B.; Zhao, H.; Fan, X.; Jiao, W.; Cao, J.; Jiang, W. Near freezing point temperature storage inhibits chilling injury and enhances the shelf life quality of apricots following long-time cold storage. *J. Food Process. Preserv.* **2019**, *43*, e13958. [CrossRef]

Article

Under Natural Field Conditions, Exogenous Application of Moringa Organ Water Extract Enhanced the Growth- and Yield-Related Traits of Barley Accessions

Nawroz Abdul-razzak Tahir [1,*], Djshwar Dhahir Lateef [2], Kamil Mahmud Mustafa [2] and Kamaran Salh Rasul [1]

1 Horticulture Department, College of Agricultural Engineering Sciences, University of Sulaimani, Sulaimani 46001, Iraq
2 Biotechnology and Crop Sciences Department, College of Agricultural Engineering Sciences, University of Sulaimani, Sulaimani 46001, Iraq
* Correspondence: nawroz.tahir@univsul.edu.iq

Abstract: Barley (*Hordeum vulgare* L.) is the preferred crop in arid regions, particularly for farmers with limited agricultural resources and low income. Typically, it is utilized for human consumption, animal feed, and malting. The discovery of natural (organic) sources of biostimulants has attracted a great deal of interest for crop productivity enhancement. Using a randomized complete block design with three main blocks, it was our aim to investigate the effects of foliar moringa (*Moringa oleifera* L.) organ extract (MOE) on the growth and yield components of a collection of barley accessions grown in Iraq. As indicated by the obtained results, almost all traits associated with barley growth and yield productivity were significantly enhanced by MOE application, relative to the respective control condition. The majority of barley accessions responded positively to the MOE treatment based on all studied traits (with the exception of 1000-kernel weight). According to the results of principal component analysis (PCA), the distribution of accessions on the two components under the MOE application was distinct from the distribution of accessions under control conditions, indicating that accessions responded differently to the MOE application. In addition, the distribution pattern of traits under MOE treatment was comparable to the distribution pattern of traits under the control condition, with the exception of two traits: total yield and 1000-kernel weight. AC5 and AC18 responded positively to the MOE application by possessing the highest total yield and harvest index values. The total yield trait registered the highest increasing value index (37.55%) based on the trait response index, followed by the straw weight (22.29%), tillering number per plant (21.44%), and spike number per plant (21.36%), while the spike length trait registered the lowest increasing value index (0.45%), compared to the traits under control conditions. So far, the results indicate that foliar application of MOE can be utilized effectively as a natural growth promoter to increase the growth and yield productivity of grown barley accessions.

Keywords: bostimulation; plant extract; *Hordeum vulgare*; growth performance; production components

Citation: Tahir, N.A.-r.; Lateef, D.D.; Mustafa, K.M.; Rasul, K.S. Under Natural Field Conditions, Exogenous Application of Moringa Organ Water Extract Enhanced the Growth- and Yield-Related Traits of Barley Accessions. *Agriculture* **2022**, *12*, 1502. https://doi.org/10.3390/agriculture12091502

Academic Editors: Daniele Del Buono, Luca Regni and Primo Proietti

Received: 2 September 2022
Accepted: 15 September 2022
Published: 19 September 2022

1. Introduction

The greatest challenge facing modern agriculture science is maintaining food production, in order to meet the needs of a growing global population without jeopardizing future generations' access to natural resources. The current level of agricultural intensification has reached a tipping point, with far-reaching, irreversible effects on the global environment and a significant decline in the range of ecosystem services once provided by nature [1]. Barley (*Hordeum vulgare* L.) is the preferred crop in arid regions, particularly among farmers with limited agricultural resources and low income. It is commonly used for human consumption, animal feed, and malting [2]. As a result of their low glycemic index and high nutritional value, barley crops are currently the subject of worldwide interest [3]. In comparison to other small grains such as wheat and rice, it is generally believed that barley yields are less susceptible to weather fluctuations [4]. In Iraq, barley is primarily grown in areas with limited precipitation and rain-fed conditions. As a result of climatic changes, crops such as wheat and rice are competing for a shrinking amount of arable land. Farmers believe that barley can replace some wheat cultivars, so they are cultivating it instead, particularly during this dry year [5]. In order to improve the yield and growth parameters of barley, agronomists, crop physiologists, and other researchers must urgently discover sustainable techniques and new innovations. Despite the fact that barley is considered to be stress-resistant [6], its productivity in harsh environmental conditions is negatively impacted by a number of factors, including water limitation, agronomic practices, heat stress, and so on [7].

Pesticides, nitrates (from nitrogen-rich fertilizers), and phosphorus are the most serious agricultural pollutants. Chemical fertilizers can certainly increase agricultural yield, and they are regarded as a crucial factor influencing the final quality of barley products [8]. Unfortunately, the use of this type of fertilizer comes at a terrible price, as it degrades the soil and pollutes the environment, in addition to being expensive. The greatest opportunity for expanding food production is to improve yields and quality through the strategic application of mineral and organic fertilizers, plant protection products, and water supply. It is critical that this process is carried out in a manner that is safe for both the environment and consumers [9]. In recent years, there has been a great deal of focus on identifying various natural (organic) sources of biostimulants for enhancing crop productivity and achieving sustainable agriculture [10,11]. There are numerous sources of biostimulants used frequently in agriculture, including humic acid [12], chitosan and chitin derivatives [13], seaweed extracts [14], and plant extracts [15].

Stimulants are beneficial, but they cannot replace chemical fertilizer in long-term agricultural output. Plant extracts of moringa can either inhibit (at high concentrations) or stimulate (at low concentrations) plant development and growth [16,17]. Moringa leaf extract (MLE) obtained from moringa (*Moringa oleifera*) is one of the most popular plant biostimulants that can be used as a substitute and natural source of mineral nutrition and fertilizer, because it contains stimulant compounds, such as cytokinins such as zeatin, antioxidants such as ascorbic acid, flavonoid, amino acids, vitamins C and A, and phenolics, and micro- and macronutrients [18]. Besides, Yasmeen et al. [19] showed that the leaf extraction of such a plant can also provide the balance among nutrients, phytohormones, and antioxidants. Zeatin is the main hormone detected in MLE, and, so far, its concentration is thousands of times higher compared to the most studied plants [20]. To improve the productivity and growth of many plants grown in normal conditions, this kind of natural biostimulant has been applied as a foliar application [21–23]. To our knowledge, few small-scale experiments have been conducted to investigate the effect of moringa mlant extract on barely crops for productivity [24]. At field scale, no study has been reported using a combined application of moringa (leaf, root, and seed) extract. Recently, an investigation into a collection of barley accessions grown in Iraq by our own research group stated different patterns of response at early stages, phenotypically, physiologically, and biochemically, for drought tolerance [25]. In this regard, the current investigation was planned to study the effects of moringa organ extract on the growth and yield of a collection of barley accessions.

2. Materials and Methods

2.1. Study Area and Experimental Layout

In this study, 59 barley accessions collected throughout Iraq were grown in the field under rain-fed conditions at the Faculty of Agricultural Sciences-University of Sulaimani Research Station (35°34′17.5″ N 45°22′01.0″ E) during the 2019–2020 growing season (Table 1). The annual precipitation was 417 mm, and average temperatures ranged from 1 to 35 °C during the growing season. The experiment was set up using a two-way analysis of variance and a randomized complete block design with two main groups. Group 1 was designated as the control or untreated group (WOM), while Group 2 represented the treated group with MOE (WM). Each group was made up of three blocks, each with 30 plants.

Plants and plots were separated by 30 and 50 cm, respectively. Seeds of the tested barley accessions were planted in early November. Each replicate was thinned to 21 plants, after the plants reached a reasonable growth stage at the start. In the field, standard agricultural practices were carried out, including hand weed control.

Table 1. Code, origin, and name of 59 barley accessions included in this study.

Accession Code	Origin	Accession Name	Accession Code	Origin	Accession Name
AC1	South of Iraq	Shoaa	AC31	Middle of Iraq	Scio/3
AC2	South of Iraq	Boraak	AC32	Middle of Iraq	Victoria
AC3	South of Iraq	Radical	AC33	Middle of Iraq	Black-Bhoos-B
AC4	South of Iraq	Arivat	AC34	Middle of Iraq	Irani
AC5	South of Iraq	16 HB	AC35	Middle of Iraq	A1
AC6	South of Iraq	Furat 9	AC36	Middle of Iraq	MORA
AC7	South of Iraq	Al-warka	AC37	Middle of Iraq	ABN
AC8	South of Iraq	Numar	AC38	Middle of Iraq	Arabi aswad
AC9	South of Iraq	Al-amal	AC39	Middle of Iraq	Clipper
AC10	South of Iraq	Rafidain-1	AC40	Middle of Iraq	Bhoos-H1
AC11	South of Iraq	Al-khayr	AC41	Middle of Iraq	BN2R
AC12	South of Iraq	BN6	AC42	Middle of Iraq	BA4
AC13	South of Iraq	IBAA-99	AC43	North of Iraq	Qala-1
AC14	North of Iraq	Saydsadiq	AC44	North of Iraq	Black-kalar
AC15	Middle of Iraq	Bhoos-244	AC45	North of Iraq	White-kalar
AC16	Middle of Iraq	IBAA-265	AC46	North of Iraq	Black-Akre
AC17	North of Iraq	White-Akre	AC47	North of Iraq	Black-Garmiyan
AC18	North of Iraq	Black-Bhoos Akre	AC48	North of Iraq	Black-Chiman
AC19	North of Iraq	Black-Zaxo	AC49	North of Iraq	Ukranian-Zarayan
AC20	North of Iraq	White-Zaxo	AC50	North of Iraq	White-Zarayan
AC21	South of Iraq	Bhoos-912	AC51	North of Iraq	Abrash
AC22	North of Iraq	White-Halabja	AC52	North of Iraq	Bujayl 1-Shaqlawa
AC23	South of Iraq	Samr	AC53	North of Iraq	Bujayl 2-Shaqlawa
AC24	South of Iraq	GOB	AC54	North of Iraq	Bujayl 3-Shaqlawa
AC25	South of Iraq	Abiad	AC55	South of Iraq	Rehaan
AC26	South of Iraq	CANELA	AC56	South of Iraq	Sameer
AC27	South of Iraq	MSEL	AC57	South of Iraq	Warka-B12
AC28	South of Iraq	Acsad strain	AC58	South of Iraq	Al-Hazzar
AC29	South of Iraq	Acsad-14	AC59	South of Iraq	IBAA-995
AC30	South of Iraq	Gk-Omega			

2.2. *Soil Analysis*

During growing seasons, the experimental site's soil texture was silty clay with electrical conductivity (EC) of 0.62 dS m^{-1}, pH 7.25, organic matter of 22.77 g Kg^{-1}, total nitrogen of 1.2 g Kg^{-1}, available phosphorus of 6.18 mg Kg^{-1}, organic matter of 23.0 g Kg^{-1}, and exchangeable potassium of 0.13 mmole L^{-1}.

2.3. *Preparation of Moringa Organ Extract (MOE) and Its Application*

Moringa (*Moringa oleifera* L.) plant parts (leaves, roots, and seeds) were harvested from young full-grown trees and planted in the nursery of the Faculty of Science, University of Sulaimani, courtesy of Jamal Saeed Rashid. MOE was made by drying a sample of moringa parts in shad and then grinding to a fine powder with a blender. Following grinding, the extract was made by combining 20 g of each part and macerating it in 1 L of distilled water. For 24 h, the mixture was shaken. The mixture was then centrifuged for 30 min at $8000\times g$. The supernatant was then filtered through Whatman filter paper to remove any residue. To achieve the required foliar spray concentrations, the supernatant was diluted 30 times with distilled water [26]. Before dusk, foliar sprays were done to promote the best penetration into leaf tissues and inhibit evaporation. During the stages of fully emerging leaves, flag leaf growth, and seed filling, the foliar MOE treatments were sprayed. At the same time, distilled water was sprayed on control plants. No synthetic fertilizer was used.

2.4. *Plant Measurements*

Three weeks after the last foliar application, the total chlorophyll measurement (TCC in Spad) was achieved using a portable chlorophyll meter CCM-200 (SPAD meter: Minolta Camera Co., Osaka, Japan), on three fully developed leaves near the plant apex of five plants. Leaf area (LA in cm^2) was also recorded at the same time. The leaf area was determined by the following formula: leaf length × leaf width × constant (0.64) [27].

Barley plants in each treated and untreated group were harvested at the end of the growing season, and parameters such as plant height (PH in cm), leaf area (LA in cm^2), total chlorophyll content (TCC in SPAD), number of the tillers per plant (TNP), number of the spikes per plant (SNP), spike length (SL in cm), awan length (AL in cm), spike weight (SW in cm), number of seeds per spike (SNS), seed weight per spike (SWS in g), 1000-kernel weight (1000-KW in g), total yield per plot (TY in g), and straw weight per plot (STW in g) were recorded.

2.5. *Statistical Analysis*

All of the recorded growth and yield parameters were statistically analyzed and assessed using the XLSTAT version 2020.3.1 and JMP version 14 statistical packages. A Duncan's Multiple-Range Test ($p \leq 0.05$) was used to compare mean values across treatments using the XLSTAT version 2020.3.1 statistical package. The dendrogram was created using JMP version 14 software. The principal component analysis (PCA) was calculated based on the mean data by using the XLSTAT version 2020.3.1 statistical package. Correlation analysis was performed by Q Research software. The radar, bar, and pie charts were created using Microsoft Excel version 2019. The percentage of trait index (PTI) was computed using the following formula:

$$\text{PTI (\%)} = [(\text{treated plants with MOE} - \text{untreated plants})/\text{untreated plants}] \times 100.$$

3. Results

3.1. *Performance of Growth Traits under the Application of MOE*

The results of a two-way analysis of variance indicate that accessions, foliar MOE application, and their interaction contribute significantly ($p \leq 0.01$) to all growth and yield component traits studied, with the exception of spike length in the foliar MOE treatment. A maximum F-value was observed for foliar MOE application (3933.46) and accessions (3177.67) based on seeds per spike (SNS), followed by 2162.29 for foliar MOE application and 877.77 for accessions based on straw weight (STW) (Table 2). The individual outcomes are displayed as follows.

The analysis of variance and mean pairwise comparison (Duncan) of all studied traits revealed statistically significant differences between treated (WM) and untreated (WOM) plants (Tables 2–4 and Table S1 in Supplementary Materials). The plants with the greatest height (90.95 cm) were those treated with moringa organ extract (WM) (Table 3). In addition, mean pairwise comparison analysis revealed significant variation among accessions for all studied traits (Table 4). AC57 recorded the greatest length at 110.03 cm, followed by AC56 at 109.30 cm and AC55 at 104.03 cm. In contrast, the barley accessions AC1, AC6, and AC5 were considered the shortest barley accessions, with respective values of 52.17, 62.93, and 65.90 cm. The plant height was positively affected by the exogenous application of moringa plant parts, as shown by the mean analysis and pairwise comparison in Table S1, which represents the interaction between MOE treatment and accessions. The AC57 accession recorded the greatest length (117 cm) under foliar application of MOE (AC57 * WM), followed by AC29 under foliar application of MOE (AC29 * WM), with a value of 113.53 cm. The barley accession AC1 under the control conditions (AC1 * WOM) is the shortest of the 59 barley accessions, with a length of 51.27 cm.

Table 2. Summary of analysis of variance of different studied traits.

	Accessions		Foliar MOE Application		Replications		Accessions * Foliar MOE Application	
Traits	**F**	**Pr > F**	**F**	**Pr > F**	**F**	**Pr > F**	**F**	**Pr > F**
PH (cm)	275.81 **	<0.0001	734.81 **	<0.0001	6.14 **	0.00	49.32 **	<0.0001
LA (cm^2)	31.17 **	<0.0001	298.34 **	<0.0001	1.14 ^{ns}	0.32	7.16 **	<0.0001
TCC (SPAD)	10.45 **	<0.0001	18.73 **	<0.0001	0.69 ^{ns}	0.50	4.02 **	<0.0001
TNP	14.38 **	<0.0001	28.57 **	<0.0001	3.76 ^{ns}	0.02	2.96 **	<0.0001
SNP	12.08 **	<0.0001	19.73 **	<0.0001	2.89 ^{ns}	0.06	2.75 **	<0.0001
SL (cm)	24.03 **	<0.0001	2.25 ^{ns}	0.13	3.45 *	0.03	3.59 **	<0.0001
AL (cm)	28.35 **	<0.0001	47.22 **	<0.0001	0.19 ^{ns}	0.83	9.28 **	<0.0001
SW (g)	143.52 **	<0.0001	70.28 **	<0.0001	30.78 **	<0.0001	12.39 **	<0.0001
SNS	3177.67 **	<0.0001	3933.46 **	<0.0001	39.11 **	<0.0001	183.40 **	<0.0001
SWS (g)	775.22 **	<0.0001	501.80 **	<0.0001	28.52 **	<0.0001	61.80 **	<0.0001
1000-KW (g)	57.82 **	<0.0001	71.49 **	<0.0001	57.73 **	<0.0001	10.59 **	<0.0001
TY (g)	20.62 **	<0.0001	49.56 **	<0.0001	0.04 *	0.96	1.78 **	0.00
STW (g)	869.77 **	<0.0001	2162.29 **	<0.0001	0.04 *	0.96	79.23 **	<0.0001

PH: plant height, LA: leaf area, TCC: total chlorophyll content, TNP: number of the tillers per plant, SNP: number of the spikes per plant, SL: spike length, AL: awan length, SW: spike weight, SNS: number of seeds per spike, SWS: seed weight per spike, 1000-KW: 1000-kernel weight, TY: total yield per plot, and STW: straw weight per plot. *: indicates a significant difference at the 0.05 level, **: indicates a highly significant variation at the 0.01 level. NS: denotes a non-significant variation.

Table 3. Pairwise comparisons (Duncan) of different studied characteristics under treatment with MOE (WM) versus control conditions (WOM).

Characteristics	Foliar Application	Mean ± Standard Error
PH (cm)	WOM	86.30 b ± 0.89
	WM	90.95 a ± 0.89
LA (cm^2)	WOM	10.62 b ± 0.18
	WM	12.31 a ± 0.18
TCC (SPAD)	WOM	11.12 b ± 0.31
	WM	12.21 a ± 0.31
TNP	WOM	15.71 b ± 0.55
	WM	17.94 a ± 0.56
SNP	WOM	12.29 b ± 0.41
	WM	13.78 a ± 0.42
SL (cm)	WOM	5.88 b ± 0.08
	WM	5.97 a ± 0.11
AL (cm)	WOM	11.15 b ± 0.14
	WM	11.62 a ± 0.11
SW (g)	WOM	1.89 b ± 0.05
	WM	2.01 a ± 0.05
SNS	WOM	33.50 b ± 0.99
	WM	37.06 a ± 0.89
SWS (g)	WOM	1.57 b ± 0.05
	WM	1.68 a ± 0.04
1000-KW (g)	WOM	47.52 a ± 0.47
	WM	45.82 b ± 0.52
TY (g)	WOM	111.61 b ± 6.16
	WM	138.83 a ± 5.21
STW (g)	WOM	352.99 b ± 11.78
	WM	412.92 a ± 11.05

PH: plant height, LA: leaf area, TCC: total chlorophyll content, TNP: number of the tillers per plant, SNP: number of the spikes per plant, SL: spike length, AL: awan length, SW: spike weight, SNS: number of seeds per spike, SWS: seed weight per spike, 1000-KW: 1000-kernel weight, TY: total yield per plot, STW: straw weight per plot. WOM: control (without application of moringa organ extract), WM: with application of moringa organ extract. Any values of means with the same letter in the same column are not significant according to Duncan's multiple range test at $p \leq 0.05$.

Table 4. Mean pairwise comparison between 59 barley accessions for three growth traits under treatment with MOE and control condition.

Accessions	PH (cm)	LA (cm^2)	TCC (SPAD)	Accessions	PH (cm)	LA (cm^2)	TCC (SPAD)
AC1	52.17 ae ± 0.46	14.42 cde ± 0.62	15.17 c–h ± 1.29	AC31	89.13 p–s ± 1.21	11.55 l–r ± 1.15	14.67 d–i ± 0.87
AC2	78.60 y ± 4.07	10.78 p–u ± 0.62	15.47 c–e ± 1.36	AC32	94.13 i–l ± 0.47	10.53 q–v ± 0.75	12.12 g–p ± 1.26
AC3	74.90 z ± 2.61	11.15 m–t ± 0.31	10.58 l–t ± 1.14	AC33	94.13 jkl ± 0.40	9.93 t–w ± 0.44	8.62 q–v ± 0.82
AC4	81.30 wx ± 3.36	12.15 i–o ± 0.78	10.40 l–t ± 0.65	AC34	96.23 ghi ± 2.07	10.02 s–w ± 0.55	8.88 p–v ± 0.79
AC5	65.90 ac ± 2.41	12.18 i–n ± 0.47	10.75 k–t ± 1.91	AC35	90.70 n–q ± 1.86	10.41 r–v ± 1.27	17.80 abc ± 0.90
AC6	62.93 ad ± 4.68	12.30 i–m ± 0.51	13.48 d–l ± 1.41	AC36	87.57 stu ± 0.60	6.91 z ± 0.93	9.25 o–v ± 0.69
AC7	93.67 jkl ± 3.95	15.07 cd ± 0.50	13.10 d–m ± 1.09	AC37	89.50 p–s ± 0.50	8.38 xy ± 0.27	8.63 q–v ± 1.32
AC8	67.83 ab ± 0.28	12.89 f–k ± 0.70	12.45 f–o ± 0.96	AC38	93.03 klm ± 1.65	11.04 m–t ± 0.61	7.40 tuv ± 0.67
AC9	88.40 rst ± 1.62	10.90 n–t ± 0.70	12.02 g–p ± 1.09	AC39	79.97 xy ± 2.91	12.30 i–m ± 0.48	16.23 bcd ± 4.45
AC10	80.27 xy ± 3.97	11.77 k–q ± 0.58	11.22 j–r ± 0.62	AC40	89.23 p–s ± 0.66	9.58 uvw ± 0.36	12.05 g–p ± 1.26
AC11	89.13 p–s ± 6.12	16.18 ab ± 0.69	11.90 h–q ± 1.11	AC41	72.73 aa ± 1.74	13.61 e–h ± 0.26	12.03 g–p ± 0.79
AC12	89.60 p–s ± 1.44	10.14 s–w ± 0.28	11.33 i–q ± 0.73	AC42	74.40 zaa ± 2.10	12.84 g–k ± 0.49	8.97 p–v ± 0.48
AC13	92.63 lmn ± 0.31	10.86 o–t ± 0.41	11.13 k–r ± 0.89	AC43	89.57 p–s ± 1.96	9.57 uvw ± 0.36	19.80 a ± 2.75
AC14	92.90 klm ± 0.65	11.26 m–s ± 0.65	12.93 e–m ± 0.97	AC44	91.13 m–p ± 1.89	8.34 xy ± 0.23	10.67 i–t ± 1.16
AC15	98.43 ef ± 1.69	15.36 bc ± 0.97	11.15 k–r ± 0.73	AC45	94.93 ijk ± 1.73	9.29 vwx ± 0.26	9.37 n–v ± 1.26
AC16	100.10 de ± 2.41	14.91 cd ± 1.12	12.87 e–m ± 1.66	AC46	82.83 w ± 2.02	11.16 m–t ± 0.78	10.90 k–s ± 0.94
AC17	95.77 g–j ± 0.57	11.28 m–s ± 0.95	15.37 c–g ± 1.51	AC47	88.50 rst ± 2.17	10.25 s–w ± 0.89	7.85 r–v ± 0.63
AC18	85.27 v ± 3.63	10.93 n–t ± 0.26	7.47 tuv ± 0.80	AC48	92.50 l–o ± 2.90	12.13 i–o ± 0.70	8.62 q–v ± 0.67
AC19	74.40 zaa ± 0.40	14.09 def ± 0.54	10.27 l–u ± 0.55	AC49	92.17 o ± 2.28	11.14 m–t ± 0.27	8.63 q–v ± 1.25
AC20	75.00 z ± 0.64	10.50 q–v ± 0.55	7.62 s–v ± 1.20	AC50	99.43 e ± 1.59	9.07 wx ± 0.82	12.13 g–p ± 0.81
AC21	102.60 bc ± 3.23	12.64 h–l ± 0.66	13.12 d–m ± 0.46	AC51	97.20 fgh ± 3.46	16.61 a ± 0.93	14.12 d–k ± 1.20
AC22	86.60 tuv ± 0.70	11.86 j–p ± 0.53	12.68 f–n ± 1.06	AC52	92.23 l–o ± 0.73	11.17 m–t ± 0.45	12.07 g–p ± 0.86
AC23	82.37 w ± 3.57	8.22 xy ± 0.25	7.60 s–v ± 0.64	AC53	86.03 uv ± 0.61	10.54 q–v ± 0.59	8.85 p–v ± 1.16
AC24	95.43 g–j ± 0.32	7.80 yz ± 0.35	7.02 uv ± 1.13	AC54	88.90 qrs ± 0.52	7.59 yz ± 0.21	8.60 q–v ± 0.21
AC25	100.00 de ± 0.81	11.90 j–p ± 0.91	11.43 i–q ± 1.62	AC55	104.03 b ± 0.35	9.39 vwx ± 0.41	16.10 b–e ± 0.56
AC26	86.57 tuv ± 1.36	9.92 t–w ± 1.33	9.77 m–v ± 1.23	AC56	109.30 a ± 0.53	13.33 e–i ± 0.31	14.53 d–j ± 1.69
AC27	89.20 p–s ± 2.81	10.78 p–u ± 0.59	6.67 v ± 0.43	AC57	110.03 a ± 4.36	13.96 d–g ± 0.37	12.42 f–o ± 1.26
AC28	97.30 fg ± 2.67	12.01 j–p ± 1.59	20.27 a ± 1.84	AC58	90.47 o–r ± 1.98	13.53 e–h ± 0.38	9.78 m–v ± 0.68
AC29	101.60 cd ± 5.34	11.73 k–q ± 0.45	10.53 l–t ± 1.01	AC59	92.87 klm ± 1.47	13.09 f–j ± 0.12	12.57 f–o ± 0.44
AC30	95.17 hij ± 0.62	13.31 e–I ± 0.98	18.78 ab ± 1.34				

PH: plant height, LA: leaf area, TCC: total chlorophyll content. According to Duncan's multiple range test at $p \leq 0.05$, any mean values with a common letter in the same column are not considered significant. The values are represented by the mean ± standard error.

Analysis of variance and mean comparisons between the WM and WOM groups revealed that the WM group had the greatest leaf area (12.31 cm^2) (Table 3). The barley accessions with the highest measurement of this trait were AC51, with a value of 16.61 cm^2, AC11, with a value of 16.18 cm^2, and AC15, with a value of 15.36 cm^2. AC36 was the least productive barley accession (6.91 cm^2), followed by AC54 (7.59 cm^2) and AC24 (7.80 cm^2). In the presence of foliar application of moringa, the interaction results revealed that AC51 under MOE application (AC51 * WM) produced the highest measurement (18.48 cm^2), followed by AC11 and AC15 under MOE treatment (AC11 * WM and AC15 * WM), with values of 17.43 and 17.40 cm^2, respectively, while AC36 under control conditions (AC36 * WOM) produced the lowest measurement (4.85 cm^2) for this trait (Table S1).

The SPAD meter CCM-200 was used to measure the total chlorophyll content (TCC) of barley flag leaves in our study. Significant differences were observed for this trait between WM and WOM (Table 2). The value of TCC was greatest at the WM plant (12.21 SPAD). The mean pairwise comparison analysis of significant variation among barley accessions, as expressed in Table 3, revealed that barley accession AC28 had the highest TCC value (20.27 SPAD) compared to the other barley accessions studied, whereas barley accession AC27 had the lowest TCC value (6.67 SPAD) (Table 4). As shown in Table S1, significant interaction effects were observed for the investigated trait between the treatment applications of moringa and barley accessions. This trait contributed significantly more to the interaction of AC39 under MOE application (AC39 * WM, 26.0 SPAD) than AC20 under control conditions (AC20 * WOM, 5.07 SPAD).

3.2. Contributing Yield Traits' Performance in the Presence of Moringa Organ Extract

In our experiment regarding the foliar effect of moringa plant extract (MOE), significant differences were observed between the untreated and treated groups for all yield-related traits except for the 1000-kernel weight (1000-KW) trait (Table 2, Table 3 and Table 6, and Supplementary Materials). The highest value (17.94) of the number of tillers per plant (TNP) was stated by the plants treated by the MOE (Table 3). Furthermore, the mean pairwise comparison between barley accessions revealed significant variation for all studied characteristics (Table 5). The results demonstrated that the barley accession AC47 had the greatest number of tillers (30.83). In contrast, barley accession AC7 had the lowest number of tillers per plant (5.67). According to Table S2 in Supplementary Materials, the mean pairwise comparison for the interaction of accession and MOE treatment revealed that AC47 had the highest tiller number per plant (37.33) in the presence of moringa application (AC47 * WM). In contrast, the AC7 accession recorded the lowest value (5.33), when the same treatment was not applied (AC7 * WOM).

Table 5. Mean pairwise comparison between 59 barley accessions for number of the tillers per plant, number of the spikes per plant, spike length, awan length, and spike weight traits after application of moringa plant extract.

Accessions	TNP	SNP	SL (cm)	AL (cm)	SW (g)
AC1	7.67 x–ab ± 0.33	5.67 r–u ± 0.49	4.42 v–y ± 0.20	10.86 n–u ± 0.19	1.94 jk ± 0.09
AC2	6.83 z–ab ± 0.31	6.00 r–u ± 0.58	6.85 d–i ± 0.27	12.62 d–g ± 0.50	2.30 e ± 0.37
AC3	6.17 aaab ± 0.75	4.00 u ± 0.52	3.79 y ± 0.07	12.43 d–h ± 0.45	1.84 l ± 0.12
AC4	9.50 w–ab ± 1.06	6.67 q–u ± 1.20	6.22 h–n ± 0.31	12.19 e–j ± 0.71	2.30 e ± 0.07
AC5	7.00 z–ab ± 0.68	5.17 stu ± 0.79	4.69 t–x ± 0.36	13.14bcd ± 0.45	1.61 rs ± 0.09
AC6	7.50 y–ab ± 0.67	5.33 stu ± 0.76	4.89 s–w ± 0.20	13.13bcd ± 0.65	1.64 qr ± 0.07
AC7	5.67 ab ± 0.71	4.33 tu ± 0.49	5.69 m–r ± 0.52	10.19 r–x ± 0.40	1.53 t ± 0.11
AC8	11.33 r–aa ± 2.04	9.17 n–s ± 2.18	4.78 t–x ± 0.18	12.37 d–i ± 0.53	1.67 pq ± 0.08
AC9	10.00 u–ab ± 1.29	7.83 p–u ± 1.40	4.26 wxy ± 0.12	12.10 e–k ± 0.26	2.07 i ± 0.07
AC10	10.67 t–ab ± 0.80	7.83 p–u ± 0.60	5.85 l–r ± 0.46	11.02 m–r ± 0.07	2.15 g ± 0.23
AC11	9.83 v–ab ± 1.54	7.33 p–u ± 1.48	4.90 s–w ± 0.17	12.67 d–g ± 0.41	1.83 lm ± 0.07
AC12	11.00 s–ab ± 1.00	8.33 o–t ± 0.71	5.13 q–v ± 0.23	11.11 l–q ± 0.25	1.78 mn ± 0.06
AC13	12.17 o–z ± 2.24	10.50 l–q ± 1.82	4.72 t–x ± 0.09	12.28 d–j ± 0.39	2.63 c ± 0.08
AC14	11.33 r–aa ± 1.09	8.33 o–t ± 1.26	4.91 s–w ± 0.11	12.43 d–h ± 0.18	2.21 f ± 0.06
AC15	14.33 l–w ± 1.28	11.67 h–p ± 1.05	5.29 p–u ± 0.51	14.10 a ± 0.88	1.96 jk ± 0.17
AC16	12.67 n–y ± 1.28	11.33 i–p ± 1.17	5.70 m–r ± 0.24	13.87 ab ± 0.36	2.11 hi ± 0.06
AC17	17.67 h–o ± 1.78	15.33 c–i ± 1.82	4.65 t–x ± 0.24	11.70 h–n ± 0.19	2.04 j ± 0.07
AC18	16.33 i–s ± 1.67	10.50 i–q ± 0.56	4.18 wxy ± 0.17	9.54 x–ab ± 0.20	1.85 l ± 0.17
AC19	13.00 m–x ± 2.21	9.83 m–r ± 1.62	4.58 u–x ± 0.17	10.99 m–s ± 0.16	1.98 jk ± 0.06
AC20	16.67 h–r ± 2.40	12.50 g–o ± 1.84	6.63 f–l ± 0.22	11.81 g–m ± 0.28	1.07 yz ± 0.06
AC21	11.67 q–z ± 1.71	10.83 j–q ± 1.72	7.51 bcd ± 0.24	14.49 a ± 0.27	2.71 b ± 0.08
AC22	11.67 q–z ± 0.71	8.83 o–s ± 0.87	4.09 xy ± 0.08	12.20 e–j ± 0.16	1.94 jk ± 0.09
AC23	18.17 g–n ± 1.14	13.17 f–n ± 1.25	7.28 c–f ± 0.28	9.33 y–ab ± 0.11	1.03 z ± 0.06
AC24	16.17 i–s ± 1.74	12.67 g–o ± 0.99	6.66 f–k ± 0.22	9.92 w–aa ± 0.27	1.17 x ± 0.06
AC25	20.83 d–j ± 2.07	16.00 b–h ± 1.84	7.35 b–f ± 0.31	13.94 a ± 0.86	1.44 u ± 0.08
AC26	19.83 d–l ± 1.17	13.83 d–m ± 1.45	6.44 g–m ± 0.19	10.93 n–t ± 0.58	1.04 yz ± 0.06
AC27	24.83 bcd ± 2.87	17.67 b–f ± 2.46	6.34 g–n ± 0.20	11.66 h–n ± 0.36	1.21 wx ± 0.06
AC28	21.67 c–i ± 3.56	19.17 abc ± 3.38	6.22 h–h ± 0.16	12.11 e–k ± 1.00	1.43 u ± 0.07
AC29	20.17 d–k ± 1.66	15.17 c–j ± 1.45	7.36 b–f ± 0.36	12.79 def ± 0.78	1.56 st ± 0.17
AC30	17.17 h–q ± 2.30	15.17 c–j ± 1.85	8.50 a ± 0.06	13.72 abc ± 0.55	1.63 qr ± 0.07
AC31	16.00 j–t ± 1.39	13.83 d–m ± 1.35	5.87 k–q ± 0.13	11.41 j–o ± 0.62	1.32 v ± 0.06
AC32	17.00 h–q ± 1.39	15.00 c–k ± 0.93	8.03 ab ± 0.21	10.45 p–w ± 0.43	1.25 w ± 0.08
AC33	19.50 d–l ± 1.26	15.17 c–j ± 1.08	6.28 g–n ± 0.14	9.96 v–aa ± 0.27	0.89 ab ± 0.06
AC34	26.33 abc ± 3.66	15.67 c–i ± 2.40	7.96 abc ± 0.16	10.07 t–z ± 0.48	1.09 y ± 0.06
AC35	17.50 h–p ± 2.40	14.33 d–l ± 1.23	6.03 j–p ± 0.35	10.46 p–w ± 0.92	2.14 g ± 0.10
AC36	19.67 d–l ± 3.76	17.50 b–f ± 2.97	5.79 m–r ± 0.15	10.40 q–x ± 0.35	1.44 u ± 0.07
AC37	24.00 b–f ± 3.01	20.17 ab ± 2.18	6.35 g–n ± 0.25	10.71 o–w ± 0.17	1.22 wx ± 0.12
AC38	26.17 abc ± 2.20	22.33 a ± 1.41	6.39 g–n ± 0.31	10.01 u–aa ± 0.21	1.10 y ± 0.10
AC39	20.00 d–l ± 1.15	18.00 b–e ± 0.89	6.13 i–o ± 0.03	11.95 f–l ± 0.42	1.22 wx ± 0.06

Table 5. *Cont.*

Accessions	TNP	SNP	SL (cm)	AL (cm)	SW (g)
AC40	15.17 k–v ± 1.96	13.67 e–m ± 1.56	6.41 g–n ± 0.08	12.43 d–h ± 0.36	1.33 v ± 0.06
AC41	19.83 d–l ± 2.50	16.33 b–g ± 1.82	7.42 b–e ± 0.40	11.31 k–p ± 0.48	1.05 yz ± 0.09
AC42	19.33 e–l ± 1.74	15.83 b–h ± 1.49	7.03 d–g ± 0.26	10.85 n–v ± 0.23	1.32 v ± 0.07
AC43	17.00 h–q ± 1.73	15.67 c–i ± 1.17	4.20 wxy ± 0.13	10.17 r–y ± 0.21	2.18 fg ± 0.06
AC44	28.00 ab ± 2.89	18.33 a–d ± 1.80	6.14 h–o ± 0.16	9.19 aaab ± 0.10	0.85 ab–ac ± 0.06
AC45	18.33 g–m ± 1.36	13.83 d–m ± 1.17	7.04 d–g ± 0.21	9.29 z–ab ± 0.23	1.19 wx ± 0.08
AC46	28.17 ab ± 1.70	19.17 abc ± 1.47	6.91 d–h ± 0.23	10.73 o–w ± 0.49	0.96 aa ± 0.10
AC47	28.50 ab ± 4.70	20.17 ab ± 2.87	5.09 r–v ± 0.14	9.29 z–ab ± 0.29	0.66 a–d ± 0.06
AC48	30.83 a ± 1.89	18.33 a–d ± 0.92	5.62 n–s ± 0.12	10.75 o–w ± 0.74	0.70 a–d ± 0.06
AC49	15.00 k–v ± 1.63	10.67 k–q ± 0.92	4.19 wxy ± 0.06	10.73 o–w ± 0.25	2.12 gh ± 0.06
AC50	19.17 e–l ± 1.80	15.00 c–k ± 1.91	5.19 q–u ± 0.12	10.46 p–w ± 0.54	1.70 op ± 0.06
AC51	12.00 p–z ± 1.15	8.50 o–t ± 0.62	6.04 j–p ± 0.27	12.59 d ± 0.25	1.91 k ± 0.11
AC52	24.33 b–d ± 2.70	16.50 b–g ± 1.06	6.73 e–j ± 0.33	9.46 y–ab ± 0.24	0.91 aaab ± 0.07
AC53	23.33 b–g ± 2.30	16.67 b–g ± 1.15	5.77 m–r ± 0.20	9.92 w–aa ± 0.28	0.82 ac ± 0.10
AC54	18.83 f–l ± 1.82	14.83 c–l ± 1.14	7.86 abc ± 0.54	10.12 s–z ± 0.12	1.31 v ± 0.10
AC55	17.00 h–q ± 1.06	14.33 d–l ± 0.88	6.08 i–o ± 1.09	11.50 i–o ± 0.32	2.45 d ± 0.20
AC56	22.00 c–h ± 1.63	17.00 b–g ± 1.93	5.63 n–s ± 0.38	10.99 m–s ± 0.15	1.76 no ± 0.12
AC57	15.33 j–u ± 0.88	12.50 g–o ± 0.76	6.38 g–n ± 0.12	8.95 ab ± 0.17	1.97 jk ± 0.14
AC58	16.67 h–r ± 1.84	14.50 d–l ± 1.65	5.77 m–r ± 0.09	12.87 de ± 0.24	2.31 e ± 0.06
AC59	18.17 g–n ± 2.61	15.17 c–j ± 1.96	5.36 o–t ± 0.15	12.98 cde ± 0.16	3.22 a ± 0.14

TNP: number of the tillers per plant, SNP: number of the spikes per plant, SL: spike length, AL: awan length, SW: spike weight. Any values of means containing the same letter in the same column are insignificant. The values are depicted by the mean ± standard error.

The number of spikes per plant (SNP) in this experiment varied between 12.29 (WOM) and 13.78 (WM) (Table 3). The SNP for barley accessions AC38 and AC3 was 22.33 and 4, respectively (Table 5). On the other hand, the same method of mean pairwise comparison was used to observe the effect of moringa treatment and determine its interaction with the examined barley accessions (Table S1). The barley accession AC28 had a higher SNP (25.33) under MOE treatment (AC28 * WM) than AC3, which had only three spikes per plant under control conditions (AC3 * WOM).

The maximum value (5.97 cm) of the spike length (SL) was recorded under the application of MOE (WM), as determined by a mean comparison between the levels of foliar treatment (Table 3). The mean comparison of the evaluated barley accessions revealed that AC30, with a length of 8.50 cm, had the highest SL, followed by AC32 and AC34, with lengths of 8.03 cm and 7.96 cm, respectively. In contrast, AC3, AC22, and AC18 are the shortest barley accessions for the SL trait, with respective values of 3.79, 4.09, and 4.18 cm (Table 5). The mean pairwise comparison of the SL between foliar application of moringa and accessions revealed a significant improvement in their interactions (Table S2). The barley accession with the best performance was AC54 under control conditions (AC54 * WOM), with a length of 8.97 cm, followed by AC30 under MOE application (AC30 * WM), with a length of 8.53 cm. As shown in Table S2, barley accession AC3 had the shortest SL in both levels (with and without MOE treatment), with values of 3.78 and 3.81 cm, respectively.

According to our analysis, the greatest awn length (AL) was found when MOE was applied (Table 3). In terms of AL, accessions AC21 (14.49 cm), AC15 (14.10 cm), and AC25 (13.94 cm) performed better than the remaining examined accessions. AC57 had the shortest length measurement (8.95 cm), followed by AC44 (9.19 cm), AC45, and AC47, with the same value (9.29 cm) (Table 5). Similarly, the mean pairwise comparison of the interaction between barley accessions and treatment status revealed that accessions AC15, AC25, and AC21 exhibited the highest levels of AL under MOE application, with values of 15.94, 15.76, and 15.02 cm, respectively. In the absence of moringa application, AC35 had the shortest AL, followed by AC57 and AC47, with values of 8.46, 8.71, and 8.81 cm, respectively (Table S2).

The application of moringa organ extract (WM) increased spike weight by 2.01 g in comparison to the control group (WOM), which is another significant finding of this study

(Table 3). AC59, with a value of 3.74 g, had the heaviest spikes, followed by AC13 (3.27 g) and AC21 (3.27 g). The mean pairwise comparison between the studied barley accessions for this investigated trait, as presented in (Table 5), exhibited that AC13 (3.27 g) and AC21 (3.27 g) had the lowest spike weight. In contrast, the three barley accessions AC47, AC48, and AC53 performed the worst in terms of spike weight, with respective values of 0.77, 0.81, and 0.97 g. As documented in Table S2, the mean pairwise comparison for detecting the influence of MOE applications and their interaction with the studied barley accessions revealed significant impacts. Barley accession AC59 (4.03 g) with moringa application (WM) performed the best, followed by AC2 (3.57 g) and AC55 (4.03 g) (3.45 g). In contrast, the absence of MOE application (WOM) significantly decreased the weight of this trait. For instance, AC47, AC53, and AC48, with respective values of 0.77, 0.79, and 0.81g, had the lowest spike weights.

The application of MOE had a substantial influence on the number of seeds per spike (SNS). The WM group demonstrated the greatest statistical significance for SNS (37.06) (Table 3). The mean pairwise comparison between barley accessions varied considerably, as demonstrated in Table 6. Barley accessions AC59, AC21, and AC13 had greater seed numbers, with values of 65.72, 56.83, and 54.83, compared to barley accessions AC47, AC52, and AC48, which all had lower grain per spike, with values of 18.22, 19.78, and 19.98, respectively. As shown in Table S2, when the data for this trait were analyzed to observe the interaction between foliar application and barley accessions, significant differences were discovered. In both conditions (WOM and WM), the AC59 produced the most seeds per spike, with values of 72.11 and 59.33, respectively. In addition to AC59, the barley accession AC13 with foliar treatment displayed a high seed number per spike, of 57.78. In contrast, the absence of MOE treatment significantly reduced the number of seeds per spike, as shown in Table 6 for barley accessions AC53, AC47, and AC48, with values of 17.22, 17.78, and 19.22, respectively.

Table 6. Mean pairwise comparison between 59 barley accessions for number of seeds per spike, seed weight per spike, 1000-kernel weight, total yield per plot, and straw weight per plot traits after application of moringa plant extract based on Duncan's multiple range test at $p \leq 0.05$.

Accessions	SNS	SWS (g)	1000-KW (g)	TY (g)	STW (g)
AC1	38.28 r ± 1.68	2.60 ef ± 0.08	50.60 f–i ± 0.68	45.73 u–z ± 7.26	160.26 ae ± 11.94
AC2	40.83 p ± 5.64	2.75 de ± 0.27	57.30 b ± 1.56	72.64 q–y ± 6.93	232.70 ab–ac ± 6.71
AC3	42.72 n ± 0.66	2.05 k–n ± 0.03	43.17 t–w ± 0.52	15.83 z ± 1.49	72.52 ag ± 12.04
AC4	51.89 e ± 0.68	2.84 d ± 0.06	44.30 q–u ± 0.65	55.97 s–z ± 7.50	191.55 ad ± 5.95
AC5	36.06 s ± 3.61	2.17 jk ± 0.13	45.13 o–t ± 1.18	27.30 yz ± 6.85	136.53 af ± 9.53
AC6	40.06 q ± 2.17	1.97 l–o ± 0.03	41.44 v–y ± 1.83	37.68 v–z ± 5.08	127.74 af ± 3.03
AC7	41.89 o ± 1.16	1.93 mno ± 0.05	36.59 ab–ac ± 0.56	30.36 w–z ± 6.13	127.80 af ± 11.48
AC8	38.89 r ± 2.05	1.93 mno ± 0.04	43.40 s–v ± 1.51	86.25 n–v ± 13.62	230.76 ac ± 22.30
AC9	48.33 h ± 1.07	2.42 ghi ± 0.05	42.84 t–x ± 0.48	118.05 j–r ± 9.36	292.31 yz ± 16.57
AC10	43.44 lm ± 3.48	2.57 fg ± 0.20	49.18 g–m ± 0.87	79.82 o–x ± 11.02	246.47 ab ± 18.27
AC11	38.44 r ± 0.73	2.22 jk ± 0.03	47.83 j–n ± 1.62	97.26 l–t ± 20.22	274.39 aa ± 37.48
AC12	41.06 p ± 2.47	2.21 jk ± 0.04	43.96 r–v ± 1.93	29.66 xyz ± 6.21	126.24 af ± 12.93
AC13	54.83 c ± 1.33	3.27 b ± 0.03	48.12 i–n ± 0.94	198.89 def ± 21.13	452.80 lmn ± 9.33
AC14	46.06 j ± 3.36	2.50 fgh ± 0.04	48.95 h–m ± 2.83	133.30 i–n ± 8.46	410.97 st ± 3.97
AC15	44.00 kl ± 0.29	2.40 ghi ± 0.06	44.61 p–u ± 1.69	129.55 i–o ± 17.29	367.80 vw ± 37.26
AC16	48.00 hi ± 0.41	2.45 f–i ± 0.04	44.04 r–u ± 0.54	203.69 def ± 20.39	484.32 j ± 7.02
AC17	44.28 k ± 1.04	2.49 f–i ± 0.08	46.01 n–r ± 0.94	209.53 cde ± 28.96	464.93 l ± 12.39
AC18	47.67 i ± 0.54	2.10 klm ± 0.12	38.62 z–ab ± 2.18	51.22 t–z ± 1.00	227.47 ac ± 43.00
AC19	53.44 d ± 1.16	2.20 jk ± 0.02	37.06 aa–ac ± 0.81	94.34 l–u ± 7.42	300.77 xy ± 10.50
AC20	23.39 xy ± 0.31	1.28 uvw ± 0.02	45.92 n–s ± 0.94	104.01 l–s ± 11.16	310.36 x ± 3.70
AC21	56.83 b ± 2.17	3.23 b ± 0.11	47.67 k–o ± 0.37	202.32 def ± 21.99	479.92 jk ± 17.12
AC22	41.22 p ± 0.87	2.19 jk ± 0.06	47.02 l–p ± 0.64	140.83 g–m ± 12.52	376.85 uv ± 5.25
AC23	19.98 ac ± 0.54	1.34 u ± 0.08	51.64 efg ± 4.53	86.46 n–v ± 15.12	360.14 w ± 33.58

Table 6. *Cont.*

Accessions	SNS	SWS (g)	1000-KW (g)	TY (g)	STW (g)
AC24	23.83 x ± 0.23	1.34 u ± 0.02	48.96 h–m ± 0.94	104.27 l–s ± 23.13	313.80 x ± 47.55
AC25	22.56 zaa ± 0.41	1.80 op ± 0.06	63.66 de ± 1.89	229.29 cd ± 21.18	726.23 b ± 18.39
AC26	23.67 xy ± 0.22	1.15 vwx ± 0.02	44.04 r–u ± 0.97	129.26 i–o ± 9.95	377.34 uv ± 14.31
AC27	23.06 yz ± 0.22	1.39 tu ± 0.03	52.37 def ± 1.17	158.37 f–k ± 14.82	409.17 st ± 37.01
AC28	25.17 w ± 0.47	1.72 pq ± 0.04	56.94 b ± 0.99	188.16 d–g ± 18.60	537.90 fg ± 7.03
AC29	27.50 u ± 2.07	1.89 no ± 0.12	56.59 b ± 1.02	184.63 d–g ± 25.17	539.63 f ± 30.43
AC30	31.83 t ± 0.56	2.00 lmn ± 0.05	51.13 e–h ± 0.86	255.06 bc ± 18.02	630.47 d ± 4.43
AC31	23.44 xy ± 0.78	1.69 pqr ± 0.02	56.63 b ± 1.63	189.66 d–g ± 16.10	509.07 h ± 17.00
AC32	24.94 w ± 0.61	1.61 qrs ± 0.04	49.89 g–k ± 0.93	167.30 e–j ± 24.20	440.33 nop ± 10.31
AC33	21.06 ab ± 0.25	1.05 xy ± 0.02	42.36 u–x ± 1.03	117.74 j–r ± 7.58	458.23 lm ± 16.21
AC34	23.78 x ± 0.73	1.26 uvw ± 0.03	46.01 n–r ± 1.04	122.67 j–q ± 21.88	467.58 kl ± 14.16
AC35	44.44 k ± 0.26	2.38 hi ± 0.04	48.22 i–n ± 1.01	134.22 i–n ± 15.98	499.93 hi ± 16.57
AC36	28.06 u ± 0.28	1.51 st ± 0.02	51.29 e–h ± 0.82	71.85 q–y ± 10.66	241.77 ab–ac ± 23.96
AC37	26.22 v ± 0.68	1.56 q–t ± 0.04	46.64 m–q ± 0.90	128.35 i–p ± 17.63	369.46 vw ± 45.64
AC38	22.11 aa ± 0.59	1.32 uv ± 0.07	49.41 g–l ± 1.91	119.29 j–r ± 14.87	415.43 rst ± 4.43
AC39	22.00 aa ± 0.26	1.52 rst ± 0.02	55.36 bc ± 1.12	180.20 d–h ± 39.55	485.06 j ± 68.21
AC40	24.61 w ± 0.31	1.68 p–s ± 0.02	54.07 cd ± 1.04	80.56 o–w ± 5.44	301.92 xy ± 1.74
AC41	23.61 xy ± 0.80	1.31 uv ± 0.06	44.41 q–u ± 1.46	114.09 k–r ± 21.49	423.04 qrs ± 22.18
AC42	26.56 v ± 0.63	1.60 qrs ± 0.05	49.56 g–l ± 0.93	101.83 l–t ± 9.89	314.77 x ± 21.00
AC43	49.83 g ± 0.25	2.45 f–i ± 0.06	43.64 r ± 1.14	212.27 cde ± 10.84	525.90 fg ± 7.04
AC44	21.00 ab ± 0.78	0.98 xy ± 0.03	40.81 w–z ± 1.29	83.43 n–v ± 10.48	437.39 opq ± 6.34
AC45	25.22 w ± 1.51	1.41 tu ± 0.04	47.59 k–o ± 1.68	144.53 g–l ± 16.38	427.37 pqr ± 32.25
AC46	22.50 zaa ± 0.90	1.13 wxy ± 0.07	42.52 u–x ± 1.76	92.38 m–u ± 13.76	449.60 mno ± 19.91
AC47	18.22 ad ± 0.29	0.77 z ± 0.02	36.46 ab–ac ± 1.25	69.57 r–y ± 12.16	368.86 vw ± 20.63
AC48	19.94 ac ± 0.39	0.81 z ± 0.02	35.01 ac ± 1.23	77.61 p–x ± 10.51	402.90 t ± 8.17
AC49	43.28 mn ± 0.31	2.40 ghi ± 0.02	49.07 h–m ± 0.58	155.21 f–k ± 16.41	490.05 ij ± 2.67
AC50	41.94 o ± 0.56	1.95 mno ± 0.02	40.65 x–z ± 0.69	213.84 cde ± 11.80	572.49 e ± 12.35
AC51	48.61 h ± 0.99	2.13 kl ± 0.05	39.21 y–aa ± 0.44	68.88 r–y ± 11.85	291.12 yz ± 5.14
AC52	19.78 ac ± 0.26	1.06 xy ± 0.03	46.06 n–r ± 2.06	70.26 r–y ± 7.23	367.29 vw ± 32.83
AC53	21.94 aa ± 2.12	0.97 y ± 0.07	37.90 aaab ± 1.39	61.91 s–z ± 9.42	371.47 vw ± 12.27
AC54	24.56 w ± 0.33	1.56 q–t ± 0.06	53.31 cde ± 3.09	92.70 m–u ± 11.22	284.98 zaa ± 9.37
AC55	51.39 e ± 1.38	3.03 c ± 0.19	47.30 k–o ± 2.45	337.47 a ± 42.52	741.59 a ± 16.77
AC56	40.83 p ± 0.52	2.21 jk ± 0.03	43.05 t–x ± 1.33	84.16 n–v ± 17.77	378.81 u ± 44.64
AC57	50.67 f ± 2.44	2.32 ij ± 0.12	38.67 z–ab ± 0.72	128.16 i–p ± 7.11	387.33 u ± 13.44
AC58	46.11 j ± 2.25	3.07 c ± 0.08	50.32 f–j ± 0.99	176.44 e–i ± 20.98	523.77 g ± 62.66
AC59	65.72 a ± 2.87	3.74 a ± 0.12	49.04 h–m ± 0.49	291.98 b ± 35.86	656.98 c ± 37.16

SNS: number of seeds per spike, SWS: seed weight per spike, 1000-KW: 1000-kernel weight, TY: total yield per plot, and STW: straw weight per plot. Any means values in the same column with the same letter are not significant. The mean ± standard error is used to represent the values.

With a value of 1.68 g, the seed weight per spike of all barley accessions significantly increased in the presence of MOE (Table 3). The mean comparison of tested barley accessions revealed that AC59 and AC47 had the highest and lowest values for this trait, respectively (3.49 and 0.66 g) (Table 6). In terms of interaction with MOE, the seed weight per spike of barley accessions studied exhibited the same response pattern as the barley accessions. The values ranged from 3.49 to 0.66 g for the AC59 and AC47 samples, respectively (Table S2).

Analysis of the data showed significant negative effects of foliar MOE application for 1000-kernel weight (1000-KW) across all barley accessions (Table 3). In the absence of MOE application, the highest value of 1000-KW (47.52 g) was measured. Among the evaluated barley accessions, AC25 (63.66 g), AC2 (57.30 g), and AC28 (56.94 g) had the highest value for the studied trait, followed by AC48 (35.01 g), AC47 (36.46 g), and AC7 (36.59 g) (Table 6). The interaction between the evaluated barley accessions and the MOE application revealed, with a value of 67.12 g, that AC25, with the absence of MOE application, performed better than the other accessions for the trait under study. The AC48 accession under MOE treatment, on the other hand, had a minimum 1000-KW value (33.82 g) (Table S2).

In response to the MOE application, positive significant results were obtained in the analysis of total yield per plot (TY) and straw weight per plot (STW) performances (Table 3).

The foliar MOE application yielded the highest value of TY, measuring 138.83 g. For TY, AC55 outperformed the other barley accessions, followed by AC59 and AC30, with values of 337.47, 291.98, and 255.06 g, respectively. Barley accession AC3 had the lowest value (15.83 g), followed by AC5 (27.29 g) and AC12 (29.66 g) (Table 6). In the case of the treatment interaction (MOE application) for the same studied trait, AC59 had the highest variability in TY for its mean among the investigated accessions, with a value of 363.27 g under MOE application. In contrast, AC5 record the lowest value of TY with a value of 12.24 g under control conditions (WOM) (Table S2). Similarly, the greatest value of STW was stated under MOE application across all accessions (Table 3). In the case of studying the STW, AC3 had the lowest value (72.51 g), followed by AC12 and AC6, with values of 126.24 and 136.54 g, respectively. AC55, on the other hand, verified the highest record with a value of 741.59 g. It was followed by the AC25 and AC59 barley accessions, with values of 726.23 and 656.98 g, respectively (Table 6). The interaction of barley accessions with MOE for STW, on the other hand, revealed a wide range of variability, as stated in Table S2. In the absence of a MOE application, AC3 had the lowest record of 45.62 g. Meanwhile, when compared to the other barley accessions, AC25 had the highest value (764.32 g) in the presence of MOE treatment.

3.3. Relationship among Various Accessions and Traits under Untreated and Treated Conditions

A heat map of pairwise correlations (two-side dendrogram) based on mean values obtained from all measured traits in the presence and absence of moringa plant extract was constructed to gain a better understanding of the relationships between studied barley accessions and studied morphological traits (Figure 1). Despite the fact that six groups were estimated in both cases, the barley accessions studied behaved and grouped differently. The majority of barley accessions associated with studied traits clustered together in group 5 under control conditions, indicating that these barley accessions shared the same linkage for the majority of the studied traits. Group 3 is considered the smallest group among constructed clades because only three genotypes were clustered in this group (AC51, AC56, and AC57), demonstrating that these barley accessions share similar associations with investigated traits and are distinct from the remaining barley accessions studied. The remaining barley accessions were classified into four groups (Figure 1A). However, a different arrangement was observed in the case of foliar application of MOE, with significant responses to this treatment by the studied barley accessions and its impacts on selected morphological parameters. The largest group in the MOE treatment (Figure 1B) included 21 barley accessions. This group (Group 2) reacted similarly to the characteristics under consideration. While two distinct barley accessions (AC59 and AC55) grouped together and demonstrated a positive relationship with studied traits, they formed a distinct cluster distinct from all other accessions, whereas the remaining barley accessions fell into other distinct clusters.

To display the correlations between the various plant parameters, principal component analysis (PCA) was performed on the experimental dataset for multifactorial comparison. PCA was used to analyze all 13 measured morphological traits in both normal and treated conditions. In both the control and treated conditions, PCA revealed that 59 different barley accessions were clustered into four clades (Figure 2). Under normal conditions, the first two factorial axes (F1, F2) account for 62.93% of the variance in the data. In the current MOE foliar application, it represented 56.97% of the data variance. Under normal circumstances, all measured traits were divided into two major clusters. Cluster 1 (upper left quarter) included SNP, TNP, SL, PH, and 1000-KW, whereas Cluster 2 (upper right quarter) included TY, TCC, AL, SWS, SW, LA, and SNS (Figure 2A). In relation to the distribution of barley accessions, the PCA plot classified 59 barley accessions into four clades. Clade 1 (upper left quarter) consisted of 11 barley accessions that are predominantly grown in the south of Iraq, whereas clade 4 (lower right quarter), with a performance that differed from clade 1, consisted of 14 barley accessions. In addition to these two clades, 16 accessions of studied barley were distributed in clade 2 (upper right quarter). The traits studied contributed more positively to this clade, suggesting that this component reflected the yield potential of each barley accession in this clade.

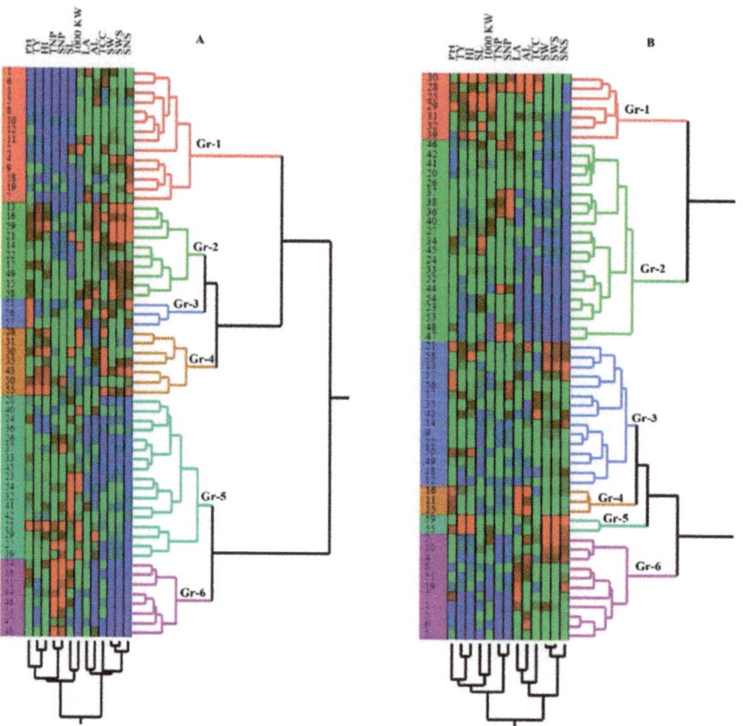

Figure 1. Dendrogram showing association among the 59 studied barley accessions and 13 measured morphological traits under both untreated (control) condition (**A**) and moringa plant extract foliar application (**B**). PH: plant height, LA: leaf area, TCC: total chlorophyll content, TNP: number of the tiller per plant, SNP: number of the spike per plant, SL: spike length, AL: awan length, SW: spike weight, SNS: number of seeds per spike, SWS: seed weight per spike, 1000-KW: 1000-kernel weight, TY: total yield per plot, STW: straw weight per plot. The numbers (1–59) denote the barley accessions. The number of formed groups ranges from Gr-1 to Gr-6.

In addition, the remaining 18 barley accessions, as shown in Figure 2A, belong to clade 3 (bottom left quarter). This determines the genetic differences between those groups that can be selected for crossing in the future breeding program, particularly in the case of AC59 and AC47, in clades 2 and 3, and AC25 and AC1, in clades 1 and 4, respectively.

Regarding the analysis of PCA in the presence of foliar application of moringa extract, distinct distribution patterns of barley accessions and studied traits can be observed when compared to the untreated condition. As depicted in Figure 2B, nearly half of the studied barley accessions are separated and evenly distributed between clades 1 and 2, with 13 accessions for each clade. In addition, clade 3 contained 17 barley accessions, in contrast to clade 1, and the remaining 16 barley accessions remained in clade 4. The present study found that studied traits had a strong correlation with barley accessions distributed across clades 1 and 2, indicating that greater emphasis should be placed on these barley accessions in order to boost final productivities in the presence of moringa plant extract. The attributed differences between these accessions may be partially attributable to their distinct genetic backgrounds and varied responses to the utilized application.

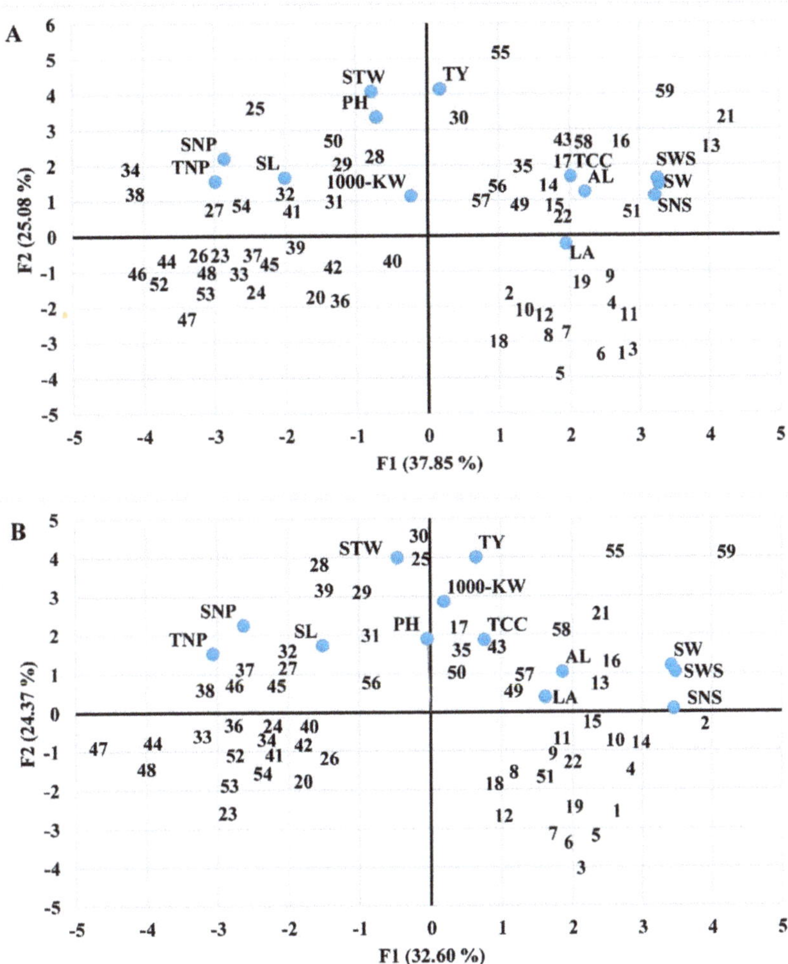

Figure 2. Biplot diagram of principal component analyses based on the first and second components for 59 barley accessions obtained from 13 morphological parameters under both control (**A**) and foliar application of moringa organ extract (**B**). PH: plant height, LA: leaf area, TCC: total chlorophyll content, TNP: number of the tiller per plant, SNP: number of the spike per plant, SL: spike length, AL: awan length, SW: spike weight, SNS: number of seeds per spike, SWS: seed weight per spike, 1000-KW: 1000-kernel weight, TY: total yield per plot, STW: straw weight. The numbers (1–59) represent the barley accessions.

The correlation coefficients measure the degree of similarity and dissimilarity between two characteristics or variables, and the nature of the association between studied parameters can be evaluated. From these mean values, Pearson correlations (r) of the studied traits under control and MOE application conditions are calculated and displayed (Figure 3). Under control conditions, a strong positive significant correlation (r = 0.97, $p < 0.0001$) was observed between SW and SWS traits, followed by TNP and SNP (r = 0.94, $p < 0.0001$) and SNS and SWS (r = 0.93, $p < 0.0001$), while weak positive significant associations were observed between AL and TY (r = 0.26 *, $p = 0.05$), SW and TY (r = 0.30 *, $p = 0.02$), and LA and TCC (r = 0.30 *, $p = 0.02$). A negative significant relationship (r = −0.27 *, $p = 0.04$)

was observed between SNS and 1000-KW (Figure 3A). Concerning the correlations between investigated parameters following foliar application of moringa plant part extract. Positive correlations among studied traits for the r value ranged between 0.98 and 0.27, corresponding to the association between SW and SWS and SL and SNP. As depicted in Figure 3B, a very robust positive significant association was found between SW and SWS yield-related characteristics (r = 0.98, $p < 0.0001$), followed by the association between SNS and SWS (r = 0.93, $p < 0.0001$) and TNP and SNP (r = 0.91, $p < 0.0001$), whereas a weak positive linkage was found with a nearly identical pattern between SL and SNP (r = 0.28 *, $p = 0.03$). A negative correlation between AL and TNP was observed (r = −0.37, $p = 0.003$).

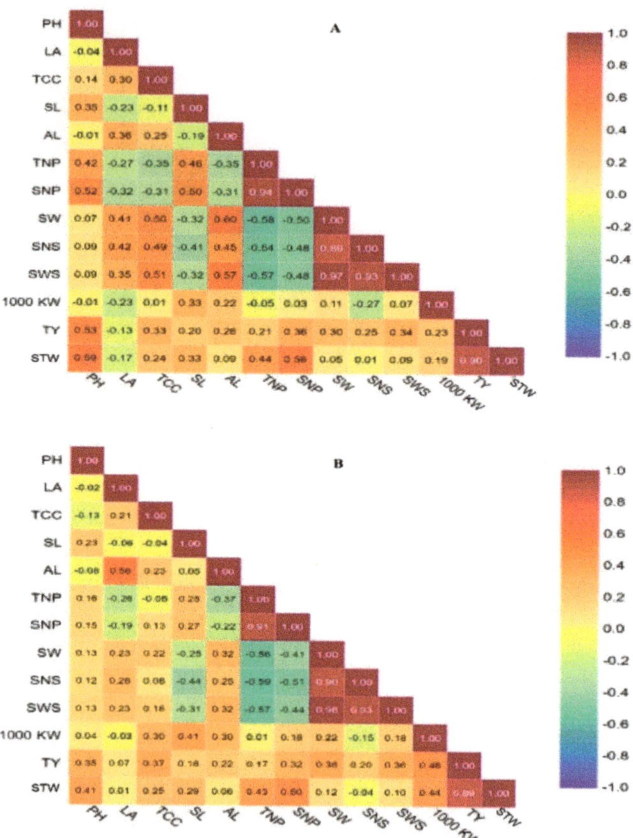

Figure 3. Pearson correlation (r) between 13 morphological parameters in the untreated (**A**) and treated (**B**) conditions with moringa organ extract. PH: plant height, LA: leaf area, TCC: total chlorophyll content, TNP: number of the tillers per plant, SNP: number of the spikes per plant, SL: spike length, AL: awan length, SW: spike weight, SNS: number of seeds per spike, SWS: seed weight per spike, 1000-KW: 1000-kernel weight, TY: total yield per plot, STW: straw weight per plot.

3.4. Percentages of Increasing (Positive Value) and Decreasing (Negative Value) Index of Various Traits among the Barley Accessions Utilized in This Study

A range of growth and yield traits are positively and negatively affected by MOE application, with the ranges varying from −14.89% to 39.90%, −20.68% to 85.04%, −44.67% to 302.06%, −38.78% to 137.14%, −34.92% to 37.75%, −32.14% to 44.25%, −45.90% to 192.86%, −26.21% to 85.84%, −15.68% to 89.37, −28.91% to 71.03%, −29.00% to 28.25%, −57.19% to 246.12%, and −54.11% to 146.17%, for PH, LA, TCC, TNP, SL, AL, SNP, SW,

SNS, SWS, 1000-KW, TY, and STW, respectively (Figure 4). As shown in Figure 4, the barley accessions responded differently to MOE. The highest scores for PH, LA, TCC, TNP, SL, AL, SNP, SW, SNS, SWS, 1000-KW, TY, and STW were, respectively, AC6, AC36, AC39, AC36, AC5, AC28, AC8, AC2, AC2, AC2, AC18, AC5, and AC18. Among the growth and yield traits studied, 1000-KW (−3.04%) was the most severely affected trait and was decreased in the majority of barley accessions (Figure 5). The most significant increase was in the TY (3.7.55%), which was followed by increases in the STW (22.29%), TNP (21.44%), and SNP (21.36%).

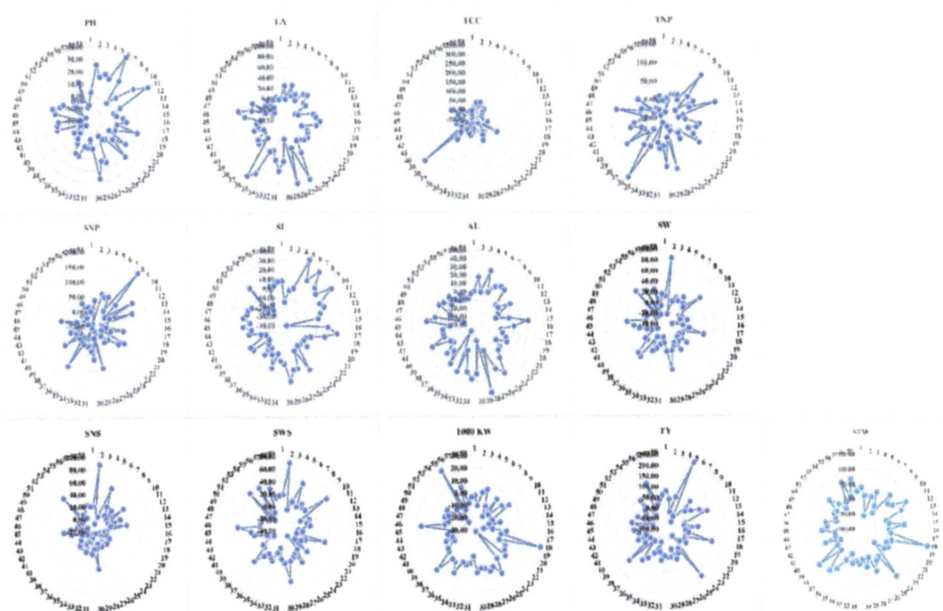

Figure 4. The radar graph depicts the responses of various accessions to MOE treatment based on growth and yield characteristics. PH: plant height, LA: leaf area, TCC: total chlorophyll content, TNP: number of the tillers per plant, SNP: number of the spikes per plant, SL: spike length, AL: awan length, SW: spike weight, SNS: number of seeds per spike, SWS: seed weight per spike, 1000-KW: 1000-kernel weight, TY: total yield per plot, STW: straw weight per plot. The numbers (1–59) represent the barley accessions.

Figure 6 depicts a PCA analysis of the studied characteristics used to establish a preliminary insight into the main distinction between barley accessions in relation to MOE response. The PCA explained a total of 42.53% of the variance, with the first axis (F1) accounting for 25.08% of the variation, and the second axis (F2) accounting for 17.45% of the variation. The PCA biplot demonstrated clearly that accessions react differently to MOE application. The PCA diagram classified all accessions into four distinct categories. The first groups (upper-right quarter) and fourth group (lower-right quarter) comprised the accessions that responded positively to MOE treatment and were deemed to have the best performance under MOE application. In contrast, the accessions in the second group (upper-left quarter) and third group (lower-left quarter) were deemed to have the lowest performance under MOE treatment. Based on the TY and STW traits, AC5 and AC18 accessions demonstrated the best performance, whereas AC2 and AC10 accessions demonstrated the best performance for the SNS, SW, and SWS traits.

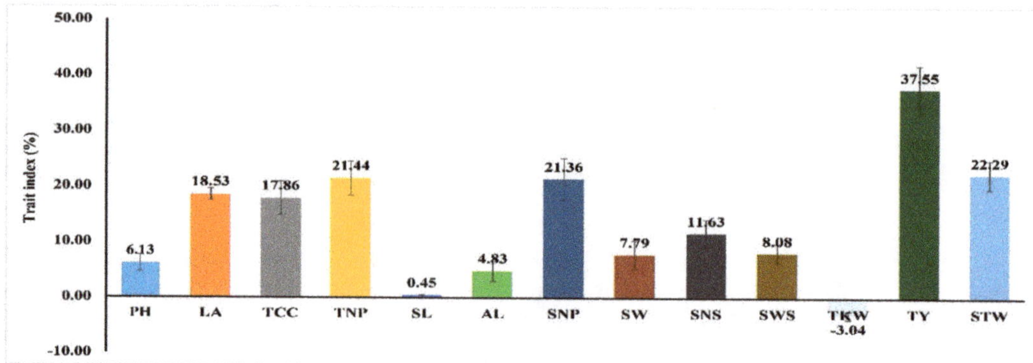

Figure 5. Percentage responses (increasing or decreasing index) of various studied characteristics across all accessions of barley under MOE application compared to control plants. PH: plant height, LA: leaf area, TCC: total chlorophyll content, TNP: number of the tillers per plant, SNP: number of the spikes per plant, SL: spike length, AL: awan length, SW: spike weight, SNS: number of seeds per spike, SWS: seed weight per spike, 1000-KW: 1000-kernel weight, TY: total yield per plot, STW: straw weight per plot. Positive and negative scores on the bars reflect the values of increasing and decreasing traits, respectively.

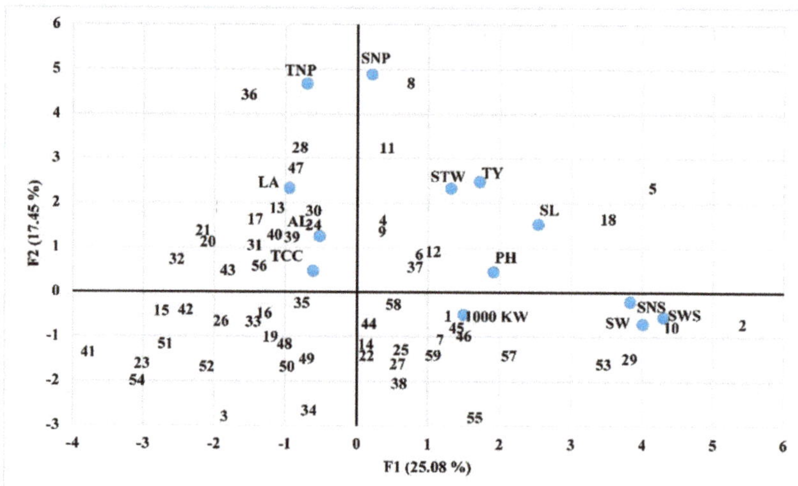

Figure 6. PCA plot depicts the distribution of barley accessions and investigated characteristics based on the percentage of responses under MOE application. PH: plant height, LA: leaf area, TCC: total chlorophyll content, TNP: number of the tillers per plant, SNP: number of the spikes per plant, SL: spike length, AL: awan length, SW: spike weight, SNS: number of seeds per spike, SWS: seed weight per spike, 1000-KW: 1000-kernel weight, TY: total yield per plot, STW: straw weight per plot. The numbers (1–59) represent the barley accessions.

3.5. Percentage of Positive and Negative Effects of MOE Application on the Studied Traits, Based on the Responses of 59 Barley Accessions

Different patterns of responses by the barley accessions under the foliar application of moringa plant parts were detected. As shown in Figure 7, the results confirmed that the application used in our investigation increased all morphological studied parameters, especially the yield traits, with the only exception of the 1000-KW trait. Regarding the analysis for displaying the percentage response by all barley accessions in the case of conducting such a foliar application, the overall view of responses for plant height trait, as shown in Figure 4, indicated the huge impact of such a treatment on the studied barley accession, in which 68% of the accession responded positively in their height to MOE application. Similarly, 83% of accessions demonstrated a positive effect of MOE on leaf area. Regarding the positive responses by barley accessions in total chlorophyll content (TCC), 64% of accessions were detected. Sixty-three percent of barley accessions were documented as having positive awn length (AL) responses. In addition, more than half of the barley accessions responded positively to moringa extract application for the spike length (SL) parameters. A progressive response was observed for the other three traits, namely tiller number per plant (TNP), total yield (TY), and straw weight (STW), in which three-quarters of the tested barley accessions responded positively to foliar application of moringa. In addition to the previous parameters, the other two traits, spike number per plant (SNP) and seed weight/spike (SWS), responded positively in 69% of barley accessions. For 1000-KW traits, 31% of barley accessions responded positively to foliar application.

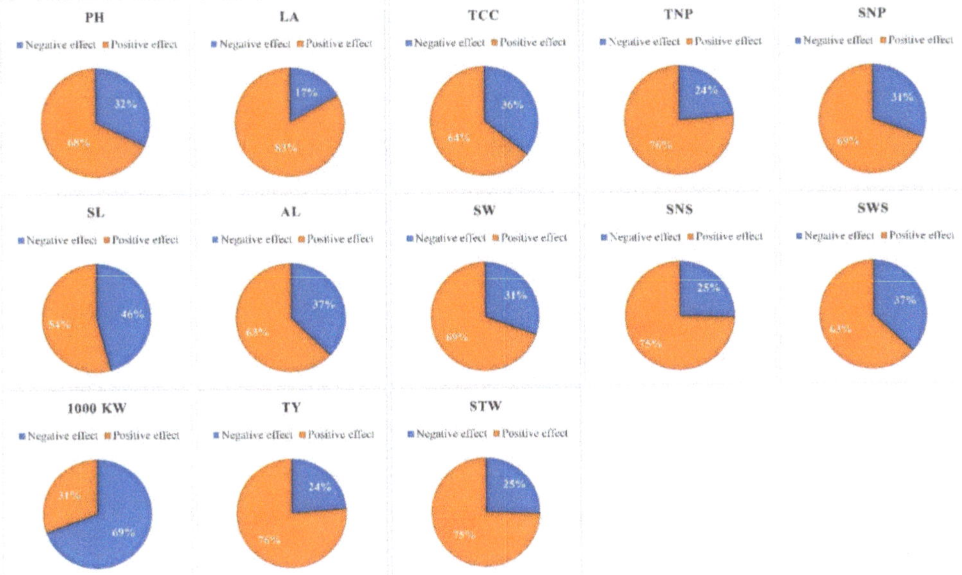

Figure 7. Illustration of the proportion of positive and negative effects of MOE application on the studied traits, based on the responses of 59 barley accessions.

4. Discussion

Plant scientists are now focusing on biostimulants and how to use them in their research to increase crop yields. Plant stimulants have been shown to improve plant health and yield quality by increasing nutrient uptake, changing plant physiology, and making plants more resilient to stress [19,22]. The ultimate goal of any breeding strategy is to increase barley and other cereal yields. To increase the yield of contributing factors, several strategies are being implemented. Crop yield in cereals is primarily determined by measuring the most important traits that are strongly related to the final yield product, such as spike length, spike number per plant, seed weight per spike, 1000-kernel weight, seed number per spike, spike weight, total yield, and harvest index. Foliar application of moringa plant extract is well-documented because it is important in improving yield contributing parameters in many plant species [21,23,28,29]. Moringa leaf extract is measured as one of the essential plant biostimulants due to the presence of phenols, antioxidants, essential nutrients, phytohormones, and ascorbates [22]. When compared to the respective control conditions, exogenous application of moringa plant part extract had a positive impact on these parameters in our experiment.

In this study, MOE had a significant positive impact on plant height. As previously stated, the moringa plant possesses an abundance of phytohormones, including gibberellin [30]. Gibberellin's metabolism and signaling are both essential for controlling plant height. The presence of this phytohormone enhanced internode elongation, leading to an increase in cell division and cell elongation [31]. Similar to our study, Rehman et al. [32] discovered a significant increase in wheat height due to the use of moringa extract.

Some traits, such as leaf area [33], awn length [34], and chlorophyll content [35], have been shown to play a major role in increasing photosynthesis under normal and stressful conditions. The primary organ, which takes a huge portion of the energy in photosynthesis, is the flag leaf. The characteristics of flag leaf are considered essential selection criteria for high grain yields in barley [36]. For this reason, the lower leaves are mostly covered by the upper plant parts and, therefore, do not directly take part in absorbing the radiation of solar energy. After the application of moringa plant extract, a significant increase in leaf area was observed, probably due to the presence of the critical phytohormones in their nature. Several phytohormones with an obvious portion were detected in moringa leaf extract by Ali et al. [30], including gibberellins, auxin, and cytokinins. It is well-documented that gibberellins improve plant height, while auxins improve the elongation of cells and promote the growth of stems, and cytokinins play a critical role in the promotion of cell division and modification of apical dominance [37]. In accordance with our findings, Chattha et al. [38] found similar outcomes in the case of using this type of extract on the wheat plant. Additionally, Ali et al. [39] showed a significant increase in the measurement of this trait on wheat varieties, after conducting the same exogenous application of moringa leaf extract.

Since chlorophyll is required to convert light energy into stored chemical energy, crop growth and yield are directly affected by chlorophyll content [40]. Correlations between leaf area, chlorophyll content, and yield were shown by many studies for barley cultivars [41–43]. This is probably due to capturing lighter chloroplast, while including a denser chloroplast. New opportunities to predict total chlorophyll content (TCC) at the various crop growth stages have been provided with the development of remote sensing equipment (SPAD), which is widely accepted by researchers [44,45]. In the regulation of photosynthesis and many physiological processes, salicylic acid (SA) plays a main role under stress conditions in maintaining these regulations within plant cells [37]. Until very recently, for barley genotypes, a foliar application of combination gibberellic acid and (SA) with a concentration of (110 mg/l and 1.5 mM) showed a significant increase in different plant physiological properties, including total chlorophyll content [46]. Many essential developmental processes are modulated by the presence or absence of cytokinins, including leaf development in the last phase, well-known as senescence, which is associated with the breakdown of chlorophyll and photosynthetic collapse. All of these undesirable changes can be slowed by cytokinins [47]. Taking all the phytohormones present in moringa leaf extract into account, it is possible to conclude that a strong direct correlation is present between the total chlorophyll content and those phytohormones. For all the above reasons, these traits (leaf area and total chlorophyll content) could be used as growth morphological markers for the selection of barley accessions having higher photosynthetic activity.

Cereals have at least two types of tillers (fertile and non-fertile). The first, also known as the productive tiller, causes the formation of spikes and is, thus, necessary for seed yield. The first type depletes the plant's mineral resources. Since this type of tiller rarely survives until the end of the plant's life, it cannot produce a yield [48]. To assess the final productivity of studied cereals, it is critical to measure the fertile tiller number per plant at this point. The obvious increase in tiller number in our results was most likely due to the presence of a cytokinin growth regulator in the moringa plant extract [30]. As a result, the trait of tiller number can be carefully chosen for studying the application of moringa plant extract. Afzal et al. [49] and Rehman et al. [32] reported that the application of moringa leaf extract increased the studied yield traits, including tiller number, in wheat, which is consistent with our findings. In addition to these findings, Koprna et al. [50], from Palack University Olomouc, stated that cytokinin application has a positive effect on the tiller number of barley varieties.

The number of spikes per plant is one of the most important yield characteristics. The selection of barley genotypes based on the number of spikes per plant may eventually lead to the selection of accessions with better yielding performance among tested accessions. This trait has a significant impact on barley genotype yield [51]. This could be due to these barley accessions' ability to respond strongly to this management. In our study, the variation in the number of spikes per plant can be attributed to the genetic potential of barley accessions and their diverse responses to foliar moringa application. In the current foliar treatment, the three barley accessions, AC28, AC47, and AC36, with values of 25.33, 25.00, and 23.67, respectively, had greater potential to produce a large SNP. The current study's findings are consistent with the findings of another group that investigated the effect of moringa extract on this specific yield trait. Afzal et al. [49] investigated the effects of three different foliar applications, moringa leaf, sorghum water extract, and salicylic acid, at concentrations of 3%, 0.075%, and 0.01%, respectively, on wheat plants under current heat stress. They applied the foliar application three times in one month, beginning with the tillering stage. Among the tested foliar applications, moringa extract and salicylic acid significantly improved this trait's performance. Similarly, Khan et al. [22] demonstrated a significant impact of moringa leaf extract alone and in combination with other plant growth promoters such as ascorbic acid and salicylic acid for this trait on wheat, by administering this treatment twice during the tillering and flowering stages.

The presence of various phytohormones and secondary metabolites in moringa plant parts may be linked to the longer spike length in the current study [30]. Similarly, Khan et al. [22] observed a significant increase in spike length on the wheat plant in the field, as a result of using the same application method. Furthermore, Zaheer et al. [52] studied wheat cultivars using various foliar applications, including cytokinins at 25 mg L^{-1} concentrations, used under drought stress conditions at three different growth stages (tiller formation, flowering, and grain filling). The longevity of spikes in their study was significantly improved in the presence of this application. As a result, it is perfectly reasonable to apply moringa foliar to increase spike length.

In our study of the awn length trait, a significant increase was observed when moringa extract was applied foliarly. As a result, increasing awn length could eventually lead to increased barley crop productivity. After the flag leaf, the awns of barley are the most important photosynthetic organs. This organ is the closest plant part to the developing grains in spikelets within the spike, acting as a source of assimilation for grain formation. The photosynthesis of barley spike organs (including awn) accounts for more than 75% of the accumulation of kernel dry weight [53]. It has been long-established that under normal growth conditions in barley, the awns organ can achieve more than 90% of spike photosynthesis [54]. As a result, these plant parts can significantly increase the proportion of net photosynthesis, resulting in a higher value of grain dry matter. Awn removal in barley genotypes had a significant effect on grain yield performance, transpiration rate, and net photosynthetic rate, all of which were reduced [55].

The increased spike weight of plants sprayed with moringa organ extract in our research was due to increased spike length, number of seeds per spike, and other yield-contributing factors previously described. A cheap, rich, and natural source of important secondary metabolic products and plant phytohormones plays an important role in barley trait improvement. The application of the moringa plant part as a foliar spray significantly increased the studied parameters in our study due to the phenomenon of remaining green for a longer period of time during grain filling. This could be due to the high concentration of cytokinin hormone in moringa extract, which is the most general coordinator between senescence and remaining green traits, ultimately improving final yield productivities.

Foliar application of moringa plant part extract had a positive effect on the trait of seed number per spike in tested barley accessions. The grain number and final yield are thought to be positively correlated with the dry weight of the spike during the spike growth phase, possibly due to improved photosynthetic capacity [56]. Zhang et al. [57] used CRISPR/Cas9 gene-editing techniques to determine the roles of cytokinin oxidase and dehydrogenase in rice among eleven candidate CKXs families for their effects on grain number, leaf senescence, and regulating the source of leaf and sink of grain. They discovered that OsCKX11 knockout significantly increased cellular cytokine levels, resulting in a delayed leaf senescence phenotype. Furthermore, the mutant OsCKX11 showed a significant increase in grain number, when compared to the wild type. It is possible that OsCKX11 regulates both grain number and photosynthesis. Previous research, as mentioned above, demonstrated the positive regulation of cytokinin in increasing the number of seeds per spike. The significant findings in our study for this trait may be linked to the presence of these essential phytohormones in moringa organ extract. As a consequence, the higher the cytokinin content, the greater the number of seeds detected in this study. It is quite clear that the combination of the activity of particular phytohormones as well as the nutritional condition of the reproductive meristem both have significant effects on final grain number [48].

MOE application positively affected seed weight per spike in the majority of barley accessions, indicating efficient nutrient use by the plant and translocation of these substrates into reproductive plant parts [58]. Similar to our results, a considerable increase in the seed weight per pod in pea plants [59], seeds in maize kernel [60], and snap bean [61] was detected due to the treatment of moringa leaf extract.

The majority of barley accessions reacted negatively to the MOE application. This reduction in 1000-KW was caused by producing a large number of seeds with small kernels that were less dense and had a low amount of food reserves, because embryo size and reserved nutrient quantity determine the quantity and quality of the seed [62].

The study of total yield and straw weight performance for its production is dependent on the genetic characteristics of the cereal crop, the nutrient status of the soil texture, the exogenous application of growth promoters, and the environmental conditions of the crop plants [22]. Under MOE, we discovered statistically significant positive values for nearly all of the explored yield traits. This is likely due to the presence of cytokinins in moringa leaves, which stimulate carbohydrate metabolism [29,63]. In addition, this characteristic creates a new sink source, leading to an increase in dry matter content. From accession to accession, the total yield of cereal grain and the values of its constituents vary. These differences in yield are strictly correlated with variation in grain number and must, therefore, rely on variation in shoot number, which produces more spikes [64]. In a similar vein, a team of researchers led by Brockman and Brennan [21] discovered significant results in grain yield and dry biomass when moringa leaf extract was applied to a greenhouse-grown wheat cultivar. In addition to total yield, straw weight is an important trait for plant breeding because it reveals the plant's capacity to allocate biomass to reproductive plant parts. It is related to grain yield and biomass in accordance with the multiplicative yield component, wherein grain yield is a product of yield biomass and harvest index [65]. This study's hypothesis, that moringa plant extract is a significant plant growth enhancer, is supported by the numerous MOE compositions discovered by other researchers as well as by the growth and productivity characteristics exhibited by plants treated with moringa plant extract.

5. Conclusions

To increase the development and productivity of barley, foliar application of moringa aqueous extract (MOE) during the crucial growth stage can be used to control the growth and productivity of barley crop plants, as demonstrated by our inquiry into increasing nearly all investigated attributes. It is possible to shed light on this unusual plant for the purposes of further research. Exogenous application of MOE positively affected all the characteristics, with the exception of 1000-kernel weight. MOE treatment exhibited the most favorable effects on overall yield and straw weight per plot. The outcome of this investigation documented that the barley accessions behaved differentially to MOE treatment. Accessions AC8 and AC18 demonstrated the largest enhancement in total yield and straw weight per plot. Further, this form of treatment may be utilized as an alternative biostimulant to conventional plant growth hormones, especially when the objective is to build an organic agricultural system. From this point, it is feasible to shed light on this amazing plant for future research programs.

Supplementary Materials: The following supporting information can be downloaded at: https://www.mdpi.com/article/10.3390/agriculture12091502/s1, Table S1. Mean pairwise comparisons of growth trait interactions between accessions and plant treated by MOE. According to the Multiple Range Duncan's test at $p \leq 0.05$, any mean values with a common letter are not considered significant; Table S2. Mean pairwise comparisons of the interaction between 59 barley accessions and foliar MOE application with yield contributing traits, based on the Multiple Range Duncan's test at $p \leq 0.05$. Any values of means holding common letter are not significant.

Author Contributions: Conceptualization, N.A.-r.T. and K.M.M.; methodology, D.D.L. and K.S.R.; data curation, D.D.L. and K.S.R.; formal analysis, N.A.-r.T.; writing—original draft preparation, N.A.-r.T. and D.D.L.; writing—review and editing, N.A.-r.T., K.M.M., K.S.R. and D.D.L. All authors have read and agreed to the published version of the manuscript.

Funding: This research received no external funding.

Institutional Review Board Statement: Not applicable.

Informed Consent Statement: Not applicable.

Data Availability Statement: The article and supplementary files contain all data.

Acknowledgments: The authors would like to thank the College of Agricultural Engineering Sciences staffs of the University of Sulaimani for their assistance and support during this work.

Conflicts of Interest: The authors declare no conflict of interest.

References

1. Giller, K.E.; Delaune, T.; Silva, J.V.; Descheemaeker, K.; van de Ven, G.; Schut, A.G.T.; van Wijk, M.; Hammond, J.; Hochman, Z.; Taulya, G.; et al. The future of farming: Who will produce our food? *Food Secur.* **2021**, *13*, 1073–1099. [CrossRef]
2. Grando, S.; Macpherson, H.G. Food barley: Importance, uses and local knowledge. In Proceedings of the International Workshop on Food Barley Improvement, Hammamet, Tunisia, 14–17 January 2002.
3. Zheng, B.; Zhong, S.; Tang, Y.; Chen, L. Understanding the nutritional functions of thermally-processed whole grain highland barley in vitro and in vivo. *Food Chem.* **2020**, *310*, 125979. [CrossRef] [PubMed]
4. Cossani, C.M.; Slafer, G.A.; Savin, R. Do barley and wheat (bread and durum) differ in grain weight stability through seasons and water–nitrogen treatments in a Mediterranean location? *Field Crops Res.* **2011**, *121*, 240–247. [CrossRef]
5. Wang, J.; Vanga, S.; Saxena, R.; Orsat, V.; Raghavan, V. Effect of climate change on the yield of cereal crops: A review. *Climate* **2018**, *6*, 41. [CrossRef]
6. Kebede, A.; Kang, M.S.; Bekele, E. Chapter Five—Advances in mechanisms of drought tolerance in crops, with emphasis on barley. In *Advances in Agronomy*; Sparks, D.L., Ed.; Academic Press: Cambridge, MA, USA, 2019; pp. 265–314. ISBN 0065-2113.
7. Fahad, S.; Bajwa, A.A.; Nazir, U.; Anjum, S.A.; Farooq, A.; Zohaib, A.; Sadia, S.; Nasim, W.; Adkins, S.; Saud, S.; et al. Crop production under drought and heat stress: Plant responses and management options. *Front. Plant Sci.* **2017**, *8*, 1147. [CrossRef] [PubMed]
8. Alemu, G.; Desalegn, T.; Debele, T.; Adela, A.; Taye, G.; Yirga, C. Effect of lime and phosphorus fertilizer on acid soil properties and barley grain yield at Bedi in Western Ethiopia. *AJAR* **2017**, *12*, 3005–3012.
9. Nardi, S.; Pizzeghello, D.; Schiavon, M.; Ertani, A. Plant biostimulants: Physiological responses induced by protein hydrolyzed-based products and humic substances in plant metabolism. *Sci. Agric.* **2016**, *73*, 18–23. [CrossRef]
10. Parađiković, N.; Teklić, T.; Zeljković, S.; Lisjak, M.; Špoljarevi, M. Biostimulants research in some horticultural plant species—A review. *Food Energy Secur.* **2019**, *8*, e00162. [CrossRef]
11. Nephali, L.; Piater, L.A.; Dubery, I.A.; Patterson, V.; Huyser, J.; Burgess, K.; Tugizimana, F. Biostimulants for Plant Growth and Mitigation of Abiotic Stresses: A Metabolomics Perspective. *Metabolites* **2020**, *10*, 505. [CrossRef]
12. Li, Y.; Fang, F.; Wei, J.; Wu, X.; Cui, R.; Li, G.; Zheng, F.; Tan, D. Humic Acid Fertilizer Improved Soil Properties and Soil Microbial Diversity of Continuous Cropping Peanut: A Three-Year Experiment. *Sci. Rep.* **2019**, *9*, 12014. [CrossRef]

13. Ur Rahman, M. The Multifunctional Role of Chitosan in Horticultural Crops; A Review. *Molecules* **2018**, *23*, 872.
14. Nabti, E.; Jha, B.; Hartmann, A. Impact of seaweeds on agricultural crop production as biofertilizer. *Int. J. Environ. Sci. Technol.* **2017**, *14*, 1119–1134. [CrossRef]
15. Drobek, M.; Frąc, M.; Cybulska, J. Plant Biostimulants: Importance of the Quality and Yield of Horticultural Crops and the Improvement of Plant Tolerance to Abiotic Stress—A Review. *Agronomy* **2019**, *9*, 335. [CrossRef]
16. Tahir, N.A.; Qader, K.O.; Azeez, H.A.; Rashid, J.S. Inhibitory allelopathic effects of *Moringa oleifera* Lamk plant extracts on wheat and *Sinapis arvensis* L. *Allelopath. J.* **2018**, *44*, 35–48. [CrossRef]
17. Tahir, N.A.; Majeed, H.O.; Azeez, H.A.; Omer, D.A.; Faraj, J.M.; Palani, W.R.M. Allelopathic plants: 27. Moringa species. *Allelopath. J.* **2020**, *50*, 35–48. [CrossRef]
18. Nouman, W.; Siddiqui, M.; Basra, S. *Moringa oleifera* leaf extract: An innovative priming tool for rangeland grasses. *Turk. J. Agric. For.* **2012**, *36*, 65–75. [CrossRef]
19. Yasmeen, A.; Basra, S.M.A.; Farooq, M.; Rehman, H.U.; Hussain, N.; Athar, H.U.R. Exogenous application of moringa leaf extract modulates the antioxidant enzyme system to improve wheat performance under saline conditions. *Plant Growth Regul.* **2013**, *69*, 225–233. [CrossRef]
20. Anwar, F.; Latif, S.; Ashraf, M.; Gilani, A.H. *Moringa oleifera*: A food plant with multiple medicinal uses. *Phytother. Res.* **2007**, *21*, 17–25. [CrossRef]
21. Brockman, H.G.; Brennan, R.F. The effect of foliar application of Moringa leaf extract on biomass, grain yield of wheat and applied nutrient efficiency. *J. Plant Nutr.* **2017**, *40*, 2728–2736. [CrossRef]
22. Khan, S.; Basra, S.M.A.; Nawaz, M.; Hussain, I.; Foidl, N. Combined application of moringa leaf extract and chemical growth-promoters enhances the plant growth and productivity of wheat crop (*Triticum aestivum* L.). *S. Afr. J. Bot.* **2020**, *129*, 74–81. [CrossRef]
23. Khan, S.; Basra, S.M.A.; Afzal, I.; Nawaz, M.; Rehman, H.U. Growth promoting potential of fresh and stored *Moringa oleifera* leaf extracts in improving seedling vigor, growth and productivity of wheat crop. *Environ. Sci. Pollut. Res.* **2017**, *24*, 27601–27612. [CrossRef] [PubMed]
24. Ahmed, M.M. Effect OF *Moringa oleifera* leaf extract on growth, metabolites and antioxidant system of barley plants. *J. Environ. Stud. Res.* **2016**, *6*, 260–271. [CrossRef]
25. Lateef, D.; Mustafa, K.; Tahir, N. Screening of Iraqi barley accessions under PEG-induced drought conditions. *All Life* **2021**, *14*, 308–332. [CrossRef]
26. Arif, M.; Kareem, S.H.S.; Ahmad, N.S.; Hussain, N.; Yasmeen, A.; Anwar, A.; Naz, S.; Iqbal, J.; Shah, G.A.; Ansar, M. Exogenously applied bio-stimulant and synthetic fertilizers to improve the growth, yield and fiber quality of cotton. *Sustainability* **2019**, *11*, 2171. [CrossRef]
27. Sesták, Z.; Catský, J.; Jarvis, P.G. *Plant Photosynthetic Production. Manual of Methods*; Dr. W. Junk NV: The Hague, The Netherlands, 1971.
28. Abd El-Mageed, T.A.; Semida, W.M.; Rady, M.M. Moringa leaf extract as biostimulant improves water use efficiency, physio-biochemical attributes of squash plants under deficit irrigation. *Agric. Water Manag.* **2017**, *193*, 46–54. [CrossRef]
29. Rady, M.M.; Mohamed, G.F. Modulation of salt stress effects on the growth, physio-chemical attributes and yields of *Phaseolus vulgaris* L. plants by the combined application of salicylic acid and *Moringa oleifera* leaf extract. *Sci. Hortic.* **2015**, *193*, 105–113. [CrossRef]
30. Ali, E.F.; Hassan, F.A.S.; Elgimabi, M. Improving the growth, yield and volatile oil content of *Pelargonium graveolens* L. Herit by foliar application with moringa leaf extract through motivating physiological and biochemical parameters. *S. Afr. J. Bot.* **2018**, *119*, 383–389. [CrossRef]
31. Gao, S.; Chu, C. Gibberellin metabolism and signaling: Targets for improving agronomic performance of crops. *Plant Cell Physiol.* **2020**, *61*, 1902–1911. [CrossRef]
32. Rehman, H.U.; Basra, S.M.A.; Rady, M.M.; Ghoneim, A.M.; Wang, Q. Moringa leaf extract improves wheat growth and productivity by affecting senescence and source-sink relationship. *Int. J. Agric. Biol.* **2017**, *19*, 479–484. [CrossRef]
33. Allel, D.; Ben-Amar, A.; Abdelly, C. Leaf photosynthesis, chlorophyll fluorescence and ion content of barley (*Hordeum vulgare*) in response to salinity. *J. Plant Nutr.* **2018**, *41*, 497–508. [CrossRef]
34. Huang, B.; Wu, W.; Hong, Z. Genetic interactions of awnness genes in barley. *Genes* **2021**, *12*, 606. [CrossRef]
35. Bahrami, F.; Arzani, A.; Rahimmalek, M. Photosynthetic and yield performance of wild barley (*Hordeum vulgare* ssp. spontaneum) under terminal heat stress. *Photosynthetica* **2019**, *57*, 9–17. [CrossRef]
36. Xue, D.-w.; Chen, M.-c.; Zhou, M.-x.; Chen, S.; Mao, Y.; Zhang, G.-p. QTL analysis of flag leaf in barley (*Hordeum vulgare* L.) for morphological traits and chlorophyll content. *J. Zhejiang Univ. Sci. B* **2008**, *9*, 938–943. [CrossRef]
37. Taiz, L.; Zeiger, E. *Plant Physiology*, 5th ed.; Sinauer Associates: Sunderland, MA, USA, 2010.
38. Chattha, U.M.; Khan, I.; Hassan, M.U.; Chattha, M.B.; Nawaz, M.; Iqbal, A.; Khan, N.H.; Akhtar, N.; Usman, M.; Kharal, M.; et al. Efficacy of extraction methods of *Moringa oleifera* leaf extract for enhanced growth and yield of wheat. *J. Basic Appl. Sci.* **2018**, *14*, 131–135. [CrossRef]
39. Ali, M.A.; Hussain, M.; Khan, M.I.; Ali, Z.; Zulkiffal, M.; Anwar, J.; Sabir, W.; Zeeshan, M. Source-sink relationship between photosynthetic organs and grain yield attributes during grain filling stage in spring wheat (*Triticum aestivum*). *Int. J. Agric. Biol.* **2010**, *12*, 509–515.

40. Sid'ko, A.F.; Botvich, I.Y.; Pis'man, T.I.; Shevyrnogov, A.P. Estimation of the chlorophyll content and yield of grain crops via their chlorophyll potential. *Biophysics* **2017**, *62*, 456–459. [CrossRef]
41. Klem, K.; Ač, A.; Holub, P.; Kováč, D.; Špunda, V.; Robson, T.M.; Urban, O. Interactive effects of PAR and UV radiation on the physiology, morphology and leaf optical properties of two barley varieties. *Environ. Exp. Bot.* **2012**, *75*, 52–64. [CrossRef]
42. Lausch, A.; Pause, M.; Schmidt, A.; Salbach, C.; Gwillym-Margianto, S.; Merbach, I. Temporal hyperspectral monitoring of chlorophyll, LAI, and water content of barley during a growing season. *Can. J. Remote Sens.* **2013**, *39*, 191–207. [CrossRef]
43. Begović, L.; Pospihalj, T.; Lončarić, P.; Štolfa Čamagajevac, I.; Cesar, V.; Leljak-Levanić, D. Distinct accumulation and remobilization of fructans in barley cultivars contrasting for photosynthetic performance and yield. *Theor. Exp. Plant Physiol.* **2020**, *32*, 109–120. [CrossRef]
44. Donnelly, A.; Yu, R.; Rehberg, C.; Meyer, G.; Young, E.B. Leaf chlorophyll estimates of temperate deciduous shrubs during autumn senescence using a SPAD-502 meter and calibration with extracted chlorophyll. *Ann. For. Sci.* **2020**, *77*, 30. [CrossRef]
45. Shibaeva, T.G.; Mamaev, A.V.; Sherudilo, E.G. Evaluation of a SPAD-502 Plus Chlorophyll Meter to estimate chlorophyll content in leaves with interveinal chlorosis. *Russ. J. Plant Physiol.* **2020**, *67*, 690–696. [CrossRef]
46. Askarnejad, M.R.; Soleymani, A.; Javanmard, H.R. Barley (*Hordeum vulgare* L.) physiology including nutrient uptake affected by plant growth regulators under field drought conditions. *J. Plant Nutr.* **2021**, *44*, 2201–2217. [CrossRef]
47. Hönig, M.; Plíhalová, L.; Husičková, A.; Nisler, J.; Doležal, K. Role of cytokinins in senescence, antioxidant defence and photosynthesis. *Int. J. Mol. Sci.* **2018**, *19*, 4045. [CrossRef] [PubMed]
48. Sreenivasulu, N.; Schnurbusch, T. A genetic playground for enhancing grain number in cereals. *Trends Plant Sci.* **2012**, *17*, 91–101. [CrossRef] [PubMed]
49. Afzal, I.; Akram, M.W.; Rehman, H.U.; Rashid, S.; Basra, S.M.A. Moringa leaf and sorghum water extracts and salicylic acid to alleviate impacts of heat stress in wheat. *S. Afr. J. Bot.* **2020**, *129*, 169–174. [CrossRef]
50. Koprna, R.; Humplík, J.F.; Špíšek, Z.; Bryksová, M.; Zatloukal, M.; Mik, V.; Novák, O.; Nisler, J.; Doležal, K. Improvement of tillering and grain yield by application of cytokinin derivatives in wheat and barley. *Agronomy* **2021**, *11*, 67. [CrossRef]
51. Araus, J.L.; Slafer, G.A.; Royo, C.; Serret, M.D. Breeding for yield potential and stress adaptation in cereals. *Crit. Rev. Plant Sci.* **2008**, *27*, 377–412. [CrossRef]
52. Zaheer, M.S.; Raza, M.A.S.; Saleem, M.F.; Erinle, K.O.; Iqbal, R.; Ahmad, S. Effect of rhizobacteria and cytokinins application on wheat growth and yield under normal vs drought conditions. *Commun. Soil Sci. Plant Anal.* **2019**, *50*, 2521–2533. [CrossRef]
53. Abebe, T.; Wise, R.P.; Skadsen, R.W. Comparative transcriptional profiling established the awn as the major photosynthetic organ of the barley spike while the Lemma and the Palea primarily protect the seed. *Plant Genome* **2009**, *2*. [CrossRef]
54. Ziegler-Jöns, A. Gas exchange of ears of cereals in response to carbon dioxide and light. *Planta* **1989**, *178*, 84–91. [CrossRef]
55. Jiang, Q.Z.; Roche, D.; Durham, S.; Hole, D. Awn contribution to gas exchanges of barley ears. *Photosynthetica* **2006**, *44*, 536–541. [CrossRef]
56. Van de Velde, K.; Ruelens, P.; Geuten, K.; Rohde, A.; van der Straeten, D. Exploiting DELLA Signaling in Cereals. *Trends Plant Sci.* **2017**, *22*, 880–893. [CrossRef]
57. Zhang, W.; Peng, K.; Cui, F.; Wang, D.; Zhao, J.; Zhang, Y.; Yu, N.; Wang, Y.; Zeng, D.; Wang, Y.; et al. Cytokinin oxidase/dehydrogenase OsCKX11 coordinates source and sink relationship in rice by simultaneous regulation of leaf senescence and grain number. *Plant Biotechnol. J.* **2021**, *19*, 335–350. [CrossRef]
58. Protich, R.; Todorovich, G.; Protich, N. Grain weight per spike of wheat using different ways of seed protection. *Bulg. J. Agric. Sci.* **2012**, *18*, 185–190.
59. Merwad, A.-R.M.A. Using *Moringa oleifera* extract as biostimulant enhancing the growth, yield and nutrients accumulation of pea plants. *J. Plant Nutr.* **2018**, *41*, 425–431. [CrossRef]
60. Maswada, H.F.; Abd El-Razek, U.A.; El-Sheshtawy, A.-N.A.; Elzaawely, A.A. Morpho-physiological and yield responses to exogenous moringa leaf extract and salicylic acid in maize (*Zea mays* L.) under water stress. *Arch. Agron. Soil Sci.* **2018**, *64*, 994–1010. [CrossRef]
61. Elzaawely, A.A.; Ahmed, M.E.; Maswada, H.F.; Xuan, T.D. Enhancing growth, yield, biochemical, and hormonal contents of snap bean (*Phaseolus vulgaris* L.) sprayed with moringa leaf extract. *Arch. Agron. Soil Sci.* **2017**, *63*, 687–699. [CrossRef]
62. Sadeghzadeh-Ahari; Hass, M.; Kashi, A.; Amri, A.; Alizadeh, K. Genetic variability of some agronomic traits in the Iranian Fenugreek landraces under drought stress and non-stress conditions. *Afr. J. Plant Sci.* **2010**, *4*, 12–20.
63. Iqbal, M.A. Role of moringa, brassica and sorghum watereExtracts in Increasing crops growth and yield: A review. *Am.-Eurasian J. Agric. Environ. Sci.* **2014**, *14*, 1150–1158.
64. Richards, R.A.; Dennett, C.W.; Qualset, C.O.; Epstein, E.; Norlyn, J.D.; Winslow, M.D. Variation in yield of grain and biomass in wheat, barley, and triticale in a salt-affected field. *Field Crops Res.* **1987**, *15*, 277–287. [CrossRef]
65. Wnuk, A.; Górny, A.G.; Bocianowski, J.; Kozak, M. Visualizing harvest index in crops. *Commun. Biometry Crop. Sci.* **2013**, *8*, 148–159.

Article

Nitrate Increases Aluminum Toxicity and Accumulation in Root of Wheat

Yan Ma [1], Caihong Bai [2,3], Xincheng Zhang [4,5,*] and Yanfeng Ding [1,*]

1 College of Agronomy, Nanjing Agricultural University, Nanjing 210095, China
2 College of Agronomy, Yulin Normal University, Yulin 537000, China
3 Key Laboratory for Conservation and Utilization of Subtropical Bio-Resources, Education Department of Guangxi Zhuang Autonomous Region, Yulin Normal University, Yulin 537000, China
4 Crop Research Institute, Huzhou Agricultural Science and Technology Development Center, Huzhou 313000, China
5 Huzhou Key Laboratory for Innovation and Application of Agricultural Germplasm Resources, Huzhou 313000, China
* Correspondence: 0617375@zju.edu.cn (X.Z.); dingyf@njau.edu.cn (Y.D.)

Abstract: Aluminum (Al) toxicity inhibits root growth, while nitrogen is an essential nutrient for plant growth and development. To explore the effects of nitrate (N) on Al toxicity and accumulation in root of wheat, two wheat genotypes, Shengxuan 6 hao (SX6, Al-tolerant genotype) and Zhenmai 168 (ZM168, Al-sensitive genotype), were used in a hydroponic experiment with four treatments (control without N or Al, N, Al, and Al+N, respectively). The results showed that N increased the inhibition of root elongation and aluminum accumulation in root. The Al-sensitive genotype suffered more serious Al toxicity than the Al-tolerant genotype. Histochemical observation clearly showed that Al prefers binding on the root apex 7–10 mm zones, and the Al-sensitive genotype accumulated more Al in these zones. Compared with other treatments, the Al+N treatment had significantly higher O_2^-, superoxides dismutase (SOD), catalase (CAT), peroxidase (POD) activities, H_2O_2, Evans blue uptake, malondialdehyde (MDA), ascorbic acid (AsA), pectin, and hemicellulose 1 (HC1) contents in both genotypes. Under Al+N treatment, O_2^- activity, Evans blue uptake, MDA, and HC1 contents of SX6 were significantly lower than those of ZM168, but SOD, CAT, and POD activities and AsA content exhibited an opposite trend. Therefore, aluminum toxicity and accumulation in root of wheat seedlings were aggravated by nitrate.

Keywords: root elongation; aluminum toxicity; antioxidant enzyme; nitrate; wheat

Citation: Ma, Y.; Bai, C.; Zhang, X.; Ding, Y. Nitrate Increases Aluminum Toxicity and Accumulation in Root of Wheat. *Agriculture* **2022**, *12*, 1946. https://doi.org/10.3390/agriculture12111946

Academic Editors: Daniele Del Buono, Primo Proietti and Luca Regni

Received: 20 October 2022
Accepted: 16 November 2022
Published: 18 November 2022

1. Introduction

About half of the world and a quarter of China's cultivated land and potential cultivated land is characterized by acidic soil [1,2]. Unfortunately, more than 60% of acidic soils are located in developing countries, and these soils are critical for food production [3]. Aluminum (Al) is the most plentiful metallic element in the crust, usually existing in the form of non-toxic aluminosilicates and oxides in neutral soils. However, in acidic soils (PH < 5), rhizotoxic Al^{3+} is solubilized into the soil solution and directly intoxicates root systems, which results in a significant reduction in crop yield worldwide [1,3]. The most typical symptom of Al toxicity is inhibition of root growth because Al mainly exists in root [4–6]. It has been well-documented that the root apex is not only the main site for Al perception and response, but also the target of Al accumulation [7–9]. The binding affinity to cell wall of root apex causes many adverse impacts, such as plasma membrane disagglomeration, signal disturbance, and reactive oxygen species (ROS) overproduction [10,11]. These disadvantages change the fraction of the cell wall and destroy its structure, thereby reducing its elasticity and plasticity, which explains the reason for inhibiting the elongation of root cells [12]. Moreover, Liu et al. (2018) [5] reported that Al-induced changes in ROS are spatially specific, as a significant decreasing gradient is exhibited from the root apex to base.

Plants have evolved different strategies for coping with Al stress to maintain reasonable growth and yield [1]. One of the mechanisms of Al tolerance is the formation of a stabilized non-phytotoxic complex with Al by the secretion of organic acid anions from the root apex, thereby alleviating aluminum toxicity [13–15]. Another mechanism that endows Al tolerance is the enhancement of antioxidative defense capabilities [16]. Accumulating evidence supports that Al stress can alter the activity of enzymes associated with reactive oxygen species (ROS) scavenging [5,6,10].

Nitrogen is an essential nutrient for plant growth and development; the effects of nitrate on root development are well studied. Nitrate shows an inhibition effect on primary root growth but the opposite effect on lateral root [17,18]. Several vital genes involved in nitrate signaling pathways have been identified, including nitrate sensor, transcription factors, protein kinases, molecular components. Recently, Chu et al. (2021) [19] found a novel transcription factor (HBI1) that regulates nitrate signal transduction by mediating ROS homeostasis. They also found that nitrate treatment decreases the production of H_2O_2, and H_2O_2 inhibits nitrate signaling, thereby forming a feedback regulatory loop to regulate plant root development. A previous study also reported that nitrate can inhibit primary root growth by regulating the production of ROS in the root tips [17].

Al and N are important factors affecting root growth, finally impact crop yield. The uptake of nitrate accompanies the OH^- secreting from roots, which increases the number of negative charge sites on the root surface for binding of Al^{3+} [20] and simultaneously increases the pectin and hemicellulose owning to the negatively charged functional groups (e.g., COO- and -OH, respectively) that possess a high capacity for binding positively charged Al^{3+} [21,22]. Root tips (0–10 mm) are generally used to investigate Al toxicity for the root growth of plants [22–24]. However, how N affects the Al toxicity and accumulation in root tips remain unclear. Thus, this study aimed to investigate the effects of N on Al toxicity in the root growth and Al accumulation in root tips of two wheat genotypes differing in Al tolerance by analyzing the root phenotype, histochemical staining, ROS, and antioxidant enzyme activity, as well as the cell-wall fractions in root tips.

2. Materials and Methods

2.1. Plant Materials and Treatments

Two wheat cultivars differing in Al tolerance, namely Shengxuan 6 hao (SX6, Al-tolerant) and Zhenmai 168 (ZM168, Al-sensitive), were used in this study. The seeds were soaked in distilled water for 1 h, and then disinfected with 1% NaClO solution by volume for 20 min and washed three times with deionized water to remove the residual NaClO on the seeds' surfaces, and then the seeds were imbibed for 12 h at 4 °C in refrigerated Petri dishes with filter papers in darkness. After the refrigeration, seeds were germinated at room temperature in darkness for 24 h. The germinated and uniform seeds were transferred to a plastic box containing 0.5 mmol/L $CaCl_2$ solution (pH 4.3). Wheat seedlings were incubated in an artificial climate chamber with a day/night cycle of 14 h/10 h, a temperature of 25 °C/20 °C, and a light intensity of 250 μmol photons $m^{-2}s^{-1}$. The solution was renewed daily.

After 4 days of pre-treatment, four treatments were adopted for 24 h, i.e., CK (0 mM $Ca(NO_3)_2$, 0 μM $AlCl_3$), N (5 mM $Ca(NO_3)_2$), Al (25 μM $AlCl_3$), and Al+N (25 μM $AlCl_3$ + 5 mM $Ca(NO_3)_2$). After 24 h, some seedings were used to determine the root length for calculating the relative root elongation and Al content. Parts of the samples' root tips (0–10 mm) were used for observing the traits by different staining methods. The rest of the samples' root tips were used to measure the antioxidant enzyme activity and cell-wall fractions.

2.2. Determination of Al Content of Cell Wall in Root Tips

The root tips (0–10 mm) were frozen at −80 °C for 12 h and then centrifuged to remove the cell saps; the residue was washed with 70% ethanol three times. The resulting cell-wall material was subsequently immersed in 0.5 mL 2 M HCl for 24 h with occasional vortexing. The Al content in the cell wall was determined according to Osawa and Matsumoto (2001) [25].

2.3. Localization of Al in Root Tips

The localization of Al was detected by hematoxylin and morin using the methods of Wu et al. (2020) [26]. Briefly, the treated root tips were soaked in 2 g/L hematoxylin with 0.2 g/L potassium iodide for 30 min. After washing for 30 min, the root tips were placed under a stereomicroscope MZ-95 (Carl Zeiss, Jena, Germany) for observation and photography. For morin staining, the root tips were immersed in 0.01% morin solution for 20 min, and then washed with deionized water for 10 min. Subsequently, root tip filming was taken by a laser scanning confocal microscope (Carl Zeiss LSCM 780, Jena, Germany) with a green fluorescence signal at 488 nm.

2.4. Membrane Integrity Verification Assay

The root tips were washed with deionized water three times for 5 min each time, stained in 0.25% (*w*/*v*) Evans blue solution for 15 min and rinsed three times with deionized water, and then observed and photographed under a visualization microscope [27]. Four stained root tips were weighed and milled in 1% sodium dodecyl sulfate (SDS) solution, centrifuged at 10,000 r/s for 10 min, the supernatant was determined at 600 nm, and the Evans blue uptake was calculated. The MDA content was measured by thiobarbituric acid (TBA) reaction according to Heath and Packer (1968) [28].

2.5. Antioxidant Enzyme Activity and Antioxidant Determination

Fresh roots were homogenized and extracted with 1 mL of 50 mmol/L sodium phosphate buffer (pH 7.0) containing 0.1 mol/L EDTA. After centrifugation, the supernatant was immediately used to determine the activities of antioxidant enzymes. Superoxide dismutase (SOD), peroxidase (POD), catalase (CAT), O_2^- activities, H_2O_2, and ascorbic acid (AsA) contents were determined according to Liu et al. (2018) [5] with minor modification. All data were obtained by absorbance methods using a Tecan Infinite M200 Microplate Reader (Tecan Group Ltd., Männedorf, Switzerland). SOD activity was assayed by monitoring its inhibition of photochemical reduction of nitro blue tetrazolium (NBT) at 550 nm. POD activity was determined by following the change of absorption at 470 nm due to guaiacol oxidation. CAT activity was measured by following the consumption of H_2O_2 at 240 nm. O_2^- activity, H_2O_2, and AsA were determined at 550 nm, 405 nm, and 536 nm, respectively.

2.6. Cell-Wall Fraction Determination

Extraction of cell-wall materials and the subsequent fractionation of cell-wall components were carried out according to Yang et al. (2011) [22] with minor modification. Fresh root tips were thoroughly homogenized with pre-cooled 75% ethanol and the homogenates were placed on ice for 20 min. Subsequently, they were centrifuged at 8000× *g* for 10 min at 4 °C, and the residues were washed for 20 min in the order of acetone, a methanol:chloroform mixture (1:1, *v*/*v*), and methanol. The supernatant was discarded and the precipitates were freeze-dried.

The pectin fraction was extracted twice by 0.5% $(NH4)_2C_2O$ (ammonium oxalate) buffer containing 0.1% $NaBH_4$ (pH 4) in a boiling water bath for 1 h. The resulting residues were subsequently subjected to triple extractions with 4% KOH containing 0.1% $NaBH_4$ at room temperature for a total of 24 h, obtaining the hemicellulose 1 (HC1). The uronic acid content in each cell-wall fraction was calculated by a calibration standard curve generated with known concentrations of Galacturonic acid (GalA).

2.7. Statistical Analysis

The values in the figures were calculated as the mean ± SD. All data were analyzed using SPSS 17.0 software (Statistical Product and Service Solutions, IBM, Endicott, NY, USA). The statistical significance among treatments was determined through one-way ANOVA, followed by Duncan's multiple range test ($p < 0.05$), and significant differences were evaluated based on $p < 0.05$.

3. Results

3.1. Effect of N and Al on Root Elongation and Al Accumulation in Root Tips

An obvious inhibition of N and Al on root elongation was observed in both genotypes compared with CK (Figure 1A,B). For the relative root elongation, no significant difference was found between N and Al treatments in SX6, while significant differences were found in ZM168. Moreover, the relative root elongations of SX6 were markedly higher than those of ZM168 in the presence of Al (Figure 1A,B). A significantly higher Al content in the cell wall of root tips (0–10 mm) was found in both genotypes exposed to Al (Figure 1C). Furthermore, the Al content under Al+N treatment was much higher than that under Al treatment, while no difference of Al content was found under CK and N treatment. The Al content of ZM168 was significantly higher than that of SX6 when Al existed.

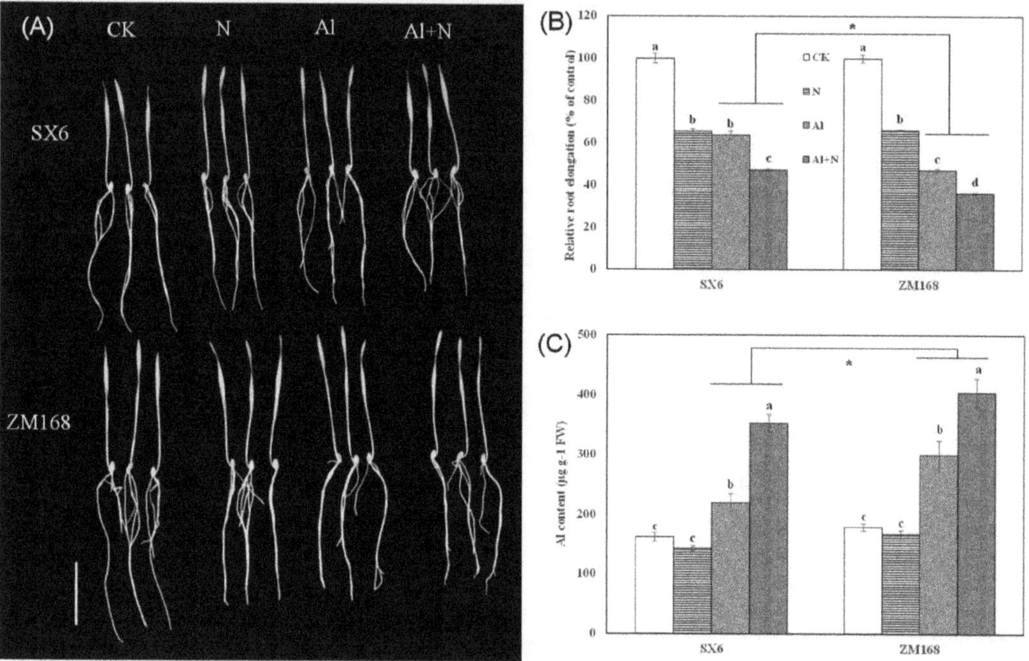

Figure 1. (**A**) Phenotypic analysis of Shengxuan 6 hao (SX6) and Zhenmai 168 (ZM168) seedlings in response to N, Al, and Al+N, scale bar = 5 cm. (**B**) Relative elongation was expressed relative to root elongation in control solutions of 0.5 mM $CaCl_2$, pH 4.3. (**C**) Al^{3+} content of the cell wall in apical 0–10 mm root segments. The values shown are means ± SD (n = 3). Different letters labeled on the columns in the same cultivar are significantly different ($p < 0.05$). * stands for a significant difference in the same treatment between two cultivars ($p < 0.05$).

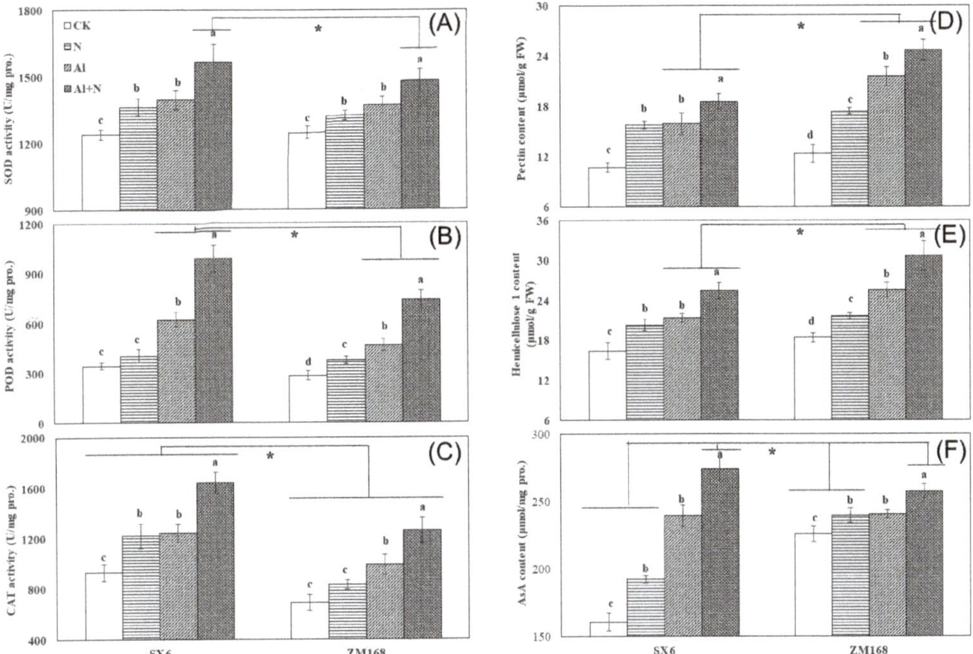

Figure 4. Effects of different treatments on antioxidant enzyme SOD (**A**), POD (**B**), CAT (**C**) activities and cell-wall fraction pectin content (**D**), HC1 content (**E**), and ascorbic acid content (**F**) in root tips. The values shown are means ± SD (n = 3). Different letters labeled on the columns in the same cultivar are significantly different ($p < 0.05$). * stands for a significant difference in the same treatment between two cultivars ($p < 0.05$).

There were significant differences on the variation amplitude of AsA content between two different genotypes. Compared with CK, the AsA content of SX6 rose steeply across treatments, increasing by 19.6%, 49.1%, and 70.6% under N, Al, and Al+N treatments, respectively, while the AsA content of ZM168 only increased by 6.2%, 6.5%, and 14.1% under N, Al, and Al+N treatments, respectively.

For the cell-wall components, compared with CK, the pectin content of SX6 increased by 48.7% and 73.9% under Al and Al+N treatments, respectively, while the pectin content of ZM168 increased by 74.8% and 100% under Al and Al+N treatments, respectively. A similar trend was found in the HC1 content, indicating that the cell-wall components changed more in the Al-sensitive genotype (ZM168) than in the Al-tolerant genotype (SX6) when root was exposed to Al or Al+N (Figure 4D,E).

4. Discussion

Root inhibition growth is a typical symptom of Al toxicity [6,29]. In this study, we found that the inhibition of root growth was the largest under Al+N treamtent in both genotypes, and nitrate promoted the accumulation of Al in root tips and an obvious synergistic inhibition occurred in the root elongation of both genotypes (Figure 1).

Moreover, the Al preferred accumulating at the root tip 0–3 mm and 7–10 mm zones, especially for the latter (Figure 2). Our findings are in close agreement with previous studies showing that Al accumulation was the highest at the 0–5 mm root apex in wheat [5] and the 0–3 mm root apex in buckwheat [30] containing the distal transition zone, which was the most Al-sensitive root apical region [9]. Though 0–2 mm zones in the root tip were considered as an indicator of genotypic sensitivity of crops to Al [9,30,31], the Al content in mature root 5–15 mm zones was four times higher than that in 0–2 mm zones under

25 μM Al^{3+} concentration [32], which supports our result that 7–10 mm zones exhibited a higher Al deposition than that of 0–3 mm zones (Figure 2). Furthermore, nitrate promoted the accumulation of Al in the root tips 0–3 mm and 7–10 mm zones, and a larger impact was found in 7–10 mm zones, especially for the Al-sensitive genotype (Figure 2). Thus, we speculated that the deposition of Al in 7–10 mm zones was a main factor in limiting the root elongation and it could be promoted by N.

ROS production and conversion play an important role in root growth [33,34]. It is well documented that the accumulation of Al induces the formation of large amounts of ROS in crop roots, which damages cell membranes and may lead to cell death [35,36], and is a key factor in inhibiting root elongation [37]. In this study, the application of Al or N significantly induced the production of ROS, including H_2O_2 and O_2^-. Furthermore, the mixture of Al and N enlarged the production of ROS compared with single Al or N treatment (Figure 3D,E). Zang et al. (2020) [17] reported that nitrate inhibited the primary root growth by reducing the H_2O_2 content in *M. truncatula*, which was the opposite of our result, probably because the H_2O_2 had an opposite effect between *T. aestivum* [6] or *A. thaliana* [34] and *M. truncatula* [17]. To investigate the contribution of ROS in lipid peroxidation and cell viability, we also examined the MDA content and Evans blue uptake, and they exhibited similar patterns to H_2O_2 (Figure 3B,C,E). This result indicated that the massive production of H_2O_2 under Al stress may play a crucial role in the triggering of lipid peroxidation and cell death [35,36]. Thus, a lower ROS content in the Al-tolerant genotype (SX6) conferred a lower root growth inhibition compared with the Al-sensitive genotype (ZM168).

To alleviate the oxidative damage caused by ROS accumulation, plants have evolved a complex defensive antioxidant system that includes a combination of enzymatic and non-enzymatic components. Reactive oxygen enzymatic scavenging systems mainly include SOD, POD, and CAT, etc., and the activities of these enzymes respond to the strength of plant resistance in different degrees [35,38,39]. In plants, AsA is an important reductant and exerts a powerful influence on plant functions [40]. In the present study, the activities of SOD, POD, CAT, and AsA were significantly increased under N or Al treatments, except for CAT content in the Al-sensitive genotype ZM168 under N treatment. The highest values of antioxidant enzyme activities in both genotypes were found under Al+N treatment, and the values in ZM168 were significantly lower than those in SX6 (Figure 4A–C). Thus, roots of both genotypes suffered heavy oxidative damage to the membranes and lipids as a result of the higher level of ROS under Al+N treatment than that under single N or Al treatment, especially for the sensitive genotype ZM168.

The cell-wall polysaccharide fraction is considered as a novel Al resistance mechanism, since cell-wall binding capacity is related to Al accumulation in plant roots [22,41]. The cell wall is a major site of Al accumulation in crops, and more than 70% of Al binds to the cell wall of wheat root [22]. Al binding may disrupt the cell-wall structure and diminish mechanical extensibility, thereby inhibiting root elongation [24,42]. Pectin and HC1 are important components of the cell wall. In this study, Al exposure increased the contents of pectin and HC1, which appeared more prominently in the Al-sensitive genotype of wheat, consistent with a previous study [6]. Moreover, N significantly increased the contents of pectin and HC1 of the cell wall in the root tips for both genotypes exposed to Al+N (Figure 4E,F) compared with N or Al treatment, especially in ZM168, resulting in a higher Al accumulation in the cell wall of root tips (Figure 1C) under Al+N treatment. Therefore, N accelerated the accumulation of Al in the wheat root and enhanced the toxicity of Al.

5. Conclusions

Nitrate significantly increased Al accumulation in the roots of wheat seedlings and thereby intensified the inhibition of root elongation by Al. Al prefers to bind on the root apex 7–10 mm zones of the roots, and Al accumulation could be promoted by N. N increased reactive oxygen species (ROS), enzyme activities from the antioxidant defense system, and cell-wall polysaccharide fraction.

Author Contributions: Conceptualization, X.Z. and Y.D.; investigation, Y.M. and X.Z.; data analysis, Y.M.; writing—original draft preparation, Y.M. and C.B.; writing—review and editing, X.Z. and Y.D.; funding acquisition, X.Z. All authors have read and agreed to the published version of the manuscript.

Funding: This research was funded by National Natural Science Foundation of China (31901443).

Institutional Review Board Statement: Not applicable.

Data Availability Statement: Not applicable.

Acknowledgments: We are grateful to thank for kindly providing the wheat seeds by Huadun Wang, who comes from Jiangsu Academy of Agricultural Sciences. We also thank Qin Chen for providing the help in microscopic observation.

Conflicts of Interest: The authors declare no conflict of interest.

References

1. Kochian, L.V.; Pineros, M.A.; Liu, J.P.; Magalhaes, J.V. Plant adaptation to acid soils: The molecular basis for crop aluminum resistance. *Annu. Rev. Plant Biol.* **2015**, *66*, 571–598. [CrossRef] [PubMed]
2. Shen, R.F.; Zhao, X.Q. The sustainable use of acid soils. *J. Agric. Food Chem.* **2019**, *9*, 16–20.
3. Liu, J.; Piñeros, M.A.; Kochian, L.V. The role of aluminum sensing and signaling in plant aluminum resistance. *J. Integr. Plant Biol.* **2014**, *56*, 221–230. [CrossRef] [PubMed]
4. Li, X.W.; Li, Y.L.; Mai, J.; Tao, L.; Qu, M.; Liu, J.Y.; Shen, R.F.; Xu, G.; Feng, Y.M.; Xiao, H.D. Boron alleviates aluminium toxicity by promoting root alkalization in transition zone via polar auxin transport. *Plant Physiol.* **2018**, *177*, 1254–1266. [CrossRef]
5. Liu, W.J.; Xu, F.J.; Lv, T.; Zhou, W.W.; Chen, Y.; Jin, C.W.; Lu, L.L.; Lin, X.Y. Spatial responses of antioxidative system to aluminum stress in roots of wheat (*Triticum aestivum* L.) plants. *Sci. Total Environ.* **2018**, *627*, 462–469. [CrossRef]
6. Sun, C.L.; Lv, T.; Huang, L.; Liu, X.X.; Jin, C.W.; Lin, X.Y. Melatonin ameliorates aluminum toxicity through enhancing aluminum exclusion and reestablishing redox homeostasis in roots of wheat. *J. Pineal Res.* **2020**, *68*, e12642. [CrossRef]
7. Ryan, P.R.; Ditomaso, J.M.; Kochian, L.V. Aluminium toxicity in roots: An investigation of spatial sensitivity and the role of the root cap. *J. Exp. Bot.* **1993**, *44*, 437–446. [CrossRef]
8. Sivaguru, M.; Horst, W.J. The distal part of the transition zone is the most aluminum-sensitive apical root zone of maize. *Plant Physiol.* **1998**, *116*, 155–163. [CrossRef]
9. Kollmeier, M.; Dietrich, P.; Bauer, C.S.; Horst, W.J.; Hedrich, R. Aluminum activates a citrate-permeable anion channel in the aluminum-sensitive zone of the maize root apex. A comparison between an aluminum-sensitive and an aluminum-resistant cultivar. *Plant Physiol.* **2001**, *126*, 397–410. [CrossRef]
10. Ma, J.F. Syndrome of aluminum toxicity and diversity of aluminum resistance in higher plants. *Int. Rev. Cytol.* **2007**, *264*, 225–252.
11. Rengel, Z.; Zhang, W.H. Role of dynamics of intracellular calcium in aluminium-toxicity syndrome. *New Phytol.* **2003**, *159*, 295–314. [CrossRef]
12. Horst, W.J.; Wang, Y.; Eticha, D. The role of root apoplast in aluminum-induced inhibition of root elongation and in aluminum resistance of plants: A review. *Ann. Bot.* **2010**, *106*, 185–197. [CrossRef] [PubMed]
13. Ma, J.F.; Ryan, P.R.; Delhaize, E. Aluminium tolerance in plants and the complexing role of organic acids. *Trends Plant Sci.* **2001**, *6*, 273–278. [CrossRef]
14. Ryan, P.R.; Tyerman, S.D.; Sasaki, T.; Furuichi, T.; Yamamoto, Y.; Zhang, W.H.; Delhaize, E. The identification of aluminium-resistance genes provides opportunities for enhancing crop production on acid soils. *J. Exp. Bot.* **2011**, *62*, 9–20. [CrossRef] [PubMed]
15. Liu, J.P.; Luo, X.Y.; Shaff, J.; Liang, C.Y.; Jia, X.M.; Li, Z.Y.; Magalhaes, J.; Kochian, L.V. A promoter-swap strategy between the *AtALMT* and *AtMATE* genes increased Arabidopsis aluminum resistance and improved carbon-use efficiency for aluminum resistance. *Plant J.* **2012**, *71*, 327–337. [CrossRef]
16. Inostroza-Blancheteau, C.; Rengel, Z.; Alberdi, M.; Mora, M.D.L.L.; Aquea, F.; Arce-Johnson, P.; Reyes-Díaz, M. Molecular and physiological strategies to increase aluminum resistance in plants. *Mol. Biol. Rep.* **2012**, *39*, 2069–2079. [CrossRef]
17. Zang, L.L.; Paven, M.C.M.L.; Clochard, T.; Porcher, A.; Satour, P.; Mojović, M.; Vidović, M.; Limami, A.M.; Montrichard, F. Nitrate inhibits primary root growth by reducing accumulation of reactive oxygen species in the root tip in *Medicago truncatula*. *Plant Physiol. Biochem.* **2020**, *146*, 363–373. [CrossRef]
18. Celis-Arámburo, T.D.J.; Carrillo-Pech, M.; Castro-Concha, L.A.; Miranda-Ham, M.D.L.; Martínez-Estévez, M.; Echevarría-Machado, I. Exogenous nitrate induces root branching and inhibits primary root growth in *Capsicum chinense* Jacq. *Plant Physiol. Biochem.* **2011**, *49*, 1456–1464. [CrossRef]
19. Chu, X.Q.; Wang, J.G.; Li, M.Z.; Zhang, S.J.; Gao, Y.Y.; Fan, M.; Han, C.; Xiang, F.N.; Li, G.Y.; Wang, Y.; et al. HBI transcription factor-mediated ROS homeostasis regulates nitrate signal transduction. *Plant Cell* **2021**, *33*, 3004–3021. [CrossRef]
20. Zhao, X.Q.; Shen, R.F.; Sun, Q.B. Ammonium under solution culture alleviates aluminum toxicity in rice and reduces aluminum accumulation in roots compared with nitrate. *Plant Soil* **2009**, *315*, 107–121. [CrossRef]

21. Chang, Y.C.; Yamamoto, Y.; Matsumoto, H. Accumulation of aluminium in the cell wall pectin in cultured tobacco (*Nicotiana tabacum* L.) cells treated with a combination of aluminium and iron. *Plant Cell Environ.* **1999**, *22*, 1009–1017. [CrossRef]
22. Yang, J.L.; Zhu, X.F.; Peng, Y.X.; Zheng, C.; Li, G.X.; Liu, Y.; Shi, Y.Z.; Zheng, S.J. Cell wall hemicellulose contributes significantly to aluminum adsorption and root growth in Arabidopsis. *Plant Physiol.* **2011**, *155*, 1885–1892. [CrossRef] [PubMed]
23. Wang, W.; Zhao, X.Q.; Chen, R.F.; Dong, X.Y.; Lan, P.; Ma, J.F.; Shen, R.F. Altered cell wall properties are responsible for ammonium-reduced aluminum accumulation in rice roots. *Plant Cell Environ.* **2015**, *38*, 1382–1390. [CrossRef] [PubMed]
24. Sun, C.L.; Lu, L.L.; Yu, Y.; Liu, L.J.; Hu, Y.; Ye, Y.Q.; Jin, C.W.; Lin, X.Y. Decreasing methylation of pectin caused by nitric oxide leads to higher aluminum binding in cell walls and greater aluminum sensitivity of wheat roots. *J. Exp. Bot.* **2016**, *67*, 97. [CrossRef]
25. Osawa, H.; Matsumoto, H. Possible involvement of protein phosphorylation in aluminum-responsive malate efflux from wheat root apex. *Plant Physiol.* **2001**, *126*, 411–420. [CrossRef]
26. Wu, L.Y.; Guo, Y.Y.; Cai, S.G.; Kuang, L.H.; Shen, Q.F.; Wu, D.Z.; Zhang, G.P. The zinc finger transcription factor ATF1 regulates aluminum tolerance in barley. *J. Exp. Bot.* **2020**, *71*, 6512–6523. [CrossRef]
27. Baker, C.J.; Mock, N.M. An improved method for monitoring cell death in cell suspension and leaf disc assays using Evans blue. *Plant Cell Tissue Organ. Cult.* **1994**, *39*, 7–12. [CrossRef]
28. Heath, R.L.; Packer, L. Photoperoxidation in isolated chloroplasts: I. Kinetics and stoichiometry of fatty acid peroxidation. *Arch. Biochem. Biophys.* **1968**, *125*, 189–198. [CrossRef]
29. Silva, S.; Rodriguez, E.; Pinto-Carnide, O.; Martins-Lopes, P.; Matos, M.; Guedes-Pinto, H.; Santos, C. Zonal responses of sensitive vs. tolerant wheat roots during Al exposure and recovery. *J. Plant Physiol.* **2012**, *169*, 760–769. [CrossRef]
30. Klug, B.; Horst, W.J. Spatial characteristics of aluminum uptake and translocation in roots of buckwheat (*Fagopyrum esculentum*). *Physiol. Plant.* **2010**, *139*, 181–191. [CrossRef]
31. Rangel, A.F.; Rao, I.M.; Horst, W.J. Spatial aluminum sensitivity of root apices of two common bean (*Phaseolus vulgaris* L.) genotypes with contrasting aluminum resistance. *J. Exp. Bot.* **2007**, *58*, 3895–3904. [CrossRef] [PubMed]
32. Samuels, T.D.; Kucukakyuz, K.; Rincon-Zachary, M. Al partitioning patterns and root growth as related to Al sensitivity and Al tolerance in wheat. *Plant Physiol.* **1997**, *113*, 527–534. [CrossRef] [PubMed]
33. Dunand, C.; Crevecoeur, M.; Penel, C. Distribution of superoxide and hydrogen peroxide in Arabidopsis root and their influence on root development: Possible interaction with peroxidases. *New Phytol.* **2007**, *174*, 332–341. [CrossRef] [PubMed]
34. Tsukagoshi, H.; Busch, W.; Benfey, P.N. Transcriptional regulation of ROS controls transition from proliferation to differentiation in the root. *Cell* **2010**, *143*, 606–616. [CrossRef] [PubMed]
35. Tabaldi, L.A.; Cargnelutti, D.; Gonçalves, J.F.; Pereira, L.B.; Castro, G.Y.; Maldaner, J.; Rauber, R.; Rossato, L.V.; Bisognin, D.A.; Schetinger, M.R.C.; et al. Oxidative stress is an early symptom triggered by aluminum in Al-sensitive potato plantlets. *Chemosphere* **2009**, *76*, 1402–1409. [CrossRef] [PubMed]
36. Delisle, G.; Champoux, M.; Houde, M. Characterization of oxalate oxidase and cell death in Al-sensitive and tolerant wheat roots. *Plant Cell Physiol.* **2001**, *42*, 324–333. [CrossRef]
37. Yamamoto, Y.; Kobayashi, Y.; Devi, S.R.; Rikiishi, S.; Matsumoto, H. Aluminum toxicity is associated with mitochondrial dysfunction and the production of reactive oxygen species in plant cells. *Plant Physiol.* **2002**, *128*, 63–72. [CrossRef]
38. Tamás, L.; Dudíková, J.; Ďurčeková, K.; Huttová, J.; Mistrík, I.; Zelinová, V. The impact of heavy metals on the activity of some enzymes along the barley root. *Environ. Exp. Bot.* **2008**, *62*, 86–91. [CrossRef]
39. Giannakoula, A.; Moustakas, M.; Syros, T.; Yupsanis, T. Aluminum stress induces upregulation of an efficient antioxidant system in the Al-tolerant maize line but not in the Al-sensitive line. *Environ. Exp. Bot.* **2010**, *67*, 487–494. [CrossRef]
40. Foyer, C.H.; Noctor, G. Ascorbate and glutathione: The heart of the redox hub. *Plant Physiol.* **2011**, *155*, 2–18. [CrossRef]
41. Zhu, X.F.; Shi, Y.Z.; Lei, G.J.; Fry, S.C.; Zhang, B.C.; Zhou, Y.H.; Braam, J.; Jiang, T.; Xu, X.Y.; Mao, C.Z.; et al. *XTH31*, encoding an in vitro XEH/ XET-active enzyme, regulates aluminum sensitivity by modulating in vivo XET action, cell wall xyloglucan content, and aluminum binding capacity in Arabidopsis. *Plant Cell* **2012**, *24*, 4731–4747. [CrossRef] [PubMed]
42. Ma, J.F.; Shen, R.; Nagao, S.; Tanimoto, E. Aluminum targets elongating cells by reducing cell wall extensibility in wheat roots. *Plant Cell Physiol.* **2004**, *45*, 583–589. [CrossRef] [PubMed]

Article

Aqueous Seaweed Extract Alleviates Salinity-Induced Toxicities in Rice Plants (*Oryza sativa* L.) by Modulating Their Physiology and Biochemistry

Kanagaraj Muthu-Pandian Chanthini [1], Sengottayan Senthil-Nathan [1,*], Ganesh-Subbaraja Pavithra [1], Pauldurai Malarvizhi [1], Ponnusamy Murugan [1], Arulsoosairaj Deva-Andrews [1], Muthusamy Janaki [1], Haridoss Sivanesh [1], Ramakrishnan Ramasubramanian [1], Vethamonickam Stanley-Raja [1], Aml Ghaith [2], Ahmed Abdel-Megeed [3] and Patcharin Krutmuang [4,5,*]

1 Sri Paramakalyani Centre for Excellence in Environmental Sciences, Manonmaniam Sundaranar University, Alwarkurichi, Tirunelveli 627 412, Tamil-Nadu, India
2 Department of Zoology, Faculty of Science, Derna University, Derna 417230, Libya
3 Department of Plant Protection, Faculty of Agriculture Saba Basha, Alexandria University, Alexandria 5452022, Egypt
4 Department of Entomology and Plant Pathology, Faculty of Agriculture, Chiang Mai University, Chiang Mai 50200, Thailand
5 Innovative Agriculture Research Center, Faculty of Agriculture, Chiang Mai University, Chiang Mai 50200, Thailand
* Correspondence: senthill@msuniv.ac.in (S.S.-N.); patcharink26@gmail.com (P.K.)

Citation: Chanthini, K.M.-P.; Senthil-Nathan, S.; Pavithra, G.-S.; Malarvizhi, P.; Murugan, P.; Deva-Andrews, A.; Janaki, M.; Sivanesh, H.; Ramasubramanian, R.; Stanley-Raja, V.; et al. Aqueous Seaweed Extract Alleviates Salinity-Induced Toxicities in Rice Plants (*Oryza sativa* L.) by Modulating Their Physiology and Biochemistry. *Agriculture* **2022**, *12*, 2049. https://doi.org/10.3390/agriculture12122049

Academic Editor: Wensheng Wang

Received: 24 August 2022
Accepted: 16 November 2022
Published: 29 November 2022

Abstract: Around the world, salinity a critical limiting factor in agricultural productivity. Plant growth is affected by salt stress at all stages of development. The contemporary investigation focused on *Chaetomorpha antennina* aqueous extracts (SWEs) to decrease the effects of salt strain on rice germination, growth, yield, and the production of key biological and biochemical characters of the rice, *Oryza sativa* L. (Poaceae). SWE improved the germination capacities of rice seedlings by promoting their emergence 36.27 h prior to those that had been exposed to saline stress. The creation of 79.647% longer radicles by SWE treatment on salt-stressed seeds which boosted the establishment effectiveness of seeds produced under salt stress longer radicles resulted in plants that were 64.8% taller. SWE treatment was effective in revoking the levels of protein (26.9%), phenol (35.54%), and SOD (41.3%) enzyme levels that were previously constrained by salinity stress. Additionally, SWE were also efficient in retaining 82.6% of leaf water content and enhancing the production of photosynthetic pigments affected by salt exposure earlier. The improvement in plant functionality was evident from the display of increase in tiller numbers/hill (62.36%), grain yield (58.278%), and weight (56.502%). The outcome of our research shows that SWEs protected the plants from the debarring effects of salinity by enhancing the plant functionality and yield by mechanistically enriching their physiological (germination and vegetative growth) and biochemical attributes (leaf RWC, photosynthetic pigments, protein, phenol, and SOD). Despite the increase in TSS and starch levels in rice grain exposed to salinity stress, SWE improved the grain protein content thus cumulatively enhancing rice nutrition and marketability. The current investigation reveals that the extracts of *C. antennina* can help alleviate rice plants from salt stress in an efficient, eco-friendly, as well as economical way.

Keywords: abiotic stress tolerance; seaweeds; protection; plant functionality; grain weight; yield

1. Introduction

Agriculture is confronting multiple issues that are escalating. Increasing food production to feed an expanding population is a major challenge. This can be attained by either expanding farmland for increased food production or by improving existing yields by applying fertilizers or using revolutionary technologies such as precision farming systems, cutting-edge irrigation, and ecologically feasible crop revolutions [1]. Abiotic stressors are major environmental restrictions reducing crop productivity globally. These abiotic stressors induce an osmotic action, specific ion effect, creation of nutritional imbalance, and oxidative damage to biomolecules and membranes [2].

Coastal areas provide ideal soil and climate conditions for agriculture, which has been practiced from time immemorial and is vital to the coastal economy. It is important to focus on coastal agriculture, making it more fruitful and attractive, and integrating it into the coastline, plans to address forthcoming encounters of food besides nutritive security for an ever-increasing human population, as well as climate change [3]. Using water resources that are low in quality such as wells and brackish surface waters causes secondary salinization [4]. Soil flooding in arid as well as humid regions causes soluble salt accumulation at the soil surface due to increased evaporation. These constraints severely limit the production of arable land, particularly in emerging nations [5]. Soil salinity is also caused by human activities such as fertilising crops. Due to the potassium in fertilisers, which can naturally create salt sylvite. Salts are a naturally occurring substance in water and soil. The ions Na+, K+, Ca^{2+}, Mg^{2+}, and Cl are in charge of salination. Normal soil pH ranges from 2.2 to 9.7, and anything beyond that causes salt content degradation in the soil. The long-term viability of irrigated farming is seriously threatened by the salinization of the soil. A major problem in agricultural sustainability is the possibility of salinization and waterlogging caused by inadequate irrigation. Globally, more than 3% of soil resources are now damaged by salt [6]. Moreover, this number is steadily increasing at a pace of two Mha annually. Numerous earlier experiments indicated that waterlogging and salinization reduce yields in a variety of crops [7]. Between 18 and 43% of agricultural productivity is lost due to salinization in arid and semiarid regions of the world. Saline soils range from salt contents of 9–18 and >18 millimhos/cm designated as moderate to highly saline soils [8]. According to Hossain et al. [9], root-zone salinization and water logging significantly reduce field crop output. According to reports, the combined impact of root-zone salinization and water logging is worse than each factor acting alone.

Plant salt sensitivity varies with developmental stage. Seed germination determines plant establishment in saltwater environments. Salinity can impair germination rates, resulting in uneven crop development and lower yields [10]. However, the plant retort to salinity at germination can be altered and sought out as a quick and reliable indication of plant establishment in salt-affected settings. To boost plant performance and protect plants from biotic and abiotic stresses, several amendments were utilized. They can be provided to germinating seeds or to plants during vegetative growth [11]. In this milieu, marine macroalgal extracts show promising effects in reducing the influence of abiotic strain on plant performances. Algae are well-known sources of plant macro- and micronutrients, as well as bioactive chemicals [12]. SWEs can be sprayed on leaves, added to hydroponic systems, or treated directly to the soil. Plants respond to their application in a variety of positive ways. SWEs have been shown to include larger bioactive compounds, including oligosaccharides and phlorotannins. Activating molecular and metabolic pathways, bioactive chemicals from SWE have been demonstrated to function as elicitor agents, promoting plant development and inducing stress responses. They are rich in growth-promoting phytohormones along with inorganic elements indispensable for plant development. The application of seaweed extracts (SWEs) as natural regulators increased crop development and yield to endure tough environmental influences [13].

O. sativa is an important grain and indispensable food for preponderance of people globally [14]. With rising rice consumption, farmers and agriculturalists are under pressure to meet demand while safeguarding the crop from diseases and pests [15]. Produce damages of up to 50–70% are stressors. As common crop, rice is cultured comprehensively in coastal areas frequently flooded by saline sea water at high tides [16]. Response of rice to salt varies greatly within species, allowing for genetic improvement for the progress of salinity stress resistant crops [17]. Salinity particularly affects the physical elements affecting rice productivity. Salinity is subsequently the most common soil concern in rice-growing countries, after drought, and it is limiting global rice output. Coastal soils are discreetly saline on the exterior, but severely saline in sub surface strata besides substrata due to a variety of environmental causes [18]. Soluble salts, particularly sodium chloride, are abundant in saline soil [19].

Poorly germinating seeds are not only a yield constrainer, but also serve as a potential host for various diseases [20]. Seeds with decreased vigour also have more difficulty responding to field conditions, causing stress [21]. Numerous seed priming approaches are used to boost seed quality and reduce yield loss [22]. Primed seeds have early emergence, a higher seed vigour index, and increased biomass along with yield [23]. Aside from enhancing disease resistance, the treated seeds may also endure abiotic stress [24]. Root-zone salinity has a noteworthy influence on yield components relevant to final grain yield. Salinity reduces the number of main branches/panicle, their dimension, and other yield-related caryopsis features. Around panicle initiation, the biggest salt effects on yield are noted, while plants recover pre-eminent from strain at the seedling stage [25]. In a harsh environment, seed priming is a meek and cost-effective technique to promote kernel germination, prompt seedling development, and yield. Seed priming with various inorganic and organic substances increased wheat salt tolerance.

The damaging effects of abiotic stress and the positive effects of organic amendments on rice have both been studied independently. However, little is known about the underlying processes that might link the capabilities of SWE in protecting the crop from salinity-induced toxicities and driving these stressed plants towards their active functioning. Henceforth, by analysing the effect of salt stress and the ameliorating effect of SWE through investigating the plant's biochemical patterns and antioxidant enzymes, this study sought to identify the effects of liquid SWE of green alga *C. antennina* on rice physiology and biochemistry exposed to salinity strain.

2. Materials and Methods

2.1. Seaweed Collection, Extraction and Extract Preparation

The seaweed, *C. antennina* was collected (Colachel beach, Kanyakumari, Tamil Nadu, India; 8°14′5168″ N and 77°14′35.209″ E; December 2020), washed, and carried forward for extraction with boiling hot water (100 g/L; 1 h) (Figure 1). The concoction was clarified and the filtrated, SWE was stored until use (4 °C). The test solutions of SWE were devised by diluting the extract with distilled water of 60, 40, and 20 mL to obtain treatment concentration of SWE 40, 60, and 80%, respectively.

2.2. Rice Seed Collection and Preparation

During the growing season, rice seeds (TN1) were collected from farming areas. Seeds of uniform size and colour were preferred for the investigation, surface sterilized with 0.1% mercuric chloride, rinsed three times in sterile distilled water, and the trial was accomplished at the biopesticide and environmental toxicology lab, Environmental Science research centre (SPKCEES) of Manonmaniam Sundaranar University, Alwarkurichi.

2.3. Preparation of Salt Solutions

For the preparation of salt treatment solutions, 58.44 g of NaCl was dissolved in 1 L of distilled water to form the stock solution, from which respective quantities (0.8, 1.0 and 1.50) were made up to 10 mL using distilled water, viz., 9.2, 1, and 8.5 mL of distilled water to obtain treatment concentrations 80, 100, and 150 mM, respectively. Since treatment concentration of 150 mM was highly significant in causing salinity stress and the SWE was able to exert a positive influence on the aftereffects of salinity stress, only S150 mM was carried forward for future experiments (treatments with salt treatments <150 mM did not produce any statistically significant results).

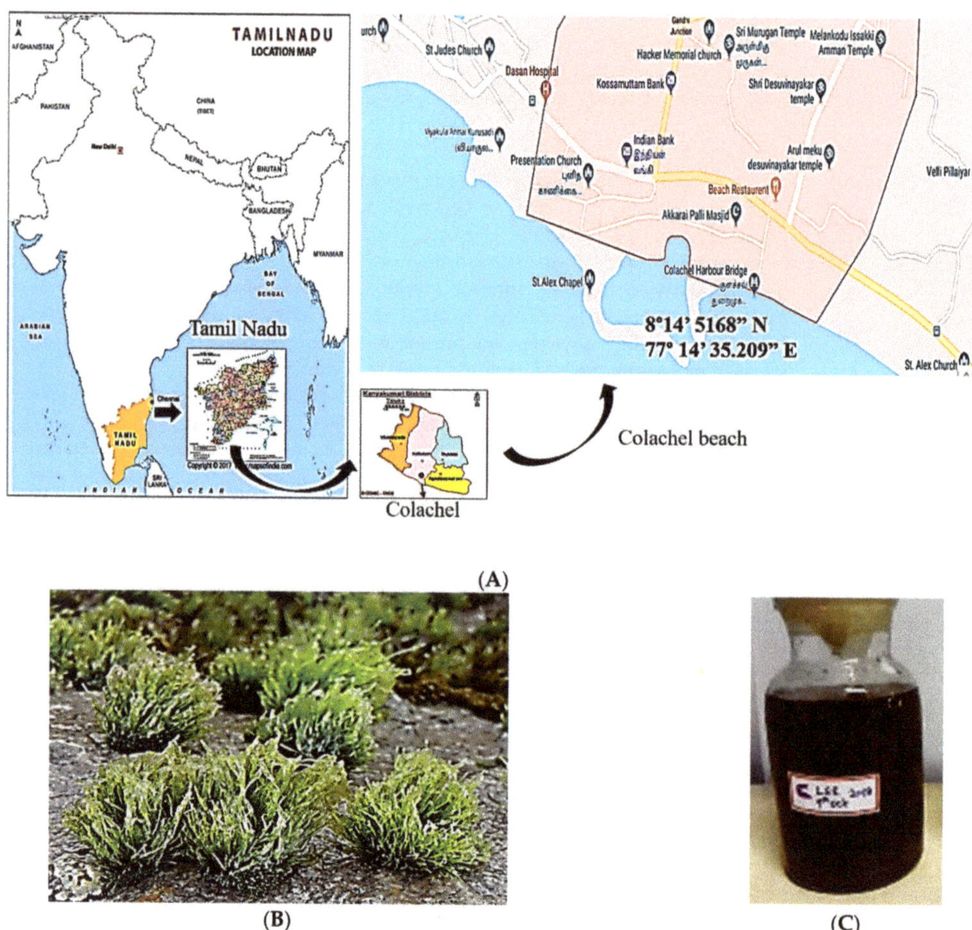

Figure 1. Sample collection site, seaweed *C. antennina*. (**A**) Sample collection site (**B**) *C. antennina* attached to rocks (**C**) liquid SWE.

2.4. Experimental Design

The experimentation consisted of different culture groups categorized by NaCl as well as SWE treatment concentrations. Starter fertilizer solution (SFS) was used as a negative control (NC) [23]. Commercial seaweed biostimulant solution (10%) was used as a positive control (PC). Both NC and PC treatments included exposure to salt (150 mM) designated as SNC and SPC. A treatment test involving a combination of SWE 80% with 150 mM of salt concentration was designated as S + SWE was also used (Table 1).

Table 1. Treatments involved in experimental design.

C	NC	SNC	PC	SPC	S150 mM	SWE80	S + SWE
Control	Negative Control	Salt 150 mM + NC	Positive control (Commercial biostimulant)	Salt 150 mM + PC	Salt treatment at 150 mM	80% SWE concentration	Salt 150 mM + SWE 80

The experiment to analyse the physiology and biochemistry of plants propagated in pots with sterile soil was brought out in randomized block design at the centre's greenhouse. The pots (15 cm diameter, 750 mL volume, and sterilized sandy soil) containing one seedling were irrigated with the appropriate treatment solutions mentioned earlier. The pots were arranged in a complete randomized design and maintained under 22–28 °C with a relative humidity of 30% and a 16 h/8 h day/night photoperiod. All the treatments were irrigated with sterile distilled water twice a day. Seeds immersed in distilled water and pots irrigated with distilled water served as control (C).

2.5. Effect of SWE on Rice Seed Germination under Salinity Stress

Rice seeds (seeds/treatment) were primed in respective solutions (10 mL; overnight), dried, and placed in correspondingly labelled sterile petri plates, and incubated (25 ± 2 °C/ consecutive 16:8 h LD). The seeds were observed for germination. Parameters associated with germination such as germination percentage (GP, %), average germination time (MGT, hours), germination energy (GE, %), and plantlet vigour index (SVI) as well as seedling growth (radicle–plumule, seedling lengths, cm) were noted [23].

$$GP = \frac{\text{Number of seeds germinated}}{\text{Total number of seeds}} \times 100 \; MGT = \frac{\sum(nT)}{\sum n}$$

where

n = number of newly germinated seeds at time T (25 °C)
T = hours from the beginning of the germination test
$\sum n$ = final germination.

$$GE = \frac{\textbf{Number of germinating seeds}}{\textbf{No. of total seeds per test post germination for 3 days}} \times 100$$

$$SVI = \text{Seedling length (cm) germination \%}$$

2.6. Effect of SWE on Rice Salt Alleviation

2.6.1. Physiology

The heights of plant, root, and shoot were measured using a ruler in all treatments, 40 days post-planting of the seedling. At the booting stage, measurement of comparative aqueous content of the shoot (RWC%) along with panicle length (cm) and tiller number (per hill) was performed following the methods described earlier [25]. The ability of SWEs on plants grown under salinity stress in terms of productivity (grain kg/hectare) and yield traits was also estimated by collecting 100 grains from the panicle of each paddy plant per treatment and determining their weight on a standard laboratory weighing scale [26].

$$RWC\ \% = \frac{(\mathbf{FW} - \mathbf{DW})}{(\mathbf{TW} - \mathbf{DW})} \times 100$$

2.6.2. Leaf Mortality

The effect of SWE on leaf mortality of rice plants was determined by counting their numbers in each plant in all treatments and the percentage of increase or increase was estimated was compared with plants growing in non-saline conditions [27].

2.6.3. Biochemistry

The effect of CA-LSE on the biochemistry of rice plants grown under salinity stress was studied using standard protocols methods to estimate whole soluble protein (TSP, mg/gFW) and phenolics (TPC, mg/gDW) [28,29]. Photosynthetic characteristics were estimated by measuring the levels of pigments. Chlorophyll (Chl a and b) and carotenoids (Car) and were extracted and estimated (mg/gFW) [30]. Beyer and Fridovich's method of super-oxide dismutase (SOD) estimation was employed and the level of the antioxidant enzyme was expressed in units (U/mg protein) [31].

Grain quality in terms of biochemistry was determined by estimating the quantities of total soluble sugars (TSS) and starch through the Anthrone method [32]. Grain protein content was also estimated [28]. The ratio of protein to starch contents was also estimated.

2.7. Statistical Analysis

To identify statistically significant variations in means between treatments, ANOVA was utilized. When statistical differences were found, the Tukey–Kramer post-hoc test was used. All experiments employed a maximum of 5 biological replicates. Tests after germination analysis employed only 8 treatments employing the highest concentrations in both salt and SWE (S150 mM and SWE 80%) for comparisons. When p was ≤ 0.05, the differences were considered statistically significant.

3. Results

3.1. Germination Parameters

The germination of rice seeds exposed to salinity stress was severely affected which was evident from their GTC (Figure 2A). Salinity stress deferred the initial germination of rice seeds by 7.73 h compared with control. SWE promoted early germination of rice seeds at 49.54 h compared with control. SWE were able to promote early germination of rice seeds under salinity stress stimulating their germination 36.27 h before that of those grown under salt stress (Figure 3). Consequently, the MGT of rice seeds exposed to salinity stress was also brought down by SWE treatment (Figure 2B). While SFS-treated salinity-exposed rice seeds (SNC) prompted MGT by only 1.89 h compared with control, 50% of S + SWE seeds emerged 23.72 h compared with control ($F_{6,28} = 126.19$; $p < 0.0001$). Seeds treated with 150 mM of salt concentration did not reach 50% of emergence to be carried for MGT calculation.

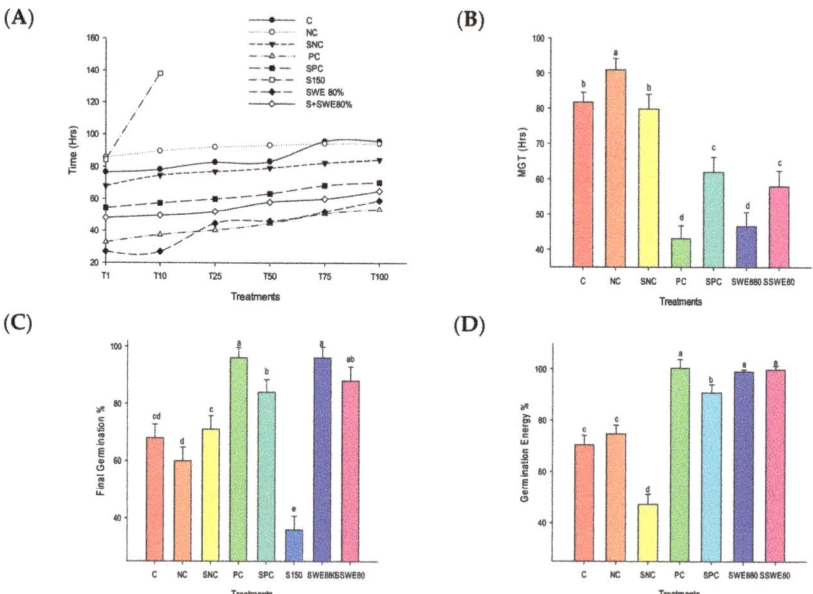

Figure 2. Effect of SWEs and salt on germination of rice seeds; (**A**) germination time course; (**B**) MGT; (**C**) final GP; (**D**) GE. Columns denoted by a different letter are significantly different at $p \leq 0.05$ in Tukey's test. Seeds exposed to salt treatment (S150 mM) did not reach 50% of germination.

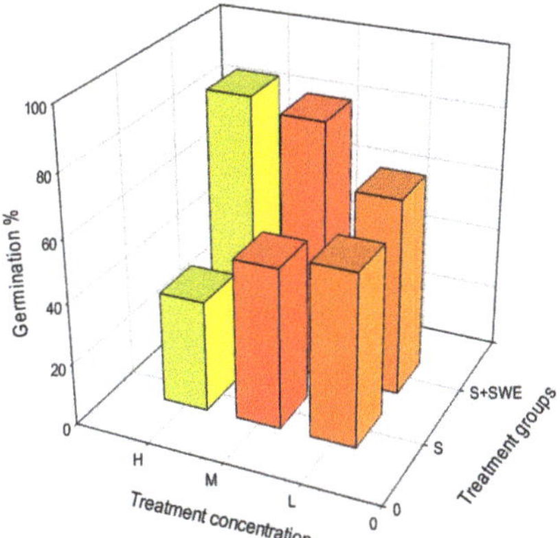

Figure 3. Efficacy of SWEs on the recovery of GP of rice seeds exposed to salt stress. the X-axis SWE Treatment concentrations (low 40% (brown), Medium 60% (red) and high 80% (yellow)) Y-axis germination%, Z-axis treatment groups.

Salt stress severely affected germination capability of seeds exhibiting 56, 52, and 36% of germination percentage at salt concentrations 80, 100, and 150 mM, respectively. SWE treatments increased GP to 80, 91, and 96% at concentrations, 40, 60, and 80, respectively. SWE were able to significantly increase the GP of seeds under salinity stress to 88 from 36‰ ($F_{6,28} = 80.35$; $p < 0.0001$) (Figure 2C). Germination energy of seeds exposed to salinity stress was severely affected ($F_{6,28} = 213.23$; $p < 0.0001$) with a corresponding impact of salinity strain on the vigour of seedlings was also observed resulting in a very lower vigour index, 133.267. SWEs were able to completely energize the seedlings (99.8%) (Figure 2D, Table 2).

Table 2. Effect of SWE and salinity stress on SVI of seedlings. Columns denoted by a different letter are significantly different at $p \leq 0.05$ in Tukey's test.

Treatments	SVI
C	333.083 ± 1.378 [g]
NC	541.5 ± 4.23 [e]
SNC	498.67± 3.98 [f]
PC	1555.65 ± 1.37 [b]
SPC	1160.17 ± 2.95 [d]
S150	133.267 ± 1.938 [h]
SWE80	1651.37 ± 3.14 [a]
SSWE80	1407.82 ± 3.74 [c]

3.2. Growth Parameters

The salt treatments drastically affected the length of radicle and plumule resulting in stunted growths (2.48 and 1.28 cm) which was 48.3 and 44.8% inferior to control ($F_{5,24} = 20.72$; $p < 0.0001$) (Figure 4A). SWE promoted the growths of radicle and plumule in seeds exposed to salinity by increasing their lengths to 6.24 and 9.54 cm from 1.28 ($F_{7,32} = 21.1$; $p < 0.0001$) and 3.72 cm ($F_{7,32} = 83.39$; $p < 0.0001$). A likely reduction in root–shoot lengths of rice plants by salt stress was also noted which was promoted by the influence of SWE (Figure 4B). Salt stress reduced the plant development by negatively influencing the development of roots and shoots decreasing them to 2.82 ($F_{7,32} = 53.78$; $p < 0.0001$) and 3 cm ($F_{7,32} = 58.27$; $p < 0.0001$) in lengths from 4.7 and 9.5 cm. However, the SWE promoted the growths of rhizome and shoots of salt stress-exposed plants to 8.6 and 13.2 cm with a parallel increase in plant height to 22.16 cm from 7.8 cm ($F_{7,32} = 317.75$;

$p < 0.0001$). SWE also increased the growth of seedlings exposed to SWE to 9.54 from 3.72 cm ($F_{7,32} = 83.39$; $p < 0.0001$) (Figure 4C).

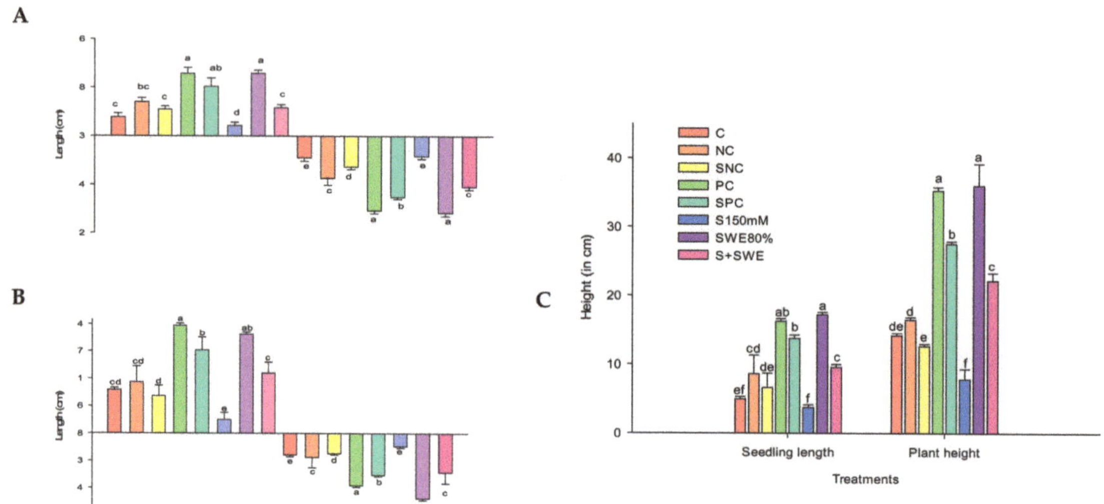

Figure 4. Effect of SWEs and salt on growth parameters of paddy (cm) (**A**) radicle–plumule length; (**B**) Root–shoot height; (**C**) seedling–plant height. Columns denoted by a different letter are significantly different at $p \leq 0.05$ in Tukey's test.

3.3. Physiological Parameters

The detrimental outcome of salt strain-induced upon RWC, panicle length, and tiller numbers of paddy plants were also positively encouraged by the treatments with SWE (Table 3). RWC of paddy leaves were increased from 40.6 to 82.6% ($F_{7,32} = 25.8$; $p < 0.0001$). An increase by SWE in panicle length and tiller numbers from 13.4 cm and 25.2 to 28.39 cm ($F_{7,32} = 25.8$; $p < 0.0001$) and 56.37 ($F_{7,32} = 168.8$; $p < 0.0001$) was also noted (Figure 5). Caryopsis qualities also enhanced by SWEs promoting the grain weight to 3.345 from 1.455 g ($F_{7,32} = 19.43$; $p < 0.0001$). SWEs also promoted yield of paddy plants exposed to salinity stress from 3230.8 to 7743.7 kg/hectare ($F_{7,32} = 10{,}138.07$; $p < 0.0001$) (Table 3).

Figure 5. Effect of SWEs and salt on panicle length (cm).

Table 3. Effect of SWE and salinity stress on RWC (%), panicle length (cm), tiller number/hill, 100 grain wt (g), grain yield (kg/hectare) of paddy. Columns denoted by a different letter are significantly different at $p \leq 0.05$ in Tukey's test.

Treatments	RWC (%)	Panicle Length (cm)	Tiller Number/Hill	100 Grain wt (g)	Grain Yield (kg/Hectare)
C	70.4 ± 4.28 c	23 ± 4.12 bc	44.8 ± 3.83 cd	2.299 ± 0.359 b	5668.8 ± 25.2 e
NC	75.2 ± 4.44 bc	26.146 ± 1.452 ab	52 ± 3.16 bcd	2.849 ± 0.353 ab	6867.9 ± 32.9 d
SNC	46 ± 23.1 d	17.81 ± 3.03 cd	38.02 ± 3.16 de	2.155 ± 0.243 bc	4842.2 ± 37.9 f
PC	95.2 ± 3.7 a	30.18 ± 3.19 a	65.63 ± 3.35 ab	3.391 ± 0.445 a	8253.6 ± 39.9 a
SPC	85.4 ± 3.65 abc	27.58 ± 4.61 ab	59.38 ± 4.16 abc	3.261 ± 0.397 a	7467.1 ± 51 c
S150	40.6 ± 3.97 d	13.4 ± 3.05 d	25.2 ± 3.96 e	1.455 ± 0.398 c	3230.8 ± 37.7 g
SWE80	93 ± 4.12 ab	30.71 ± 3.03 a	66.96 ± 2.05 a	3.502 ± 0.412 a	8174.4 ± 46.9 a
SSWE	82.6 ± 3.97 abc	28.39 ± 3.5 ab	56.37 ± 2.87 abc	3.345 ± 0.357 a	7743.7 ± 42.1 b

3.4. Biochemical Parameters

Salt stress resulted in increased rates of leaf mortality (8.4%), which was revoked by SWEs by 85.12% ($F_{7,32}$ = 30.32; p < 0.0001) (Table 4). Salt stress rigorously derailed the leaf biochemical constituents of paddy plants reducing the levels of TSP, TPC, and SOD from 1.786, 16.16, and 1.76 to 1.21 mg/g FW ($F_{7,32}$ = 4.34; p < 0.0001), 10.86 mg/g DW ($F_{7,32}$ = 39.35; p < 0.0001) and 1.3 U/mg protein ($F_{7,32}$ = 4.19; p < 0.0001), respectively. Consequently, SWEs increased the leaf biochemistry of salt-exposed paddy plants to 1.656 mg/g FW, 25.07 mg/g DW, and 3 U/mg proteins of TSP, TPC, and SOD, respectively (Table 4). Paddy leaf pigments such as chla and b along with carotenoids were also positively influenced by SWE treatments that increased from 1.1704 and 1.165 along with 1.628 mg/g FW to 2.32 ($F_{7,32}$ = 8.07; p < 0.0001), 1.535 ($F_{7,32}$ = 2.52; p < 0.0001), and 2.0274 mg/g FW ($F_{7,32}$ = 4.8; p < 0.0001), respectively (Figure 6).

Table 4. Effect of SWE and salinity stress on leaf mortality (%), TSP (mg/gFW), TPC (GAE mg/g DW), and SOD (U/mg protein) on paddy leaves. Columns denoted by a different letter are significantly different at $p \leq 0.05$ in Tukey's test.

Treatments	Leaf Mortality (%)	TSP (mg/gFW)	TPC (GAE mg/g DW)	SOD (U/mg Protein)
C	2.52 ± 0.396 c	1.786 ± 0.0397 d	16.16 ± 1.85 b	1.76 ± 0.428 ab
NC	2.44 ± 0.365 c	1.344 ± 0.0355e f	26.08 ± 2.008 a	2.3 ± 0.346 ab
SNC	5.2 ± 2.68 b	1.3 ± 0.0346d e	11.96 ± 2.24 bc	1.52 ± 0.396 b
PC	1.048 ± 0.0013 e	2.192 ± 0.0317 a	28.04 ± 3.16 a	2.454 ± 0.415 ab
SPC	1.23 ± 0.3 d	1.738 ± 0.0445b c	25.89 ± 3.09 a	2.192 ± 0.259 ab
S150	8.4 ± 1.14 a	1.21 ± 0.0351 ef	10.86 ± 2.144 c	1.3 ± 0.346 b
SWE80	1.118 ± 0.0604 d	1.93 ± 0.0383 b	27.66 ± 3.35 a	2.45 ± 0.415 ab
SSWE	1.248 ± 0.233 de	1.656 ± 0.0304 c	25.07 ± 2.3 a	3 ± 1.414 a

Grain biochemical contents was analysed in terms of carbohydrate content and protein. TSS level was found to be enhanced by salinity stress (17.85%) compared with control (Table 5). Despite the 13.69% decrease in TSS content in SWE-treated rice plants compared with salt-treated plants, TSS content was found to be present 6.14% greater than those exposed to salt stress ($F_{7,32}$ = 46.5; p < 0.0001). A similar pattern of increase in starch content in grains emerged out of plants exposed to salt stress (18.21%) and those treated with SWE was also observed (0.827%); however, the increase in starch content brought about by SWE application on salt-stressed plants was not statistically significant (p > 0.05). Protein levels in grains of salt-stressed plants was decreased by 7.54%; However, SWE treatments increased grain protein level by 23.12% ($F_{7,32}$ = 57.28; p < 0.0001). This consequently affected the protein–starch ratio. Salinity stress severely decrease protein–starch ratio by 23.6% and was elevated by SWE treatment to 0.0886 from 49.4 ($F_{7,32}$ = 53.71; p < 0.0001).

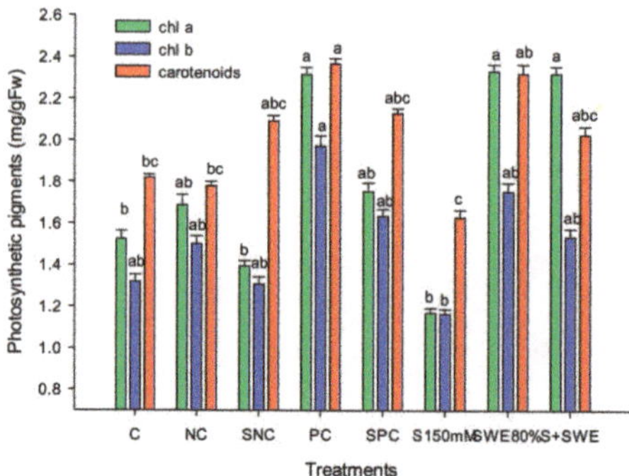

Figure 6. Effect of SWEs and salt on photosynthetic pigments, chlorophyll (a and b), and carotenoids (mg/g FW) of paddy leaves. Columns denoted by a different letter are significantly different at $p \leq 0.05$ in Tukey's test.

Table 5. Grain biochemical traits of plants on salinity stress and SWE treatments. Columns denoted by a different letter are significantly different at $p \leq 0.05$ in Tukey's test.

Treatments	Carbohydrate (mg/g DW)		Protein (mg/g DW)	Protein: Starch
	TSS	Starch		
C	138.2 ± 2.21 [e]	588 ± 3.8 [ef]	53.1 ± 1.8 [d]	0.090 ± 0.001 [c]
NC	157 ± 2. 1[c]	662.5 ± 4.03 [d]	62.4 ± 2.03 [b]	0.0941 ± 0.0001 [b]
SNC	164.2 ± 2.03 [b]	684 ± 3.85 [c]	54 ± 1.7 [d]	0.0789 ± 0.001 [f]
PC	142.1 ± 1.98 [d]	567 ± 3.81 [g]	67.36 ± 1.89 [ab]	0.118 ± 0.002 [a]
SPC	148.34 ±2.07 [d]	713 ± 4.02 [b]	59 ± 2.01 [bc]	0.082 ± 00.01 [e]
S150	168.3 ± 2.31 [b]	719 ± 4.36 [ab]	49.4 ± 2.02 [e]	0.0687 ± 0.001 [g]
SWE80	145 ± 1.9 [d]	581.23 ± 4.2 [e]	70.2 ± 1.9 [a]	0.1207 ± 0.002 [a]
SSWE	179 ± 2.13 [a]	725 ± 3.95 [a]	64.26 ± 2 [b]	0.0886 ± 0.001 [d]

4. Discussion

Salty conditions are known to limit plant development; saline soils and saline irrigation pose major issues for vegetable crop productivity [33]. With increasing salt, salt stress can delay and limit plant development differentiation, as well as lower the fresh weight of leaf, stem, and root tissue [34]. Poor germination of *O. sativa* control seeds is uncommon for a well-domesticated crop, since most crop species demonstrate quick germination and can reach 100% germination under ideal conditions [35]. A sustainable eco-friendly technique of using seaweed extracts was devised to minimize both the agricultural constraint and to increase rice output. SWE had an influence on the sprouting and development of rice.

Seed emergence is hampered by the equipoise, which is bordered by the embryo's development abilities as well as the endosperm's mechanical resistance, which needs to be weakened for germination [36]. The endosperm cap is weakened by a succession of enzymes and phytohormones, which accelerate cell development in the embryo and alter the radicle's emergence. As a result, seed germination augmentation practices such as seed instructing are increasingly being used to boost crop growth and production.

Until recently, seaweeds were being investigated as possible crop development besides produce enhancers, with the goal of replacing chemical fertilizers due to their superior efficiencies, larger action range, eco-friendliness, and cost effectiveness. Seaweeds have been observed to have remarkable plant growth encouraging potentials, including enhanced plant height, root, and shoot lengths, and are therefore classified as plant growth biostimulants, according to Craigie et al. [37]. The radicle and plumule are crucial in determining the underpinning of a plant in the field since they are the major developmental plant growth phase. Seeds with a prominent radicle and plumule germinate quickly and have increased competence [19]. Plant establishment efficiency is also specified by lengthier radicles. Seeds with a shorter radicle–plumule may have problems transporting nutrients to the embryo [38].

Seed priming effects of SWEs were instituted to ensure a favourable impact on early emergence and plant growth. SWEs not only spurred early seed emergence but also raised seed GP, whereas increasing salt treatments delayed seed emergence and lowered GP. The toxic effects of Na^+ and Cl in the germination process might be to blame for the decrease in germination [39]. Salt stress appears to influence seed germination by limiting seed water absorption, causing protein synthesis abnormalities, and causing excessive nutrient pool usage [40]. Regardless of the circumstance, greater salt concentrations inhibited more than 80% of seed germination, and SWEs increased the GP by 59.09%. SWEs are known to have beneficial impacts on the germination of a variety of crops [41].

The proportion of seeds that germinate quickly is expressed as germination energy. While salt stress depleted the germination energy completely, SWEs enhanced the GE to its maximal degree (80–90%). The relative increase in seed emergence is linked to the seed eminence, which appears to enhance by SWE treatment. Furthermore, Amabika and Sujatha [42] demonstrated that priming red gram seeds with extract of the seaweed, *Sargassum myriocystum* increased the seedling vigour index of emerged seedlings. This thereby proves the fact that higher SVI is a good indicator of healthy seedlings.

Sangare et al. [43] indicated that the true performance ability of a seed can be elucidated by comparing its SVI with that of a control. Higher SVI of SWE-treated seeds denote an increase in seed quality. Other characteristics of SWE-treated seeds showed greater seedling-plant height, radicle-root, plumule-shoot lengths, and dry-wet weight compared with control. Enhanced seed germination and growth rates of brinjal and tomato, as well as chilli, were connected with increased SVI after seed priming with SWE of *Ulva lactuca*, *Padina pavonic*, and *S. johnstonii* [44]. This beneficial impact might be accredited to various growth-regulating chemicals found in SWEs, such as kinetin, gibberellic acid, and ethylene, which have been linked to the reversal of seed dormancy [45]. The growth hormones found in SWE may serve a crucial part in initiating the creation of hydrolytic enzymes from scratch. These phytohormones may cause abscisic acid inhibitors to seep from the seeds, improving germination rates [46].

Yang and Guo [47] found that the principal response to salinity stress is a suppression of shoot and root vegetative development, which was seen in the current experiment. The same discoveries were established by Khosravinejad et al. [48], who observed a substantial reduction in shoot lengthening in barley when NaCl treatment was increased. Compared with plants in a saline state, liquid extract delivers the highest outcomes in terms of plant development under salt stress conditions by a considerable increase. Seaweed elements such as macro- and microelement nutrients, vitamins, amino acids, auxins, and cytokinins, which impact cellular metabolism in plants and lead to increased growth, can be connected to improved wheat vegetative development [49]. Researchers from all around the world have confirmed that seaweeds boost plant development [50–52].

Although salt stress reduced rice plant protein and phenolic contents, SWE enhanced both biochemical properties by 22 and 56.8%, respectively. Wheat plants treated with *Fucus spiralis* LSEs showed a similar rise in TSP and TPC [45]. All kinds of stressors cause an increase in the synthesis of hazardous oxygen derivatives. Plants have effective mechanisms for scavenging active oxygen species, which protect them from oxidative processes that are harmful to them [53]. Antioxidant enzymes show a vital part in the defensive mechanisms of the plant system. The administration of SWE enhanced the synthesis of the SOD enzyme that was previously reduced by salt stress in the current investigation. The application of extracts of *F. spiralis* to wheat plants resulted in a comparable rise in SOD activity [45]. Salinity levels have a big impact on relative water content [54]. Plants exposed to salt at higher levels had much decreased water content in their leaves. Plants under salinity stress had their RWC increased by SWEs. Bulgari et al. [55] stated that similar applications of ascorbic acid and abscisic acid, as well as seaweed extracts, have been demonstrated to protect plants against water loss and retain water. Plant growth, tiller number, length of panicle, seed weight, and caryopsis traits all decrease with increasing saline levels, according to Abdullah et al. [56], which correlates with our findings. The treatment with SWEs, which cumulatively improved panicle length and augmented the number of tillers/hills, improved all the above characteristics.

In our research, SWEs had a comparable favourable effect on grain weight, yield, and photosynthetic pigments. Reduced chlorophyll concentration, fragmentation of chloroplast membranes, besides disruption of biochemical activities, are all examples of abiotic stressors that have a noteworthy influence on plant photosynthesis. NaCl stress dramatically reduced chlorophyll content in paddy, according to the findings of this study. The seaweed extract was shown to increase plant chlorophyll content by promoting its production. Under NaCl stress, the polysaccharides of the algae *Lessonia nigrescens* (LNP) greatly boosted chlorophyll levels in plants. LNP treatment decreased lipid peroxidation and alleviated the salt-induced loss of chlorophyll content [57]. The high content of macro- and micronutrients, growth stimulators, bioactive ingredients, and unique plant growth promoting products in the alga can be accredited to the helpful outcome of algal amendment besides also their stress-dismissing influence on rice yield [58].

The varied effects of the treatments changed the biochemistry and grain weight. Despite the fact that TSS and starch levels increased, rice plants treated with SWE had higher protein contents. The total protein–starch ratio of grains treated with SWE, which previously reduced due to salt stress, rose as a result. Thitisaksakul et al. previously established that salt stress increases the starch content of rice grains [59]. However, it was suggested that genotypes affected how much biochemical material accumulated in rice grains in response to salt stress [60]. Proteins are thought to be the factors that determine the sensory quality of rice [61]. According to reports, rice grain proteins can change depending on the salinity. It is considered that the more protein a grain contains, the less likely it is to break during the milling process [62]. In light of nutrition, storage, and marketability, increased rice grain protein content encouraged by SWE treatment under saline stress is a desirable rice quality. Soil salinity has been reported to enact a negative impact on the plants in all stages of development, affecting the plant growth, development, functionality, and eventually the crop's quality and yield. Application of SWE was hence tested for its salt alleviation capacity right from germination and throughout the growth of rice plant. It was evident that the SE promoted seed germination that was severely constrained by salinity stress and eventually producing plants of greater height and vigour. This can be correlated to the enhanced biochemical attributes that contributed to increase in the synthesis of vital plant biochemical such as protein, phenols, antioxidant enzyme, and chlorophyll pigments. An increase in water retention in leaves treated with SWE exposed to salt stress and lower leaf mortality rates dues to SWE treatments clearly indicates that SWE was successful in conferring protection to rice plants against salt-induced toxicities. An increase in grain yield and weight along with grain biochemical attributes such as protein and protein:starch also indicates that SWE had exerted a profound effect in promoting the productivity of plant as a result of enhanced plant functionalities. Hence the current study proposes that SWE promotes effective use of plant resources, increases plant development, and increases tolerance to unfavourable environmental circumstances.

5. Conclusions

These discoveries propose that seaweed extracts can help rice plants cope with salt stress by boosting rice seed germination and plant development. The extracts significantly enhanced panicle length and tiller numbers, both of which were badly harmed by salt stress. SWE also improved the content of photosynthetic pigments. The treatment of SWE had a good effect on the caryopsis characteristics and yield. SWE significantly reduced leaf death rates while also boosting protein, phenolic content, and antioxidant enzyme levels. Extracts from the halotolerant macroalgae *C. antennina* can help plants deal with salt stress, according to our findings.

Author Contributions: Conceptualization, K.M.-P.C. and S.S.-N.; Methodology, K.M.-P.C. and G.-S.P.; Validation, K.M.-P.C., A.G. and S.S.-N.; Investigation, G.-S.P., P.M. (Pauldurai Malarvizhi), A.G. and P.M. (Ponnusamy Murugan); Formal analysis, A.D.-A., P.M. (Pauldurai Malarvizhi), M.J., H.S. and R.R.; Data curation, K.M.-P.C., S.S.-N., A.G. and V.S.-R.; writing—original draft preparation, K.M.-P.C.; writing—review and editing, A.A.-M., P.K. and S.S.-N. All authors have read and agreed to the published version of the manuscript.

Funding: This research was supported by the Department of Science and Technology (DST-FIST), India under FIST program (SR/FIST/LS-1/2019/522). This research work was partially supported by Chiang Mai University, Thailand.

Institutional Review Board Statement: Not applicable.

Informed Consent Statement: Not applicable.

Data Availability Statement: Data are contained within the article.

Conflicts of Interest: The authors declare no conflict of interest.

References

1. Rohr, J.R.; Barrett, C.B.; Civitello, D.J.; Craft, M.E.; Delius, B.; DeLeo, G.A.; Hudson, P.J.; Jouanard, N.; Nguyen, K.H.; Ostfeld, R.S.; et al. Emerging human infectious diseases and the links to global food production. *Nat. Sustain.* **2019**, *2*, 445–456. [CrossRef]
2. Kalaivani, K.; Kalaiselvi, M.M.; Senthil-Nathan, S. Effect of Methyl Salicylate (MeSA) induced changes in rice plant (*Oryza sativa*) that affect growth and development of the rice leaffolder, *Cnaphalocrocis medinalis*. *Physiol. Mol. Plant Pathol.* **2018**, *101*, 116–126. [CrossRef]
3. Barbier, E.B.; Hacker, S.D.; Kennedy, C.; Koch, E.W.; Stier, A.C.; Silliman, B.R. The value of estuarine and coastal ecosystem services. *Ecol. Monogr.* **2011**, *81*, 169–193. [CrossRef]
4. Yasuor, H.; Yermiyahu, U.; Ben-Gal, A. Consequences of irrigation and fertigation of vegetable crops with variable quality water: Israel as a case study. *Agric. Water Manag.* **2020**, *242*, 106362. [CrossRef]
5. Choudhary, O.P.; Grattan, S.R.; Minhas, P.S. Sustainable crop production using saline and sodic irrigation waters. In *Alternative Farming Systems, Biotechnology, Drought Stress and Ecological Fertilisation*; Springer: Berlin/Heidelberg, Germany, 2011; pp. 293–318.
6. Singh, A. Soil salinization management for sustainable development: A review. *J. Environ. Manag.* **2021**, *277*, 111383. [CrossRef] [PubMed]
7. Pulido-Bosch, A.; Rigol-Sanchez, J.P.; Vallejos, A.; Andreu, J.M.; Ceron, J.C.; Molina-Sanchez, L.; Sola, F. Impacts of agricultural irrigation on groundwater salinity. *Environ. Earth Sci.* **2018**, *77*, 197. [CrossRef]
8. FAO. Management of Salt Affected Soils: 'Soil Management' under 'FAO SOILS PORTAL'. In Food and Agriculture Organization' of the 'United Nations'. Rome. 2020. Available online: http://www.fao.org/soils-portal/soil-management/management-of-some-problemsoils/salt-affected-soils/more-information-on-salt-affected-soils/en/ (accessed on 15 November 2022).
9. Hossain, M.A.; Araki, H.; Takahashi, T. Poor grain filling induced by waterlogging is similar to that in abnormal early ripening in wheat in Western Japan. *Field Crops Res.* **2011**, *123*, 100–108. [CrossRef]
10. Luo, X.; Dai, Y.; Zheng, C.; Yang, Y.; Chen, W.; Wang, Q.; Chandrasekaran, U.; Du, J.; Liu, W.; Shu, K. The ABI4-RbohD/VTC2 regulatory module promotes reactive oxygen species (ROS) accumulation to decrease seed germination under salinity stress. *New Phytol.* **2021**, *229*, 950–962. [CrossRef]
11. Kalaivani, K.; Maruthi-Kalaiselvi, M.; Senthil-Nathan, S. Seed treatment and foliar application of methyl salicylate (MeSA) as a defense mechanism in rice plants against the pathogenic bacterium, *Xanthomonas oryzae* pv. *oryzae*. *Pest. Biochem. Physiol.* **2021**, *171*, 104718. [CrossRef]
12. Kocira, S.; Szparaga, A.; Kuboń, M.; Czerwińska, E.; Piskier, T. Morphological and biochemical responses of *Glycine max* (L.) Merr. to the use of seaweed extract. *Agronomy* **2019**, *9*, 93. [CrossRef]
13. Stirk, W.A.; Rengasamy, K.R.; Kulkarni, M.G.; van Staden, J. Plant biostimulants from seaweed: An Overview. *Chem. Biol. Plant Biostimulants* **2020**, *2*, 31–55.
14. Bohra, A.; Jha, U.C.; Jha, R.; Naik, S.J.; Maurya, A.K.; Patil, P.G. Genomic interventions for biofortification of food crops. In *Quality Breeding in Field Crops*; Springer: Berlin/Heidelberg, Germany, 2019; pp. 1–21.
15. Nakashima, K.; Yanagihara, S.; Muranaka, S.; Oya, T. Development of Sustainable Technologies to Increase Agricultural Productivity and Improve Food Security in Africa. *Jpn. Agric. Res. Q.* **2022**, *56*, 7–18. [CrossRef]
16. Francini, A.; Sebastiani, L. Abiotic stress effects on performance of horticultural crops. *Horticulturae* **2019**, *5*, 67. [CrossRef]
17. Reddy, A.M.; Francies, R.M.; Rasool, S.N.; Reddy, V.R. Breeding for tolerance stress triggered by salinity in rice. *Int. J. Appl. Biol. Pharm. Technol.* **2014**, *5*, 167–176.
18. Rasel, H.M.; Hasan, M.R.; Ahmed, B.; Miah, M.S. Investigation of soil and water salinity, its effect on crop production and adaptation strategy. *Int. J. Water Res. Environ. Eng.* **2013**, *5*, 475–481.
19. Pessarakli, M.; Szabolcs, I. Soil salinity and sodicity as particular plant/crop stress factors. In *Handbook of Plant and Crop Stress*, 4th ed.; CRC Press: Boca Raton, FL, USA, 2019; pp. 3–21.

20. Shi, W.; Yin, X.; Struik, P.C.; Xie, F.; Schmidt, R.C.; Jagadish, K.S. Grain yield and quality responses of tropical hybrid rice to high night-time temperature. *Field Crops Res.* **2016**, *190*, 18–25. [CrossRef]
21. Liu, H.; Hussain, S.; Zheng, M.; Sun, L.; Fahad, S.; Huang, J.; Cui, K.; Nie, L. Progress and constraints of dry direct-seeded rice in China. *J. Food Agric. Environ.* **2014**, *12*, 465–472.
22. Stanley-Raja, V.; Senthil-Nathan, S.; Chanthini, K.M.P.; Sivanesh, H.; Ramasubramanian, R.; Karthi, S.; Shyam-Sundar, N.; Vasantha-Srinivasan, P.; Kalaivani, K. Biological activity of chitosan inducing resistance efficiency of rice (*Oryza sativa* L.) after treatment with fungal based chitosan. *Sci. Rep.* **2021**, *11*, 20488. [CrossRef]
23. Chanthini, K.M.; Stanley-Raja, V.; Thanigaivel, A.; Karthi, S.; Palanikani, R.; ShyamSundar, N.; Sivanesh, H.; Soranam, R.; Senthil-Nathan, S. Sustainable agronomic strategies for enhancing the yield and nutritional quality of wild tomato, *Solanum Lycopersicum* (l) var *Cerasiforme* Mill. *Agronomy* **2019**, *9*, 311. [CrossRef]
24. Jisha, K.C.; Puthur, J.T. Seed priming with BABA (β-amino butyric acid): A cost-effective method of abiotic stress tolerance in *Vigna radiata* (L.) Wilczek. *Protoplasma* **2016**, *253*, 277–289. [CrossRef]
25. Radanielson, A.M.; Angeles, O.; Li, T.; Ismail, A.M.; Gaydon, D.S. Describing the physiological responses of different rice genotypes to salt stress using sigmoid and piecewise linear functions. *Field Crops Res.* **2018**, *220*, 46–56. [CrossRef] [PubMed]
26. Hasanuzzaman, M.; Fujita, M.; Islam, M.N.; Ahamed, K.U.; Nahar, K. Performance of four irrigated rice varieties under different levels of salinity stress. *Int. J. Integr. Biol.* **2019**, *6*, 85–90.
27. Thitisaksakul, M.; Tananuwong, K.; Shoemaker, C.F.; Chun, A.; Tanadul, O.U.; Labavitch, J.M.; Beckles, D.M. Effects of timing and severity of salinity stress on rice (*Oryza sativa* L.) yield, grain composition, and starch functionality. *J. Agric. Food Chem.* **2015**, *63*, 2296–2304. [CrossRef]
28. Kielkopf, C.L.; Bauer, W.; Urbatsch, I.L. Bradford assay for determining protein concentration. *Cold Spring Harb. Protoc.* **2020**, *4*, 102269. [CrossRef] [PubMed]
29. Taga, M.S.; Miller, E.E.; Pratt, D.E. Chia seeds as a source of natural lipid antioxidants. *J. Am. Oil Chem. Soc.* **1984**, *61*, 928–931. [CrossRef]
30. Manaf, H.H. Beneficial effects of exogenous selenium, glycine betaine and seaweed extract on salt stressed cowpea plant. *Ann. Agric. Sci.* **2016**, *61*, 41–48. [CrossRef]
31. Beyer, W.F.; Fridovich, I. Assaying for superoxide dismutase activity: Some large consequence of minor changes in conditions. *Ann. Biochem.* **1987**, *161*, 559–566. [CrossRef]
32. Maness, N. Extraction and analysis of soluble carbohydrates. In *Plant Stress Tolerance*; Humana Press: Totowa, NJ, USA, 2010.
33. Minhas, P.S.; Ramos, T.B.; Ben-Gal, A.; Pereira, L.S. Coping with salinity in irrigated agriculture: Crop evapotranspiration and water management issues. *Agric. Water Manag.* **2020**, *227*, 105832. [CrossRef]
34. Bonomelli, C.; Celis, V.; Lombardi, G.; Mártiz, J. Salt stress effects on avocado (*Persea americana* Mill.) plants with and without seaweed extract (*Ascophyllum nodosum*) application. *Agronomy* **2018**, *8*, 64. [CrossRef]
35. Kalaivani, K.; Kalaiselvi, M.M.; Senthil-Nathan, S. Effect of methyl salicylate (MeSA), an elicitor on growth, physiology and pathology of resistant and susceptible rice varieties. *Sci. Rep.* **2016**, *6*, 34498. [CrossRef]
36. Seethalakshmi, S.; Umarani, R. Biochemical changes during imbibition stages of seed priming in tomato. *Int. J. Chem. Stud.* **2018**, *6*, 454–456.
37. Craigie, J.S. Seaweed extract stimuli in plant science and agriculture. *J. Appl. Phycol.* **2011**, *23*, 371–393. [CrossRef]
38. Chanthini, K.M.P.; Senthil-Nathan, S.; Stanley-Raja, V.; Thanigaivel, A.; Karthi, S.; Sivanesh, H.; Sundar, N.S.; Palanikani, R.; Soranam, R. *Chaetomorpha antennina* (Bory) Kützing derived seaweed liquid fertilizers as prospective bio-stimulant for *Lycopersicon esculentum* (Mill). *Biocatal. Agric. Biotechnol.* **2019**, *20*, 101190. [CrossRef]
39. Zhumabekova, Z.; Xu, X.; Wang, Y.; Song, C.; Kurmangozhinov, A.; Sarsekova, D. Effects of Sodium Chloride and Sodium Sulfate on *Haloxylon ammodendron* Seed Germination. *Sustainability* **2020**, *12*, 4927. [CrossRef]
40. El Moukhtari, A.; Cabassa-Hourton, C.; Farissi, M.; Savouré, A. How does proline treatment promote salt stress tolerance during crop plant development? *Front. Plant Sci.* **2020**, *11*, 1127. [CrossRef]
41. Begum, M.; Bordoloi, B.C.; Singha, D.D.; Ojha, N.J. Role of seaweed extract on growth, yield and quality of some agricultural crops: A review. *Agric. Rev.* **2018**, *39*, 321–326. [CrossRef]
42. Amabika, S.; Sujatha, K. Effect of priming with seaweed extracts on germination and vigour under different water holding capacities. *Seaweed Res. Utiln.* **2015**, *37*, 37–44.
43. Sangare, S.K.; Compaore, E.; Buerkert, A.; Vanclooster, M.; Sedogo, M.P.; Bielders, C.L. Field-scale analysis of water and nutrient use efficiency for vegetable production in a West African urban agricultural system. *Nutr. Cycl. Agroecosyst.* **2012**, *92*, 207–224. [CrossRef]
44. Patel, R.V.; Pa, K.Y. Effect of hydropriming and biopriming. *Res. J. Agric. For.* **2017**, *5*, 1–14.
45. Latique, S.; Mohamed Aymen, E.; Halima, C.; Chérif, H.; Mimoun, E.K. Alleviation of salt stress in durum wheat (*Triticum durum* L.) seedlings through the application of liquid seaweed extracts of *Fucus spiralis*. *Commun. Soil Sci. Plant Anal.* **2017**, *48*, 2582–2593. [CrossRef]
46. du Jardin, P.; Xu, L.; Geelen, D. Agricultural Functions and Action Mechanisms of Plant Biostimulants (PBs) an Introduction. *Chem. Biol. Plant Biostimulants* **2020**, 1–30. [CrossRef]
47. Yang, Y.; Guo, Y. Elucidating the molecular mechanisms mediating plant salt-stress responses. *New Phytol.* **2018**, *217*, 523–539. [CrossRef] [PubMed]

48. Khosravinejad, F.; Heydari, R.; Farboodnia, T. Effect of salinity on organic solutes contents in barley. *Pak. J. Biol. Sci.* **2019**, *12*, 158–162. [CrossRef] [PubMed]
49. Di Filippo-Herrera, D.A.; Hernández-Carmona, G.; Muñoz-Ochoa, M.; Arvizu-Higuera, D.L.; Rodríguez-Montesinos, Y.E. Monthly variation in the chemical composition and biological activity of *Sargassum horridum*. *Bot. Mar.* **2018**, *61*, 91–102. [CrossRef]
50. Arioli, T.; Mattner, S.W.; Winberg, P.C. Applications of seaweed extracts in Australian agriculture: Past, present and future. *J. Appl. Phycol.* **2015**, *27*, 2007–2015. [CrossRef]
51. Battacharyya, D.; Babgohari, M.Z.; Rathor, P.; Prithiviraj, B. Seaweed extracts as biostimulants in horticulture. *Sci. Hortic.* **2015**, *196*, 39–48. [CrossRef]
52. Hernández-Herrera, R.M.; Santacruz-Ruvalcaba, F.; Zañudo-Hernández, J.; Hernández-Carmona, G. Activity of seaweed extracts and polysaccharide-enriched extracts from *Ulva lactuca* and *Padina gymnospora* as growth promoters of tomato and mung bean plants. *J. Appl. Phycol.* **2016**, *28*, 2549–2560. [CrossRef]
53. Oude Essink, G.H.; Van Baaren, E.S.; De Louw, P.G. Effects of climate change on coastal groundwater systems: A modeling study in the Netherlands. *Water Resour. Res.* **2010**, *46*, 1–16. [CrossRef]
54. Shahi, C.; Bargali, K.; Bargali, S.S. Assessment of salt stress tolerance in three varieties of rice (*Oryza sativa* L.). *Progress. Agric.* **2015**, *6*, 50–56.
55. Bulgari, R.; Franzoni, G.; Ferrante, A. Biostimulants application in horticultural crops under abiotic stress conditions. *Agronomy* **2019**, *9*, 306. [CrossRef]
56. Abdullah, Z.K.; Khan, M.A.; Flowers, T.J. Causes of sterility in seed set of rice under salinity stress. *J. Agron. Crop Sci.* **2001**, *187*, 25–32. [CrossRef]
57. Zou, P.; Lu, X.; Zhao, H.; Yuan, Y.; Meng, L.; Zhang, C.; Li, Y. Polysaccharides derived from the brown algae *Lessonia nigrescens* enhance salt stress tolerance to wheat seedlings by enhancing the antioxidant system and modulating intracellular ion concentration. *Front. Plant Sci.* **2019**, *10*, 48. [CrossRef] [PubMed]
58. Ward, F.; Deyab, M.; El-katony, T. *Biochemical Composition and Bioactivity of Dictyota from Egypt*; LAP LAMBERT Academic Publishing: Sunnyvale, CA, USA, 2017.
59. Irakoze, W.; Prodjinoto, H.; Nijimbere, S.; Rufyikiri, G.; Lutts, S. NaCl and Na_2SO_4 salinities have different impact on photosynthesis and yield-related parameters in rice (*Oryza sativa* L.). *Agronomy* **2020**, *10*, 864. [CrossRef]
60. Razzaq, A.; Ali, A.; Safdar, L.B.; Zafar, M.M.; Rui, Y.; Shakeel, A.; Shaukat, A.; Ashraf, M.; Gong, W.; Yuan, Y. Salt stress induces physiochemical alterations in rice grain composition and quality. *J. Food Sci.* **2020**, *85*, 14–20. [CrossRef] [PubMed]
61. Calingacion, M.; Laborte, A.; Nelson, A.; Resurreccion, A.; Concepcion, J.C.; Daygon, V.D.; Mumm, R.; Reinke, R.; Dipti, S.; Bassinello, P.Z.; et al. Diversity of global rice markets and the science required for consumer-targeted rice breeding. *PLoS ONE* **2014**, *9*, e85106. [CrossRef]
62. Lee, I.S.; Lee, J.O.; Ge, L. Comparison of terrestrial laser scanner with digital aerial photogrammetry for extracting ridges in the rice paddies. *Surv. Rev.* **2009**, *41*, 253–267. [CrossRef]

Article

Effects of Oak Leaf Extract, Biofertilizer, and Soil Containing Oak Leaf Powder on Tomato Growth and Biochemical Characteristics under Water Stress Conditions

Nawroz Abdul-razzak Tahir [1,*], Kamaran Salh Rasul [1], Djshwar Dhahir Lateef [2] and Florian M. W. Grundler [3]

1 Horticulture Department, College of Agricultural Engineering Sciences, University of Sulaimani, Sulaimani 46001, Iraq
2 Crop Science and Biotechnology Department, College of Agricultural Engineering Sciences, University of Sulaimani, Sulaimani 46001, Iraq
3 INRES—Molecular Phytomedicine, University of Bonn, Karlrobert-Kreiten-Str. 13, D-53115 Bonn, Germany
* Correspondence: nawroz.tahir@univsul.edu.iq; Tel.: +964-7701965517

Abstract: Drought stress is one of the most significant abiotic stresses on the sustainability of global agriculture. The finding of natural resources is essential for decreasing the need for artificial fertilizers and boosting plant growth and yield under water stress conditions. This study used a factorial experimental design to investigate the effects of oak leaf extract, biofertilizer, and soil containing oak leaf powder on the growth and biochemical parameters of four tomato genotypes under water stress throughout the pre-flowering and pre-fruiting stages of plant development. The experiment had two components. The first component represented the genotypes (two sensitive and two tolerant), while the second component represented the treatment group, which included irrigated plants (SW), untreated and stressed plants (SS), treated plants with oak leaf powder and stressed (SOS), treated plants with oak leaf powder and oak leaf extract and stressed (SOES), and treated plants with oak leaf powder and biofertilizers and stressed (SOBS). When compared with irrigated or control plants, drought stress under the treatments of SS, SOS, SOES, and SOBS conditions at two stages and their combination significantly lowered shoot length (12.95%), total fruit weight per plant (33.97%), relative water content (14.05%), and total chlorophyll content (26.30%). The reduction values for shoot length (17.58%), shoot fresh weight (22.08%), and total fruit weight per plant (42.61%) were significantly larger in two sensitive genotypes compared with tolerant genotypes, which recorded decreasing percentages of 8.36, 8.88, and 25.32% for shoot length, shoot fresh weight, and total fruit weight per plant, respectively. Root fresh weight and root dry weight of genotypes treated with SS, SOS, SOES, and SOBS, on the other hand, increased in comparison with control plants. Tomato fruits from stressed plants treated with SS, SOS, SOES, and SOBS had considerably higher levels of titratable acidity, ascorbic acid, and total phenolic compounds than irrigated plants during all stress stages. Under water stress conditions, the addition of oak leaf powder to soil, oak leaf extract, and biofertilizer improved the biochemical content of leaves in all genotypes. Furthermore, leaf lipid peroxidation was lower in plants treated with SOES and SOBS, and lower in the two tolerant genotypes than in the two susceptible genotypes. In conclusion, the application of SOS, SOES, and SOBS demonstrated a slight decrease in some morpho-physiological and fruit physicochemical traits compared with SS treatment. However, the application of oak leaf powder and oak leaf extract can be described as novel agricultural practices because they are low-cost, easy to use, time-consuming, and can meet the growing demands of the agricultural sector by providing environmentally sustainable techniques for enhancing plant resistance to abiotic stress. The usage of the combination of leaf crude extract, oak leaf powder, and arbuscular mycorrhizal fungus should be investigated further under stress conditions.

Keywords: drought; *Solanum lycopersicum*; biostimulation; plant tissue; plant response; enhancement of tolerance

Citation: Tahir, N.A.-r.; Rasul, K.S.; Lateef, D.D.; Grundler, F.M.W. Effects of Oak Leaf Extract, Biofertilizer, and Soil Containing Oak Leaf Powder on Tomato Growth and Biochemical Characteristics under Water Stress Conditions. *Agriculture* **2022**, *12*, 2082. https://doi.org/10.3390/agriculture12122082

Academic Editors: Daniele Del Buono, Primo Proietti and Luca Regni

Received: 17 November 2022
Accepted: 2 December 2022
Published: 4 December 2022

1. Introduction

The tomato (*Solanum lycopersicum* L.) belongs to the Solanaceae family, which includes nearly 2800 species, and is one of the world's most important vegetables and crops [1,2]. Its production has increased continuously, reaching nearly 186 million tons of fresh fruit in 2020 [3]. It is consumed as a fresh or processed fruit because of its high nutritional value, which includes vitamins, folate, and phytochemicals [4]. Tomatoes are also considered a perfect fleshy fruit model system because they can be easily grown under different conditions, have a short life cycle, and have simple genetics owing to their small genome and lack of gene duplication [5].

Water resources around the world have decreased as a result of climate change and global warming. Agriculture productivity is significantly impacted by water constraints around the world [6]. The plant's internal water content is affected by low soil water availability, which inhibits its physiological and biochemical functions. Despite the tomato's economic importance, it is susceptible to drought stress, especially during its blooming and fruit enlargement phases [7,8], which prevents seed germination, slows down plant development, and lowers fruit yields [9]. Additionally, little is known about the crucial role of stress-responsive genes, the processes behind their response to abiotic pressures, and the mechanisms underlying their response to biotic challenges [10].

An understanding of how plants respond to fluctuations in environmental conditions is crucial for predicting plant and ecosystem responses to climate change [11]. The plant's response to drought stress is highly dependent on the duration and severity of the stress, but is also influenced by the plant's genotype and its developmental stage [12]. The plants change their cellular activities by producing different defense mechanisms in response to water stress. Drought causes osmotic stress, which can result in turgor loss, membrane deterioration, protein degradation, and often high amounts of reactive oxygen species (ROS), which cause tissue oxidative damage [11]. The antioxidant enzyme systems are produced by some antioxidant enzymes and osmotic substances such as soluble sugars, proteins, and free prolyls, which scavenge these ROS and protect macromolecules in plant cells [13]. Plants adopt different strategies, including the accumulation of some substances with the capability to retain water, such as proline, compatible solutes, and those that evade water deficits by modifying water consumption such as root system traits and C3/C4 or CAM photosynthesis [11,14]. Stressed plants produce some important metabolites, like organic acids, polyamines, amino acids, and lipids, which moderately alleviate stress by acting as osmoregulators, antioxidants, and defense compounds [15]. Some protein kinases are turned on in most plants when they are under water stress. These include mitogen-activated protein kinases (MAPKs), calcium-dependent protein kinases (CDPKs), calcineurin B-like (CBL)-interacting protein kinases (CIPKs), and members of the sucrose non-fermenting-1 (SNF1)-related protein kinase 2 (SnRK2) family [16,17].

The addition of plant tissue to soil improves soil quality by reducing the risk of soil erosion and increasing crop yields [18]. Plant tissue application also plays a crucial role in sustaining and improving the chemical, physical, and biological properties of the soil by providing mineral nutrients and protecting the soil's water content [19] and may have an effect on plant water uptake [20]. Silicon (Si) is a nutritional mineral in the plant residue that promotes plant growth and development, particularly under dry conditions. Si ameliorates osmotic and ionic stressors associated with drought [21]. Si-treated plants maintained stomatal conductance and transpiration rate, leaf relative water content, as well as root and whole-plant hydraulic conductivity [22].

Natural biofertilizer is a product made from living microorganisms that are extracted from cultivated or root soil. It is safe for the environment and soil health, and it is essential for atmospheric nitrogen fixation and phosphorus solubilization, which leads to increased nutrient uptake and tolerance to drought and moisture stress [23]. Rhizobacteria that promote plant growth (PGPR) are a favorable interaction between microbes and plants that can speed up plant growth. One category of rhizobacteria consists of *Bacillus* species, which support plant growth, increase nutrient availability, increase the production of plant hormones, generate volatiles, and lessen the effects of drought [24,25]. Several studies analyzed the chemical profile of leaves of different oak species and confirmed the presence of several chemical elements, including phenolic, flavonoid, and terpenoid substances. In addition, they demonstrated significant radical scavenging, antibacterial, and antitopoisomerase activity [26–30]. To the best of our knowledge, no research has been conducted on the use of oak leaf extract and powder as biostimulator factors in water stress situations.

Owing to the presence of high amounts of chemical compounds related to growth and antioxidant activity, the hypothesis of this study was to test and determine the biological activity of oak tissues. The goal of this study was to determine the effects of oak leaf extract, biofertilizer, and soil incorporating oak leaf powder on the growth and biochemical traits of four tomato genotypes under water stress conditions during two stages of plant development. This research will help farmers find new ways to use oak leaf powder and extract because they are cheap, easy to use, and do not take much time. They can also meet the growing needs of the agricultural industry by providing environmentally friendly ways to make plants more resistant to abiotic stress.

2. Materials and Methods

2.1. Plant Materials

This study used two susceptible tomato genotypes, Braw and Yadgar, and two tolerant tomato genotypes, Raza Pashayi and Sandra, based on the results of in vitro tests of 64 tomato genotypes to drought stress by polyethylene glycol-MW 6000 (unpublished data). The tomato genotypes were collected from the Agricultural Research Center of the Ministry of Agriculture and Water Resources in Kurdistan, Iraq.

2.2. Experimental Design Components, Plant Treatement, and Growth Conditions

The experiment is divided into three groups. The plants in Group 1 were stressed before flowering. The plants in the second group were stressed prior to fruiting. The third category includes plants that were stressed before flowering and fruiting. To conduct this investigation, a factorial completely randomized design (CRD) with two components was applied. The first component represented tomato genotypes (two sensitive and two tolerant) and the second component represented the treatment group, which consisted of irrigated plants (SW), stressed plants (SS), stressed plants + oak leaf powder (SOS), stressed plants + oak leaf powder + oak leaf extract (SOES), and stressed plants + oak leaf powder + biofertilizers (SOBS). Seeds of four genotypes were planted in plastic trays in a plastic house. Fully developed and healthy oak leaves (*Quercus aegilops* Oliv.) were gathered at the vegetative stage on 17 May 2021, dried, and ground into powder for the SOS, SOES, and SOBS treatments. The seedlings were transplanted into the plastic pots (40 cm in height and 18 cm in diameter). The pots for SW and SS treatments contained only 10 kg of soil, whereas the pots for SOS, SOES, and SOBS contained 10 kg of soil and 80 g of oak leaf powder. Each treatment was composed of eight replications (eight plants) (Figure S1).

To make the extract of oak leaf, 60 g of powdered oak leaves were dissolved in 1 L of distilled water, shaken for 3 h, and then incubated overnight at 5 °C [31,32]. After centrifuging for 30 min at 4000 rpm, the supernatant was collected and diluted (1:29 v/v) with distilled water. This extract was applied four times by foliar spray before flowering (first stress stage) and fruiting (second stress stage) with three-day intervals. Leaf extract was sprayed before flowering on 7 June, 10 June, 13 June, and 16 June 2021 and before fruiting on 15 July, 18 July, 21 July, and 24 July 2021. For biofertilizer treatment, 40 mg per plant of Fulzyme Plus (JH Biotech.; Inc.; USA) was applied as fertigation three times in 15 days. This biofertilizer consisted of beneficial bacteria like *Bacillus subtilis* and *Pesudomonas putida* (2×10^{10} g); enzymes like protease, amylase, lipase, and chitinase; and hormones like gibberellin (0.3%) and cytokinin (0.3%). Water stress at 40% of field capacity was applied before flowering (the first stress stage) for six days and fruiting (the second stress stage) for four days [7]. The plants grew over the spring and summer sessions of 2021. The average daytime and nighttime relative humidity in the greenhouse during the experiment was 42.84/17.17% and the average temperature was 39.55/23.59 °C. Plants were kept in a regular photoperiod with 14 h of natural light per day. Weeds were physically eliminated during the plant's growing stage, and unhealthy or dried leaves were taken out.

The soil in the experiment was silty clay in texture, with an EC of 0.61 dS m^{-1}, a pH of 7.5, an organic matter content of 17.79 g kg^{-1}, a total nitrogen content of 15.56 g kg^{-1}, a phosphorus content of 4.44 mg kg^{-1}, an available potassium content of 0.16 meq L^{-1}, and an exchangeable phosphorus content of 0.2 mg kg^{-1}.

2.3. Evaluation of Morphological and Physiological Parameters

Plant morphological data from eight plants per treatment, including shoot length (SL in cm), shoot fresh weight (SFW in g), shoot dry weight (SDW in g), root length (RL in cm), root fresh weight (RFW in g), root dry weight (RDW in g), and fruit weight per plant (FWT in g), were measured at the end of the stress period. The total chlorophyll content of the leaves of eight plants (TCC in SPAD) was determined using a SPAD-meter at the end of the stress period. Using the method outlined by Lateef et al. [33], the relative water content (RWC in %) of the leaves was estimated using six leaves from eight tomato plants harvested at the end of the stress period.

2.4. Tomato Leaves' and Fruits' Collection

At the end of the stress point, fresh tomato leaves were collected, ground using liquid nitrogen, and frozen at −20 °C for use in biochemical investigations. Tomato fruits were hand-harvested at full maturity and stored at −20 °C for use in tomato fruit quality tests.

2.5. Moisture Content, Titratable Acidity, and Total Soluble Solid Measurement

The moisture content (MC) of eight plants was estimated by weighing 10 g of fresh tomato fruit and then drying the samples at 70 °C for 72 h until a consistent weight was achieved. The weight of the dry samples was determined and the MC percentage was calculated using the following equation [34,35]:

$$\text{MC }(\%) = \frac{\text{FW} - \text{DW}}{\text{DW}} \times 100$$

where MC is the moisture content of tomato fruit, FW is the fresh weight of tomato fruit, and DW is the dry weight of tomato fruit.

Titratable acidity (TA) was determined by combining 3 mL of tomato juice with two to three drops of phenolphthalein and titrating the mixture with 0.1 N NaOH [36]. TA was computed using the following formula:

$$\text{TA }(\%) = \frac{\text{Volume of titrant } \times \text{ N (NaOH)} \times \text{ Acid equivilent}}{\text{Volume of used juice } \times 1000} \times 100$$

Total soluble solids (TSSs, Brix) was determined using a digital refractometer [34,35]. Fruits of six plants from each level of treatments were subjected to this test.

2.6. Measurement of Biochemical Traits

2.6.1. Ascorbic Acid Content (ASC)

Ascorbic acid content (ASC) was determined by combining 0.4 g of powdered tomato fruit tissue with 1300 µL of 1% (*w*/*v*) HCl and vigorously shaking the mixture for 30 min. The mixture was centrifuged for 10 min at 13,000× *g* rpm and the supernatant was collected. The supernatant was mixed with 1900 µL of 1% (*v*/*v*) HCl and measured at 243 nm against a blank containing 1% (*v*/*v*) of HCl [37].

2.6.2. Carotenoid Content (CAC)

One gram of powdered tomato fruit tissue was mixed with 1000 µL of 100% methanol, and the mixture was incubated overnight at 5 °C. After centrifuging the samples for 8 min at 13,000× *g* rpm, 500 µL of the supernatant was collected and mixed with 1500 µL of 100% methanol. At 470 nm, the sample was read against a blank of 100% methanol [38].

2.6.3. Soluble Sugar Content (SSC)

Using the method described by Lateef et al. [33], the concentration of soluble sugar in fresh leaves and fruits was determined.

2.6.4. Proline Content (PC)

The proline content of the fresh leaves was determined using the method of Lateef et al. [33].

2.6.5. Total Phenolic Content (TPC)

According to Lateef et al. [33], fresh fruits and leaves were tested for their total phenolic content (TPC).

2.6.6. Antioxidant Compound Capacity (AC)

The antioxidant capacity was evaluated by combining 0.1 g of ground fresh leaves with 1 mL of 60% (*v*/*v*) acidic methanol (%99 methanol + %1 HCl). After shaking the mixture for 10 min, the sample was incubated at 5 °C overnight. The mixture was centrifuged for 15 min at 12,000× *g* rpm to collect the supernatant. Using the 1-diphenyl-2-picrylhydrazyl (DPPH) method as described by Lateef et al. [33], the antioxidant capacity of supernatant (extract) was assessed.

2.6.7. Antioxidant Enzyme Activity

The activities of guaiacol peroxidase (GPA) and catalase (CAT) were determined using the procedures reported by Lateef et al. [33].

2.6.8. Lipid Peroxidation Assays

As a biomarker of membrane oxidative damage caused by the water stress, the concentration of malondialdehyde (MDA), which is the final product of lipid peroxidation, was measured [39]. This experiment was initiated by mixing an amount of grinded powder leaves (0.4 g) with 2 mL of Tris-HCl buffer solution (pH 7.4) comprising 1.5% (*w*/*v*) of polyvinylpyrrolidone (PVP). Then, the mixture was shaken well for a duration of 10 min. Afterwards, the solution mixture was centrifuged at 10,000× *g* rpm for half an hour. All of the upper layers were then taken and transferred to a glass tube. Following that, 2 mL of 0.5% (*w*/*v*) thiobarbituric acid in 20% trichloroacetic acid (*w*/*v*) was mixed with the supernatant and boiled for 31 min at 95 °C in a water bath. After heating, the samples were immediately placed in a cold-water bath to stop the reactions, and the pinkish color appeared among the samples. The reaction mixture, after centrifugation at 4000× *g* rpm for 12 min, was measured at two different wavelengths, 532 and 600 nm. The first measurement is a true measurement of the sample, while the second is for correcting unclear turbidity by subtracting the value of absorbance at 600 nm. The concentration of lipid peroxidation (LP) was stated in nmol g^{-1} seedling fresh weight:

$$LP = \frac{AB532 - AB600 \times 1000 \times VL}{EC \times WE}$$

where AB532 is the absorbance at 532 nm, AB600 is the absorbance at 600 nm, VL is the volume of extract (mL), WE is the fresh weight of the sample (g), and EC is the extinction coefficient of 155 $mM^{-1}cm^{-1}$.

2.7. GC-MS Analysis of Oak Leaf Extract

The chemical components of oak leaf extract were identified using an Agilent 7890 B gas chromatograph and an Agilent 5977 mass spectrometer, both manufactured by MSD, USA. HP-5MS UI capillary column (30 m × 0.25 × 0.25 mm) fused with 5% phenyl methyl siloxane and a splitless injector were used in a gas chromatograph. The initial temperature in the column oven was 40 °C, held steady for 60 s, and then increased to 300 °C at a rate of 10 °C per minute. To do this, we used a constant flow rate of 1 mL/min of helium as the carrier gas and heated the injector to 290 °C. In the splitless model, the injection volume was 1 mL, the purge flow was 3 mL/min, the total flow was 19 mL/min, and the pressure was 7.0699 psi. The mass spectrometer was run with the help of the Mass Hunter GC/MS Acquisition software and the Mass Hunter qualitative program, which scanned fragments in the range of 35 m/z to 650 m/z. The interface temperature (MSD transfer line) was set at 290 °C, the ionization source temperature was set at 230 °C, and the quad temperature was set at 150 °C. The solvent cut time began at 4 min and ended between 35 and 40 min.

2.8. Statistical Data Analysis

XLSTAT version 2019.2.2 (Boston, USA) was used to run statistical analyses (two-way analysis of variance, Duncan's multiple range test, and principal component analysis (PCA)) for assessing the data obtained in this study at $p \leq 0.05$ [40]. The trait index was calculated by the following formula [41]:

$$\text{Trait index } (\%) = \frac{(\text{Mean of treated and stressed plants} - \text{Mean of irrigated plants})}{\text{Mean of irrigated plants}} \times 100$$

The values of all studied traits are represented by the mean ± standard deviation (SD). Each value is the average of three replications for physicochemical parameters and eight replications for morpho-physiological traits.

3. Results

3.1. Effect of Various Treatments on the Morpho-Physiological and Fruit Physicochemical Traits of Tomato under Water Stress

Plant development and growth are essentially the results of cell division, cell enlargement, and differentiation, and they are regulated by a variety of genetic, physiological, ecological, and morphological processes, as well as their interconnections [42]. The analysis of variance on morphological characters, relative water content (RWC), and total chlorophyll content (TCC) in the first stress stage (before flowering), the second stress stage (before fruiting), and their combinations revealed that treatments had a significant effect (Table S1 and Figures S2 and S3). When compared with control plants, all levels of treatment resulted in a significant percentage decrease in shoot length (SL), shoot fresh weight (SFW), shoot dry weight (SDW), fruit weight per plant (FWT), relative water content (RWC), and total chlorophyl content (TCC). In comparison with control plants, the stressed plant group (SS) that was not exposed to powdered oak tissue, oak leaf extract, or biofertilizer at any stage had the highest decline percentages for all traits (Table 1).

According to the results of the interaction, Braw under SOBS application resulted in the highest increasing percentages of SFW (33.35%), SDW (51.30%), and RFW (145.06%) compared with the irrigated plants (SW) during the first stress stages, while Yadgar under untreated and stressful conditions (SS) resulted in the maximum decreasing values for FWT (50.38%) and RWC (18.72%) (Table S4). The interaction results showed that, during the second stress stages, Braw under SOBS application contributed to the greatest increases in SFW (5.03%), SDW (29.64%), and RFW (258.68%) compared with the control conditions, while Yadgar (48.30%) and Sandra (48.11%) under the SOS condition caused the greatest decreases in FWT and TCC. As per Table S4, the interaction outcomes demonstrated that the Sandra genotype under SOBS application contributed to the highest increases in SDW (2.74%), and RDW (255.70%) compared with SW conditions, and that Yadgar under the

SS condition caused the greatest decreases in SL (26.52%) and FWT (63.89%) during the combination of both stress stages.

Table 1. Effect of oak leaf powder, oak leaf extract, and biofertilizer on the morpho-physiological characteristics of tomato plants at various stress stages. Positive and negative values signify increasing and declining, respectively.

Increasing and Decreasing Percentages Compared with Irrigated Plants in the First Stress Stage									
Treatment	SL (%)	SFW (%)	SDW (%)	RL (%)	RFW (%)	RDW (%)	FWT (%)	RWC (%)	TCC (%)
SOBS	−6.70 a ± 5.15	1.90 a ± 20.80	7.72 a ± 27.31	4.94 a ± 16.58	74.93 a ± 50.94	99.21 a ± 84.92	−27.30 ab ± 9.53	−11.94 ab ± 3.26	−24.30 b ± 12.18
SOES	−6.54 a ± 8.47	−5.78 b ± 7.59	−3.24 b ± 8.14	0.03 ab ± 12.43	44.00 b ± 58.76	43.76 ab ± 117.82	−21.36 a ± 16.97	−9.44 a ± 8.07	−9.80 a ± 18.67
SOS	−7.13 a ± 5.52	−8.93 b ± 6.53	−11.53 c ± 6.95	−4.57 b ± 15.55	30.05 b ± 26.34	22.51 b ± 20.64	−28.59 b ± 10.07	−14.24 bc ± 3.89	−33.34 b ± 8.43
SS	−13.52 b ± 6.56	−24.76 c ± 15.11	−22.98 d ± 14.67	−16.67 c ± 11.37	16.16 b ± 29.80	26.11 b ± 52.72	−32.58 b ± 11.58	−16.75 c ± 4.69	−31.30 b ± 10.72
Increasing and Decreasing Percentages Compared with Irrigated Plants in the Second Stress Stage									
Treatment	SL (%)	SFW (%)	SDW (%)	RL (%)	RFW (%)	RDW (%)	FWT (%)	RWC (%)	TCC (%)
SOBS	−9.28 a ± 4.57	−7.95 a ± 13.31	3.02 a ± 18.52	15.70 a ± 17.39	107.57 a ± 104.78	121.80 a ± 91.08	−28.49 a ± 13.46	−10.12 a ± 4.59	−24.18 b ± 12.07
SOES	−9.04 a ± 5.33	−14.16 b ± 10.09	−5.18 b ± 11.10	5.96 b ± 15.92	94.83 a ± 86.72	104.54 a ± 82.27	−30.57 a ± 11.03	−8.87 a ± 7.42	−16.85 a ± 12.49
SOS	−13.00 a ± 6.63	−20.21 c ± 10.32	−14.03 c ± 8.70	−11.47 c ± 14.93	30.54 b ± 39.65	36.60 b ± 26.44	−35.13 b ± 10.26	−10.88 a ± 3.99	−33.29 c ± 10.34
SS	−20.56 b ± 8.62	−29.05 d ± 15.83	−25.14 d ± 14.76	−12.84 c ± 11.43	29.59 b ± 39.99	31.48 b ± 67.26	−37.64 b ± 11.47	−18.14 b ± 6.84	−31.55 c ± 10.75
Increasing and Decreasing Percentages Compared with Irrigated Plants in the First and Second Stress Stages									
Treatment	SL (%)	SFW (%)	SDW (%)	RL (%)	RFW (%)	RDW (%)	FWT (%)	RWC (%)	TCC (%)
SOBS	−13.95 a ± 7.95	−15.22 b ± 12.10	−9.28 a ± 11.43	8.52 a ± 17.42	92.02 a ± 83.34	101.46 a ± 100.45	−41.10 a ± 13.87	−15.59 a ± 6.65	−26.22 b ± 11.54
SOES	−14.57 a ± 7.57	−9.14 a ± 12.32	−8.57 a ± 8.88	2.52 ab ± 14.18	89.63 a ± 76.02	100.71 a ± 106.78	−39.72 a ± 12.98	−13.70 a ± 6.65	−16.24 a ± 14.80
SOS	−17.50 a ± 8.64	−16.48 b ± 13.13	−13.05 b ± 10.59	−3.73 b ± 9.00	51.78 b ± 52.23	62.56 b ± 117.59	−40.04 a ± 11.37	−16.87 a ± 6.45	−32.91 c ± 8.37
SS	−23.80 b ± 9.89	−36.00 c ± 23.39	−27.49 c ± 16.15	−12.98 c ± 6.97	27.08 b ± 35.52	37.11 c ± 86.64	−45.10 b ± 13.80	−22.06 b ± 5.42	−35.64 c ± 10.40

SL: shoot length, SFW: shoot fresh weight, SDW: shoot dry weight, RL: root length, RFW: root fresh weight, RDW: root dry weight, FWT: fruits weight per plant, RWC: relative water content, TCC: total chlorophyl content, SS: stressed plants that had not been treated, SOS: stressed plants that had been treated with oak leaf powder, SOES: stressed plants that had been treated with oak leaf powder and oak leaf extract, SOBS: stressed plants that had been treated with oak leaf powder and biofertilizers. Duncan's multiple range test at $p \leq 0.05$ indicates that any mean values sharing the same letter in the same column are not statistically significant. The value is represented by trait index ± standard deviation (SD). Each value is the average of eight measurements.

The analysis of variance (ANOVA) of the data reported a significant influence of the treatment on the fruit's physicochemical properties (Table S2). As stated in Table 2, the titratable acidity (TA). ascorbic acid content (ASC), and total pheolic content (TPC) responded positively to different levels of treatment in all stages of growth. In the first stress stage, the highest increasing percentages of TA, ASC, and TPC were obtained by the treatments SS (11.23%), SOBS (23.50%), and SOES (11.10%), respectively. The TA, ASC, and TPC responded favorably to various treatments during the second stress stage. The highest increasing TA (12.63%), ASC (18.49%), and TPC (12.21%) values were seen in the treatments SS, SOES, and SOBS, respectively. Similarly, when two stress measures were combined, the same results were found. In the SS and SOES applications, the highest percentage increases in TA (19.05%), ASC (13.11%), and TPC (10.42%) were shown. Under all stress conditions, a decreasing amount was also observed in the moisture content (MC), total soluble solids (TSSs), and carotenoid content (CAC). The SS application showed the largest decline in percentage in MC, TSS, and CAC. With the first stress stage, the soluble sugar content (SSC) decreased by 3.27 and 2.78% under SOBS and SOES conditions, respectively. The SSC responded favorably to the SOBS and SOES applications during the second stress stage, increasing by 1.68 and 2.73%, respectively. Under all levels of treatment (SS, SOS, SOES, and SOBS), the SSC values for both stress stages together decreased.

Table 2. Influence of oak leaf powder, oak leaf extract, and biofertilizer on the fruit physicochemical parameters of tomato plants at different stress stages. Increasing and decreasing are labeled by a positive and negative value, respectively.

Increasing and Decreasing Percentages Compared with Irrigated Plants in the First Stress Stage							
Treatment	**MC**	**TA**	**TSS**	**ASC**	**CAC**	**SSC**	**TPC**
SOBS	−0.67 a ± 0.45	2.64 bc ± 6.43	−1.73 a ± 4.35	23.50 a ± 12.40	−3.02 b ± 4.43	3.27 a ± 7.73	8.55 b ± 12.38
SOES	−0.59 a ± 0.35	1.03 c ± 6.81	−2.08 a ± 3.60	22.10 b ± 14.06	−1.80 a ± 4.25	2.78 a ± 8.68	11.10 a ± 10.93
SOS	−0.87 b ± 045	5.82 b ± 5.12	−4.43 b ± 4.29	15.92 c ± 11.81	−5.35 c ± 6.19	−2.34 b ± 7.72	9.06 b ± 10.41
SS	−1.28 c ± 0.83	11.23 a ± 6.53	−6.67 c ± 4.37	6.41 d ± 10.22	−7.64 d ± 7.61	−8.28 c ± 11.84	5.06 c ± 10.63
Increasing and Decreasing Percentages Compared with Irrigated Plants in the Second Stress Stage							
Treatment	**MC**	**TA**	**TSS**	**ASC**	**CAC**	**SSC**	**TPC**
SOBS	−0.65 a ± 0.046	3.45 b ± 7.54	−1.59 a ± 3.23	17.22 b ± 18.89	−0.03 a ± 6.67	1.68 b ± 10.98	12.21 a ± 14.70
SOES	−0.55 a ± 0.039	2.12 b ± 7.50	−2.42 a ± 3.26	18.49 a ± 16.69	−2.40 b ± 7.74	2.73 a ± 9.89	11.37 a ± 14.15
SOS	−0.87 b ± 0.44	6.00 b ± 5.91	−4.21 b ± 4.64	13.40 c ± 17.44	−4.70 c ± 8.46	−2.71 c ± 9.85	9.37 b ± 12.41
SS	−1.23 c ± 0.78	12.63 a ± 11.80	−6.51 c ± 5.14	2.29 d ± 12.86	−7.80 d ± 10.15	−8.35 d ± 13.56	4.68 c ± 9.02
Increasing and Decreasing Percentages Compared with Irrigated Plants in the First and Second Stress Stages							
Treatment	**MC**	**TA**	**TSS**	**ASC**	**CAC**	**SSC**	**TPC**
SOBS	−0.90 a ± 0.58	6.65 c ± 6.14	−3.86 a ± 4.53	12.73 a ± 16.38	−2.58 a ± 9.90	−1.49 a ± 10.12	9.12 b ± 13.75
SOES	−0.86 a ± 0.59	7.19 c ± 5.68	−4.27 a ± 4.50	13.11 a ± 17.24	−5.12 b ± 9.91	−1.69 a ± 9.79	10.42 a ± 14.22
SOS	−1.21 b ± 0.74	11.47 b ± 5.94	−5.88 b ± 5.32	7.04 b ± 17.51	−6.92 c ± 10.56	−6.41 b ± 10.34	7.65 c ± 12.05
SS	−1.55 c ± 1.01	19.05 a ± 11.13	−8.54 c ± 5.74	−3.73 c ± 14.17	−10.10 d ± 11.68	−12.31 c ± 13.16	1.78 d ± 10.46

MC: moisture content, TA: titratable acidity, TSS: total soluble solids, ASC: ascorbic acid content, CAC: carotenoid content, SSC: soluble sugar content, TPC: total phenolics content, SS: stressed plants that had not been treated, SOS: stressed plants that had been treated with oak leaf powder, SOES: stressed plants that had been treated with oak leaf powder and oak leaf extract, SOBS: stressed plants that had been treated with oak leaf powder and biofertilizers. Duncan's multiple range test at $p \leq 0.05$ indicates that any mean values sharing the same letter in the same column are not statistically significant. The value is represented by trait index ± standard deviation (SD). Each value is the average of three measurements.

A multivariate analytic technique called principal component analysis (PCA) is used to evaluate the similarity between the levels of treatment. Additionally, it is also used to determine the relationship between attributes. In total, 16 determined variables concerning the morpho-physiological and fruit physicochemical traits under four levels of treatment were subjected to a principal component analysis. Based on an eigenvalue > 1, we extracted a total of two first components with a cumulative distribution of 95.63% (85.05% for the first component and 11.59% for the second component), 96.53% (90.26% for the first component and 6.27% for the second component), and 97.04% (92.95% for the first component and 4.09% for the second component) for the first, second, and their combination stress stages, respectively (Figure 1). Different distributions of studied traits and treatments were observed on the PCA plot. Under first stress stage, the most notable contributors to the observed variance along PC1 were SL, SFW, RL, RWC, MC, TA, TSS, ASC, CAC, and SSC. However, the greatest amount of variance along PC2 was caused by SDW, RFW, RDW, FWT, TCC, and TPC (Figure 1A). The most noteworthy contributions to the observed variance along PC1 during the second stress stage were SL, SFW, SDW, FWT, MC, TSS, CAC, SSC, and TPC. Nevertheless, RL, RFW, RDW, RWC, TCC, MC, TA, and ASC were responsible for the bulk of the variation along PC2 (Figure 1B). Under both stress stages, the SL, SDW, RFW, RDW, RWC, TCC, MC, TA, TSS, ASC, SSC, and TPC were the major contributors to the observed variance along PC1. SFW, RL, FWT, and CAC, on the other hand, were responsible for the majority of the variation along PC2 (Figure 1C).

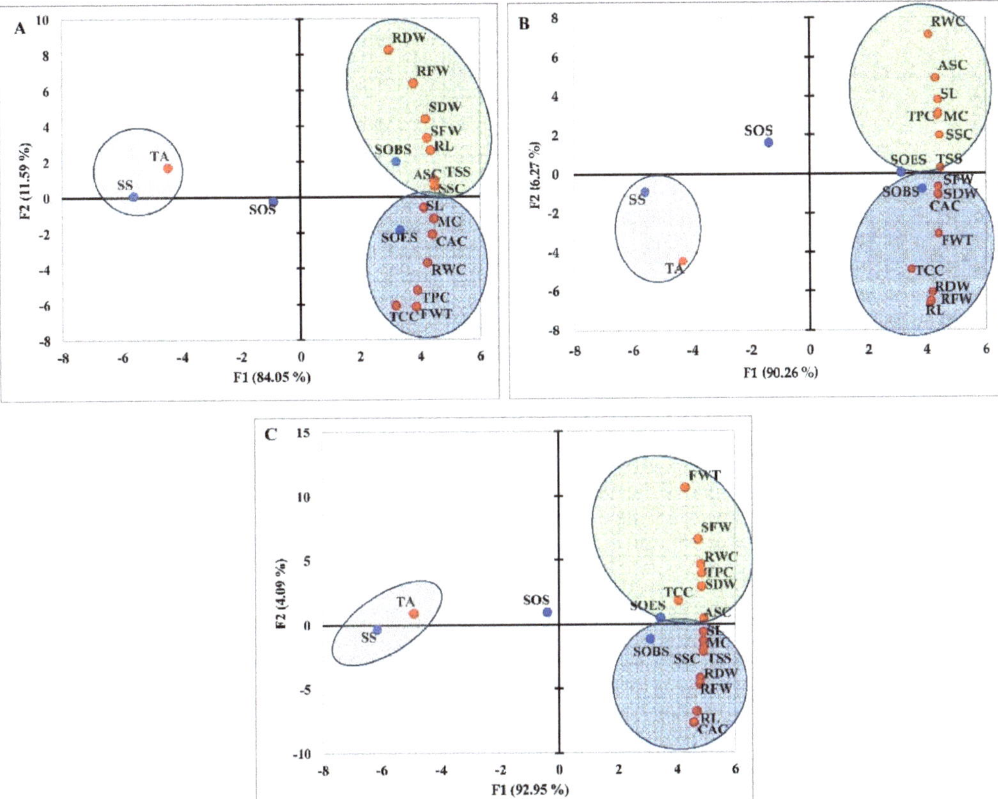

Figure 1. PCA plot showing the distribution of various morpho-physiological and fruit physico-chemical traits and treatments under first (**A**), second (**B**), and both (**C**) stress conditions. SL: shoot length, SFW: shoot fresh weight, SDW: shoot dry weight, RL: root length, RFW: root fresh weight, RDW: root dry weight, FWT: fruits weight per plant, RWC: relative water content, TCC: total chlorophyl content, MC: moisture content, TA: titratable acidity, TSS: total soluble solids, ASC: ascorbic acid content, CAC: carotenoid content, SSC: soluble sugar content, TPC: total phenolic content, SS: stressed plants that had not been treated, SOS: stressed plants that had been treated with oak leaf powder, SOES: stressed plants that had been treated with oak leaf powder and oak leaf extract, SOBS: stressed plants that had been treated with oak leaf powder and biofertilizers. F1 and F2 represent the first and second components, respectively.

The application of powdered oak leaf, leaf oak extract, and biofertilizers reduced titratable acidity (TA) in fruit in all stress stages compared with the untreated plant under stress conditions and formed the first group in the left of the PCA plot (brown outline). During the first stress stage, the characteristics of the plants treated with SOBS with high percentage values of RL, SFW, SDW, RFW, RDW, TSS, ASC, and SSC were included in the second group on the upper right quadrant (green outline) of the PCA plot. The third group in the lower right quadrant (blue outline) of the PCA plot is made up of attributes in SOES-treated plants with high SL, RWC, TCC, FWT, MC, CAC, and TPC values. Under the second stress stage, the characteristics of the plants treated with SOES that had high values of SL, TSS, MC, SSC, TPC, RWC, and ASC formed the second group in the upper right quadrant (green circle) of the PCA plot. Furthermore, the traits in the plants treated with SOBS with high percentage values of RL, RFW, RDW, SFW, SDW, FWT, TCC, and CAC were included in the third group on the lower right quadrant (blue circle) of the PCA plot. In the combination of both stress stages, traits with high percentage values of SFW, SDW, FWT, RWC, TCC, TPC, and ASC in plants treated with SOES comprised the second group in the upper right quadrant (green circle) of the PCA plot. The third group was in the lower right quadrant (blue outline) of the PCA plot. It was made up of plants treated with SOBS and having high values of SL, RL, RFW, RDW, MC, CAC, TSS, and SSC.

3.2. *Influence of Genotypes on the Morpho-Physiological and Physicochemical Characteristics of Tomato Fruit under Application of SS, SOS, SOES, and SOBS*

Under conditions of water stress, analysis of variance (ANOVA) revealed highly significant genotype effects on the morpho-physiological traits of the first stress stage (before blooming), the second stress stage (before fruiting), and their combinations (Table S1). Shoot length (SL), shoot fresh weight (SFW), shoot dry weight (SDW), fruit weight per plant (FWT), relative water content (RWC), and total chlorophyll content (TCC) were all significantly lower in all genotypes as compared with control plants. SL (13.13%), SFW (17.96%), SDW (17.83%), FWT (42.63%), and RWC (15.68%) exhibited the largest decreasing percentages in the stressed Yadgar genotype. In all stress stages, the tolerant genotypes (Raza Pashayi and Sandra) had lower decreasing amounts of SL and FWT than the sensitive genotypes (Braw and Yadgar) (Table 3). Root fresh weight (RFW) and root dry weight (RDW) demonstrated high increasing percentages in four genotypes for all stress levels under water stress circumstances.

Table 3. Impact of tomato genotypes treated with oak leaf powder, oak leaf extract, and biofertilizer at different stress stages on the morpho-physiological traits. Increasing and declining percentages are represented by positive and negative values, respectively.

Increasing and Decreasing Percentages Compared with Irrigated Plants during the First Stress Stage									
Genotypes	**SL (%)**	**SFW (%)**	**SDW (%)**	**RL (%)**	**RFW (%)**	**RDW (%)**	**FWT (%)**	**RWC (%)**	**TCC (%)**
Raza Pashayi	−4.98 a ± 5.24	−5.75 a ± 2.20	−3.79 b ± 1.95	−6.52 bc ± 11.01	34.19 b ± 17.88	53.76 ab ± 16.65	−19.47 a ± 6.44	−11.75 ab ± 4.63	−31.53 b ± 8.38
Sandra	−5.14 a ± 5.81	−8.03 a ± 7.99	−9.21 c ± 9.88	−5.18 b ± 7.93	36.34 b ± 60.49	98.55 a ± 146. 60	−21.26 ab ± 14.64	−10.13 a ± 8.03	−32.62 b ± 23.06
Braw	−10.65 b ± 6.62	−5.83 a ± 30.20	0.79 a ± 35.77	−14.79 c ± 13.11	66.29 a ± 64.37	24.59 b ± 41.84	−26.48 b ± 5.15	−14.81 b ± 5.28	−17.60 a ± 8.38
Yadgar	−13.12 b ± 7.06	−17.96 b ± 9.03	−17.83 d ± 9.88	10.23 a ± 19.20	28.30 b ± 25.01	14.68 b ± 16.65	−42.63 c ± 6.24	−15.68 b ± 3.01	−16.98 a ± 11.40
Increasing and Decreasing Percentages Compared with Irrigated Plants during the Second Stress Stage									
Genotypes	**SL (%)**	**SFW (%)**	**SDW (%)**	**RL (%)**	**RFW (%)**	**RDW (%)**	**FWT (%)**	**RWC (%)**	**TCC (%)**
Raza Pashayi	−8.71 a ± 4.44	−6.32 a ± 1.83	−4.32 a ± 1.17	−2.97 bc ± 17.02	43.51 b ± 27.94	76.21 b ± 24.00	−18.68 a ± 8.55	−12.57 ab ± 4.76	−31.19 c ± 7.80
Sandra	−9.35 a ± 5.40	−13.32 b ± 8.57	−7.98 a ± 10.95	−11.49 c ± 10.65	53.32 b ± 51.17	158.73 a ± 94.25	−29.87 b ± 4.25	−8.93 a ± 8.25	−39.90 d ± 8.22
Braw	−18.44 b ± 9.70	−21.72 c ± 20.98	−6.05 a ± 28.75	−0.54 b ± 20.29	150.13 a ± 110.13	52.48 b ± 66.73	−36.66 c ± 3.49	−14.55 b ± 4.13	−14.13 a ± 9.99
Yadgar	−15.39 b ± 6.74	−30.00 d ± 6.22	−22.97 b ± 7.07	12.36 a ± 20.13	15.57 b ± 22.11	6.99 c ± 11.82	−46.60 d ± 6.52	−11.96 ab ± 8.40	−20.65 b ± 7.58
Increasing and Decreasing Percentages Compared with Irrigated Plants during the First and Second Stress Stages									
Genotypes	**SL (%)**	**SFW (%)**	**SDW (%)**	**RL (%)**	**RFW (%)**	**RDW (%)**	**FWT (%)**	**RWC (%)**	**TCC (%)**
Raza Pashayi	−9.97 a ± 3.90	−6.23 a ± 1.68	−3.28 a ± 1.61	0.14 b ± 11.89	36.62 c ± 23.30	53.07 b ± 26.86	−25.90 a ± 5.29	−22.04 c ± 2.75	−33.06 c ± 7.58
Sandra	−12.01 a ± 6.33	−13.64 b ± 9.47	−10.01 b ± 10.39	−9.82 c ± 8.32	73.29 b ± 45.59	238.51 a ± 47.09	−36.77 b ± 3.01	−15.54 ab ± 6.41	−40.04 d ± 6.58
Braw	−27.55 c ± 7.28	−24.46 c ± 29.40	−18.78 c ± 18.75	−4.75 bc ± 14.41	143.07 a ± 75.82	12.30 c ± 48.49	−44.83 c ± 4.28	−18.42 bc ± 5.54	−15.80 a ± 10.58
Yadgar	−20.29 b ± 6.15	−32.50 d ± 8.00	−26.31 d ± 6.58	8.77 a ± 16.95	7.52 d ± 18.88	−2.04 c ± 17.34	−58.46 d ± 5.43	−12.23 a ± 8.06	−22.11 b ± 13.26

SL: shoot length, SFW: shoot fresh weight, SDW: shoot dry weight, RL: root length, RFW: root fresh weight, RDW: root dry weight, FWT: fruits weight per plant, RWC: relative water content, TCC: total chlorophyl content. Any mean values sharing the same letter in the same column are not statistically significant, according to Duncan's multiple range test at $p \leq 0.05$. The values are represented by the standard deviation of the trait index. Each value is the average of eight measurements. The value is represented by trait index ± standard deviation (SD).

The analysis of variance (ANOVA) of the data obtained for the fruit physicochemical traits found a significant genotype effect (Table S2). According to Table 4, all stress stages contributed to a reduction in the four genotypes' moisture content (MC). The fruit of tolerant genotypes showed higher increasing values in the ASC, CAC, and TPC characteristics than sensitive genotypes under all stress stages. Under all stages of stress, the TA was higher in sensitive genotypes than in tolerant genotypes. Additionally, the Sandra genotype showed an increase in CAC of 2.22, 5.01, and 6.95% for the first, second, and both of them together, respectively.

In accordance with Table S5, the mean pairwise comparison for the interaction of genotypes and different treatments showed that Sandra had the highest increasing percentages in CAC (3.18%) and SSC (16.32%) in the presence of the SOES application, followed by Raza Pashayi with the highest increasing percentages in ASC (36.58%) and TPC (22.43%). With the exception of TA with the treatment of SS and SOS, the Braw genotype reported the highest declining values in all physicochemical parameters under the first stress stage. The Sandra genotype registered the largest percentage increases in TSS (2.63%), ASC (38.21%), CAC (6.02%), and SSC (16.67%) compared with irrigated plants (Table S5), while the Braw genotype showed declining trends in all physicochemical measures except TA under the second stress stage. Sandra had the largest increasing percentages in TSS (1.75%), CAC (9.47%), and SSC (11.81%) with SOBS application, followed by Raza Pashayi in ASC (28.71%) and TPC (27.20%) in the presence of SOES application during both stress stages (Table S5).

Table 4. Effect of tomato genotypes treated at various stress stages with oak leaf powder, oak leaf extract, and biofertilizer on the fruit physicochemical traits. Increasing and decreasing percentages are indicated by positive and negative values, respectively.

Increasing and Decreasing Percentages Compared with Irrigated Plants during the First Stress Stage							
Genotypes	**MC (%)**	**TA (%)**	**TSS (%)**	**ASC (%)**	**CAC (%)**	**SSC (%)**	**TPC (%)**
Raza Pashayi	−0.45 a ± 0.14	−1.49 b ± 9.61	−2.06 b ± 2.35	29.29 a ± 6.52	−0.84 b ± 1.37	5.73 b ± 3.59	19.82 a ± 2.46
Sandra	−0.35 a ± 0.18	5.97 a ± 5.27	1.32 a ± 2.84	25.46 b ± 10.82	2.22 a ± 0.73	9.07 a ± 3.84	14.18 b ± 2.72
Braw	−1.46 c ± 0.68	7.46 a ± 6.65	−6.68 c ± 2.81	0.26 d ± 5.12	−8.24 c ± 3.86	−9.45 c ± 8.47	−8.43 d ± 3.12
Yadgar	−1.14 b ± 0.25	8.77 a ± 5.84	−7.49 c ± 3.07	12.93 c ± 6.40	−10.95 d ± 3.73	−9.94 c ± 4.69	8.20 c ± 1.97
Increasing and Decreasing Percentages Compared with Irrigated Plants during the Second Stress Stage							
Genotypes	**MC (%)**	**TA (%)**	**TSS (%)**	**ASC (%)**	**CAC (%)**	**SSC (%)**	**TPC (%)**
Raza Pashayi	−0.39 a ± 0.20	−3.71 c ± 5.59	−1.93 b ± 1.54	24.72 b ± 4.60	2.04 b ± 1.51	2.73 b ± 2.79	24.29 a ± 6.83
Sandra	−0.33 a ± 0.17	6.01 b ± 6.63	1.32 a ± 2.09	27.38 a ± 12.61	5.01 a ± 1.03	12.67 a ± 4.54	15.19 b ± 4.13
Braw	−1.44 c ± 0.57	14.87 a ± 7.93	−6.15 c ± 3.28	−11.16 d ± 2.62	−13.05 d ± 2.80	−15.41 d ± 7.74	−8.02 d ± 1.74
Yadgar	−1.14 b ± 0.29	7.03 b ± 5.53	−7.97 d ± 3.12	10.46 c ± 8.75	−8.92 c ± 7.82	−6.64 c ± 4.02	6.18 c ± 1.03
Increasing and Decreasing Percentages Compared with Irrigated Plants during the First and Second Stress Stages							
Genotypes	**MC (%)**	**TA (%)**	**TSS (%)**	**ASC (%)**	**CAC (%)**	**SSC (%)**	**TPC (%)**
Raza Pashayi	−0.42 a ± 0.20	3.12 c ± 5.58	−2.90 b ± 3.32	20.78 a ± 7.76	0.19 b ± 2.75	0.33 b ± 3.46	21.28 a ± 6.70
Sandra	−0.65 b ± 0.14	9.65 b ± 5.28	0.24 a ± 2.71	20.57 a ± 9.39	6.95 a ± 1.89	7.81 a ± 4.09	12.68 b ± 3.67
Braw	−2.14 d ± 0.70	17.57 a ± 10.89	−8.79 c ± 1.59	−17.15 c ± 6.30	−16.17 c ± 4.31	−18.87 d ± 7.90	−11.39 d ± 2.23
Yadgar	−1.31 c ± 0.32	14.01 a ± 5.66	−11.09 d ± 2.59	4.96 b ± 6.22	−15.68 c ± 2.75	−11.17 c ± 3.38	6.39 c ± 1.82

MC: moisture content, TA: titratable acidity, TSS: total soluble solids, ASC: ascorbic acid content, CAC: carotenoid content, SSC: soluble sugar content, TPC: total phenolics content. Any mean values sharing the same letter in the same column are not statistically significant, as determined by the Duncan's multiple range test at $p \leq 0.05$. The value is represented by trait index ± standard deviation (SD). Each value is the average of three measurements.

3.3. Impact of Various Treatments on the Biochemical Responses of the Leaves of Tomato Plants under Conditions of Water Stress

To gain a better understanding of the mechanism of tolerance in plants treated with SS, SOS, SOES, and SOBS under water deficit stress, a number of biochemical measurements were performed on the leaves of tomato plants. As shown in Table S3, significant variations were detected among different levels of treatment for all biochemical characters of the leaves of the tomato under all stress stages. The maximum values of proline content (PC), soluble sugar content (SSC), guaiacol peroxidase (GPA), and catalase (CAT) were recorded by the tomato plants treated with SOES, while the highest values of total phenolic content

(TPC) and antioxidant activity (AC) were observed by the plants treated with SOBS under the first and second stress stages. Moreover, under the combination of first and second stress stages, the plants treated with SOBES displayed the greatest values of all biochemical traits, with the exception of the LP trait. Furthermore, the control plants (SW) exhibited the minimum values of all chemical characters of the leaves of tomato under all stress stages. Low amounts of lipid peroxidation were observed by SW (5.24 nmol g^{-1} FLW), followed by SOES (7.15 nmol g^{-1} FLW) and SOBS (8.46 nmol g^{-1} FLW), under the first, second, and their combination stress stages (Table 5).

Seven different variables relating to the biochemical parameters of leaves treated with SW, SS, SOS, SOES, and SOBS were subjected to a principal component analysis (PCA). Based on an eigenvalue greater than one, the first two components displayed cumulative distributions of 93.53, 93.35, and 98.11% for the first, second, and their combined stress stages, respectively (Figure 2A–C). The biochemical characteristics and treatments were dispersed in various ways across the PCA plot throughout the first, second, and combined stages of stress. The characteristics that had the most significance in affecting the observed variance along PC1 were PC, SSC, TPC, AC, GPA, and CAT. However, the LP characteristic was the primary driver of variance along PC2. In comparison with untreated plants (SS), the application of SOBS, SOES, and SOS reduced the amount of lipid peroxidation in the leaves during all stages of stress. On the right side (blue circle) of the PCA plot, characteristics of SOBS- and SOES-treated plants with high PC, TPC, AC, SSC, GPA, and CAT values were noted throughout the first, second, and their combined stress stages. On the other hand, the plants that received SS treatment produced more LP (brown circle).

Table 5. Impact of oak leaf powder, oak leaf extract, and biofertilizer on the biochemical characteristics of the leaves of tomato plants under various stress stages.

Treatment	PC (µg g^{-1})	SSC (µg g^{-1})	TPC (µg g^{-1})	AC (µg g^{-1})	LP (nmol g^{-1})	GPA (units min^{-1} g^{-1})	CAT (units min^{-1} g^{-1})
First Stress Stage							
SOBS	1546.37 b ± 503.08	569.04 b ± 99.21	433.90 a ± 98.38	1010.20 a ± 173.44	8.46 c ± 1.13	0.26 b ± 0.06	139.61 b ± 42.49
SOES	1956.50 a ± 489.76	612.64 a ± 109.34	399.21 b ± 90.59	1006.99 b ± 175.26	7.15 d ± 0.98	0.34 a ± 0.06	160.71 a ± 56.00
SOS	1322.91 c ± 619.10	524.14 c ± 96.68	344.91 c ± 57.07	966.79 c ± 171.26	11.05 b ± 2.24	0.25 b ± 0.08	118.51 c ± 65.65
SS	1307.65 d ± 578.09	417.19 d ± 108.74	325.57 d ± 56.09	892.80 d ± 94.45	13.10 a ± 2.26	0.16 c ± 0.09	87.66 d ± 45.71
SW	1054.58 e ± 425.20	374.14 e ± 91.47	312.23 e ± 63.52	893.31 d ± 129.46	5.24 e ± 0.78	0.13 d ± 0.06	64.94 e ± 42.26
Second Stress Stage							
Treatment	**PC (µg g^{-1})**	**SSC (µg g^{-1})**	**TPC (µg g^{-1})**	**AC (µg g^{-1})**	**LP (nmol g^{-1})**	**GPA (units min^{-1} g^{-1})**	**CAT (units min^{-1} g^{-1})**
SOBS	2058.81 b ± 426.81	742.65 b ± 110.77	428.24 a ± 20.91	986.05 a ± 120.82	9.92 c ± 1.40	0.24 b ± 0.08	126.62 b ± 30.89
SOES	2534.00 a ± 433.44	782.93 a ± 89.71	402.68 b ± 29.48	974.43 b ± 159.49	8.17 d ± 1.68	0.33 a ± 0.08	159.09 a ± 40.27
SOS	1813.81 c ± 396.27	627.76 c ± 146.44	378.99 c ± 35.48	909.19 c ± 163.83	10.59 b ± 1.32	0.23 c ± 0.06	113.64 c ± 24.69
SS	1616.82 d ± 444.00	529.00 d ± 122.78	322.51 e ± 70.59	902.40 d ± 139.83	12.83 a ± 2.85	0.17 d ± 0.07	81.17 d ± 15.04
SW	1126.37 e ± 533.06	501.76 e ± 145.17	334.84 d ± 32.40	895.51 e ± 109.70	7.20 e ± 0.51	0.15 e ± 0.08	64.94 e ± 40.50
Combination of Both Stress Stages							
Treatment	**PC (µg g^{-1})**	**SSC (µg g^{-1})**	**TPC (µg g^{-1})**	**AC (µg g^{-1})**	**LP (nmol g^{-1})**	**GPA (units min^{-1} g^{-1})**	**CAT (units min^{-1} g^{-1})**
SOBS	2057.01 b ± 391.73	764.63 b ± 121.99	453.15 b ± 58.23	1029.29 b ± 96.55	10.67 c ± 1.11	0.30 b ± 0.11	155.84 b ± 63.92
SOES	2217.65 a ± 330.37	856.54 a ± 96.24	493.69 a ± 122.67	1092.47 a ± 120.92	8.94 d ± 1.21	0.44 a ± 0.19	217.53 a ± 90.38
SOS	1689.96 c ± 485.67	670.35 c ± 109.86	407.97 c ± 59.26	938.58 c ± 96.75	12.53 b ± 2.66	0.27 c ± 0.10	137.99 c ± 20.56
SS	1661.24 d ± 268.52	580.69 d ± 12.76	316.07 e ± 76.85	907.34 d ± 118.09	13.95 a ± 2.80	0.17 d ± 0.09	95.78 d ± 17.32
SW	1126.37 e ± 533.06	501.76 e ± 145.17	334.84 d ± 32.40	895.51 d ± 109.70	7.20 e ± 0.51	0.15 e ± 0.08	64.94 e ± 40.50

PC: proline content, SSC: soluble sugar content, TPC: total phenolic content, AC: antioxidant activity, LP: lipid peroxidation, GPA: peroxidase, CAT: catalase, SS: stressed plants that had not been treated, SOS: stressed plants that had been treated with oak leaf powder, SOES: stressed plants that had been treated with oak leaf powder and oak leaf extract, SOBS: stressed plants that had been treated with oak leaf powder and biofertilizers. Duncan's multiple range test at $p \leq 0.05$ indicates that any mean values sharing the same letter in the same column are not statistically significant. The value is represented by mean ± standard deviation (SD). Each value is the average of three measurements.

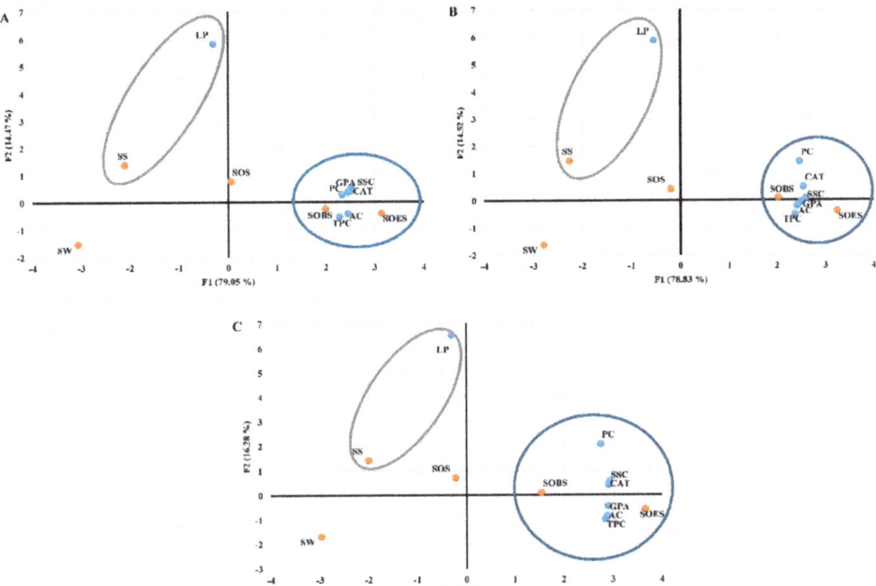

Figure 2. PCA plot illustrating the distribution of leaf biochemical characteristics and treatments under first (**A**), second (**B**), and both (**C**) stress circumstances. PC: proline content, SSC: soluble sugar content, TPC: total phenolic content, AC: antioxidant activity, LP: lipid peroxidation, GPA: peroxidase, CAT: catalase, SS: stressed plants that had not been treated, SOS: stressed plants that had been treated with oak leaf powder, SOES: stressed plants that had been treated with oak leaf powder and oak leaf extract, SOBS: stressed plants that had been treated with oak leaf powder and biofertilizers. F1 and F2 represent the first and second components, respectively.

3.4. Impact of Different Genotypes Treated with SW, SS, SOS, SOES, and SOBS on the Biochemical Responses of the Leaves of Tomato Plants under Circumstances of Water Stress

Tomato plant leaves were analyzed chemically in order to acquire a better knowledge of the mechanism of tolerance in genotypes treated with SS, SOS, SOES, and SOBS. As demonstrated in Table S3, substantial differences were identified between different genotypes for all biochemical characteristics of tomato leaves under all stress stages. The tolerant genotype Sandra had the highest values of PC, SSC, and AC during the first stress stage, whereas the tolerant genotype Raza Pashayi had the highest scores of TPC, GPA, and CAT traits. The sensitive genotype Yadgar showed the minimum values of all chemical characteristics with the exception of the LP trait. As a comparison between tolerant and sensitive genotypes, the mean values of SSC, GPA, and CAT in tolerant genotypes were higher than those obtained in sensitive plants. The highest scores of LP were found in sensitive plants (Table 6). Under the second stress stage, the tolerant genotype Sandra had the highest values of PC, TPC, AC, and CAT, while the tolerant genotype Raza Pashayi had the highest value of GPA. Except for the PC and LP features, the sensitive genotype Yadgar displayed the lowest values for all biochemical parameters (Table 6). Comparing tolerant and sensitive genotypes, the mean TPC, AC, and GPA values of tolerant genotypes were greater than those of sensitive plants. The susceptible plants (Braw and Yadgar) had the highest levels of LP. Sandra's genotype exhibited the greatest PC, AC, and CAT scores in response to both stress periods. With the exception of the LP trait, Yadgar genotypes had the lowest values for all leaf biochemical parameters (Table 6).

Table 6. Effects of oak leaf powder, oak leaf extract, and biofertilizer on the biochemical traits of the leaves of tomato plants under different levels of stress.

Genotypes	PC (μg g^{-1})	SSC (μg g^{-1})	TPC (μg g^{-1})	AC (μg g^{-1})	LP (nmol g^{-1})	GPA (units min^{-1} g^{-1})	CAT (units min^{-1} g^{-1})
First Stress Stage							
Raza Pashayi	1093.08 d ± 146.59	517.72 b ± 160.20	433.82 a ± 65.06	987.30 c ± 34.97	6.86 d ± 2.03	0.30 a ± 0.06	172.73 a ± 40.01
Sandra	2220.97 a ± 257.08	610.25 a ± 88.53	371.72 c ± 70.52	1080.95 a ± 106.77	9.39 b ± 2.34	0.26 b ± 0.12	132.47 b ± 46.10
Braw	1252.82 b ± 497.38	499.26 c ± 75.14	393.63 b ± 48.06	1020.41 b ± 103.02	9.28 c ± 3.50	0.19 c ± 0.07	93.51 c ± 59.81
Yadgar	1183.54 c ± 526.99	370.49 d ± 69.11	253.48 d ± 21.77	727.43 d ± 34.97	10.46 a ± 3.77	0.16 d ± 0.08	58.44 d ± 17.35
Second Stress Stage							
Genotypes	PC (μg g^{-1})	SSC (μg g^{-1})	TPC (μg g^{-1})	AC (μg g^{-1})	LP (nmol g^{-1})	GPA (units min^{-1} g^{-1})	CAT (units min^{-1} g^{-1})
Raza Pashayi	1620.41 c ± 526.17	606.11 c ± 181.90	378.73 b ± 51.11	1010.00 b ± 59.28	8.93 c ± 1.88	0.31 a ± 0.06	103.90 bc ± 27.77
Sandra	2278.41 a ± 491.93	658.77 b ± 100.11	414.37 a ± 27.83	1102.97 a ± 47.37	7.93 d ± 1.19	0.23 b ± 0.12	128.57 a ± 68.90
Braw	1621.13 c ± 938.94	795.56 a ± 82.71	373.60 c ± 23.28	858.92 c ± 27.47	10.77 b ± 2.48	0.19 c ± 0.07	106.49 b ± 47.70
Yadgar	1799.90 b ± 121.84	486.85 d ± 110.53	327.12 d ± 74.40	762.16 d ± 54.98	11.34 a ± 3.01	0.17 d ± 0.04	97.40 c ± 18.69
Combination of Both Stress Stages							
Genotypes	PC (μg g^{-1})	SSC (μg g^{-1})	TPC (μg g^{-1})	AC (μg g^{-1})	LP (nmol g^{-1})	GPA (units min^{-1} g^{-1})	CAT (units min^{-1} g^{-1})
Raza Pashayi	1903.64 b ± 567.18	658.83 c ± 190.89	432.28 b ± 97.33	1003.38 b ± 57.89	9.30 c ± 2.13	0.40 a ± 0.13	132.47 b ± 47.07
Sandra	2114.72 a ± 403.89	724.07 b ± 157.46	405.43 c ± 18.45	1082.30 a ± 52.67	9.01 d ± 1.28	0.29 b ± 0.21	176.62 a ± 125.44
Braw	1382.77 d ± 117.24	793.94 a ± 72.50	459.33 a ± 118.99	955.36 c ± 168.75	12.48 a ± 3.63	0.17 d ± 0.06	103.90 c ± 36.73
Yadgar	1600.67 c ± 669.52	522.35 d ± 126.03	307.53 d ± 65.85	849.50 d ± 79.51	11.85 b ± 3.14	0.20 c ± 0.06	124.68 b ± 30.28

PC: proline content, SSC: soluble sugar content, TPC: total phenolic content, AC: antioxidant activity, LP: lipid peroxidation, GPA: peroxidase, CAT: catalase. Duncan's multiple range test at $p \leq 0.05$ reveals that any mean values in the same column that share the same letter are not statistically significant. The value is represented by mean ± standard deviation (SD). Each value is the average of three measurements.

Raza Pashayi had the highest values in SSC (711.79 μg g^{-1}), TPC (518.13 μg g^{-1}), and CAT (220.78 units min^{-1} g^{-1}) in the availability of the SOES treatment, while Sandra had the highest values in PC (2446.05 μg g^{-1}) and GPA (0.42 units min^{-1} g^{-1}) under the first stress stage, as shown in Table S6. In comparison with irrigated plants during the second stress stage, the Sandra genotype recorded the highest values for AC (1151.89 μg g^{-1}), GPA (0.40 units min^{-1} g^{-1}), and CAT (214.29 units min^{-1} g^{-1}). When SOES was applied, Sandra had the highest SSC (976.60 μg g^{-1}), GPA (0.62 units min^{-1} g^{-1}), and CAT (363.64 units min^{-1} g^{-1}) scores, while Raza Pashayi had the highest PC (2571.69 μg g^{-1}) score during both stress stages.

3.5. GC/MS Analysis of Oak Leaf Extract

Table 7 displays the phytochemical composition of the extracts as determined by GC/MS analysis. The extract contained twenty-four components. The major compounds were heptasiloxane, 1,1,3,3,5,5,7,7,9,9,11,11,13,13-tetradecamethyl-(32.50%), silane, dimethoxydimethyl-(11.67), octasiloxane, 1,1,3,3,5,5,7,7,9,9,11,11,13,13,15,15-hexadecamethyl-(10.88%), 1-hexadecanol (9.37%), behenic alcohol (8.86%), 2,4-di-tert-butylphenol (7.02%), 1-octadecene (6.73%), acetic acid, chloro-, octadecyl ester (1.55%), dichloroacetic acid, 4-hexadecyl ester (1.52%), coumatetralyl isomer-2 ME (1.49%), 1-dodecanol (1.29%), chloroacetic acid, and pentadecyl ester (1.01%).

Table 7. Substances detected by GC/MS analysis and biological activity of the major compounds in the leaf water extract of *Quercus aegilops*.

Name of Compound	Retention Time (min)	Peak Area	Concentration (%)	Biological Activity of Major Compounds
Silane, dimethoxydimethyl-	5.17	7,202,705.00	11.67	Antibacterial [43]
Cyclotrisiloxane, hexamethyl-	6.93	327,535.00	0.53	
Silane, methyldimethoxyethoxy-	8.30	268,638.00	0.44	
Oxime-, methoxy-phenyl-	9.38	231,616.00	0.38	
Tetraethyl silicate	10.54	314,412.00	0.51	
1-Dodecanol	13.90	796,766.00	1.29	Antibacterial [44]
1-Hexadecanol	16.73	1,941,423.00	9.37	Reduction of evaporation [45]
Carbonic acid, decyl undecyl ester	16.84	397,832.00	0.64	
7-Tetradecene	16.90	307,446.00	0.50	

Table 7. *Cont.*

Name of Compound	Retention Time (min)	Peak Area	Concentration (%)	Biological Activity of Major Compounds
Chloroacetic acid, tetradecyl ester	17.04	250,824.00	0.41	
2,4-Di-tert-butylphenol	18.29	4,329,760.00	7.02	Antioxidant [46–48]
Carbonic acid, eicosyl vinyl ester	19.31	422,435.00	0.68	
Dichloroacetic acid, 4-hexadecyl ester	19.36	536,815.00	1.52	Antimicrobial [49]
1-Octadecene	21.45	4,150,402.00	6.73	Antioxidant and antimicrobial [50,51]
Acetic acid, chloro-, octadecyl ester	21.58	542,636.00	1.55	No activity was reported
1,2-Benzenedicarboxylic acid, bis(2-methylpropyl) ester	22.32	229,532.00	0.37	
18-Norabietane	23.10	242,753.00	0.39	
Behenic alcohol	23.48	3,132,644.00	8.86	Antifungal [52]
Chloroacetic acid, pentadecyl ester	23.58	273,527.00	1.01	No activity was reported
Coumatetralyl isomer-2 ME	23.67	918,610.00	1.49	No activity was reported
Acetic acid, chloro-, octadecyl ester	24.34	507,902.00	0.82	
Cyclotetrasiloxane, octamethyl-	27.02	268,576.00	0.44	
Heptasiloxane, 1,1,3,3,5,5,7,7,9,9,11,11,13,13-tetradecamethyl-	32.59	21,294,993.00	32.50	Insecticidal and antibacterial [53,54]
Octasiloxane, 1,1,3,3,5,5,7,7,9,9,11,11,13,13,15,15-hexadecamethyl-	38.79	6,719,421.00	10.88	

4. Discussion

Plant growth results from cell division, cell enlargement, and differentiation and is regulated by a wide range of genetic, physiological, ecological, and morphological processes, as well as the interaction between these factors [42]. Damage to physiological and biochemical processes, such as a delay in stomatal conductance, a decrease in nutrient uptake, a breakdown of leaf pigments, a decrease in photosynthesis, a stop in the rate of net assimilation and photosystem photochemical efficiency parameters, an increase in reactive oxygen species (ROS), and oxidative damage caused by water stress, reduced the morphological features [55]. The fresh weight, plant height, and productivity of the stressed tomato plants were all lower than those of the control plant (watered plants), as found by previous studies [12,56–58]. Relative water content and total chlorophyll content also decreased under SS condition. The same results were also found in tomato plants studied by Khan et al. [59], Ibrahim et al. [55], and Ullah et al. [42].

Root fresh weight (RFW) and root dry weight (RDW) under situations of water stress have shown significantly increased percentages for all degrees of treatment under all stress stages. The plant treated with SOBS and SOES had significantly higher RFW and RDW trait values than the control group (SW) during all stress stages. As a comparison among the three stress stages, the plants treated with SOBS showed the greatest increases in RWF (107%) and RDW (127.80%) in the second stress stage. The increased root surface area and root volume in plants during the search for water in the soil is mostly responsible for the higher RFW and RDW observed across all stress stages in comparison with untreated and unstressed plants. Additionally, a large number of prominent compounds found in leaf extract, including silane, heptasiloxane, and octasiloxane are thought to be silicon (Si) sources and are responsible for the increasing RL, RFW, and RDW in plants exposed to SOES at all stress levels. The leaf extract also had the compounds 2,4-Di-tert-butylphenol and 1-octadecene, which have antioxidant properties that reduce the synthesis of ROS products and membrane lipid peroxidation [46–48,50]. In addition, the leaf extract contained a 1-hexadecanol compound, which is used to reduce water evaporation in reservoirs. Si-enhanced cell-wall extensibility in the root's growth zone likely contributes to root elongation. Root density and length were both increased by Si in Purslane [60]. Sorghum's root length was found to be increased by Si, according to research by Sonobe et al. [61]. It is also likely that the higher RFW and RDW in SOES-treated plants are due to the ability of Si and 2,4-Di-tert-butylphenols to minimize ROS overproduction, which reduces membrane lipid peroxidation. On the other hand, our research showed that both SFW and SDW were lower in the SOES-treated plant. This may be because of the fact that Si controls the levels of polyamine and 1-aminocyclopropane-1-carboxylic acid in response to drought stress, which improves root growth, the ratio of roots to shoots, water uptake at the roots, and hydraulic conductance. Root endodermal silicification and suberization are also boosted by Si-mediated alterations in root growth, which help plants better retain water and tolerate the negative effects of drought [62]. In comparison with plants treated with SS and SOS, SFW and SDW in plants treated with SOES and SOBS may have increased owing to a decrease in ROS products and membrane lipid peroxidation. Furthermore, RFW and RDW increased in plants treated with SOBS throughout all stress stages, and these increases were induced by the presence of cytokinin, enzymes (lipase, amylase, protease, and chitinase), *Bacillus subtilis*, and *Pesudomonas putida*. These components of the SOBS treatment improve root area and volume by degrading organic matter and boosting phosphorus availability in the soil [63]. *Bacillus subtilis* and *Pesudomonas putida* invade plant rhizospheres and produce volatile organic chemicals that can affect plant development and root architecture in a variety of plants [64,65].

Drought stress, on the other hand, can alter the chemical composition of fruits. Organic acids (malic and citric acid) and soluble sugars are among the primary osmotic components found in ripe fruits [66]. Organic acids are stored by plants in order to reduce their osmotic potential and prevent cell turgor pressure from decreasing [67,68]. Vitamin C, also known as ascorbic acid, is found in all parts of plants. It plays a pivotal role in the development and expansion of plants. Ascorbic acid is the plant's primary antioxidant, which neutralizes the active forms of oxygen. Our results showed that the ascorbic acid content of the red fruit of the stressed plant increased owing to the water shortage. This increase in ascorbic acid may be vital for detoxifying reactive oxygen species. Antioxidant capability is determined by the phenolic contents of tomato fruits (TPC), and an increase in the TPC amount results in a decrease in oxidative alterations in cells owing to a lower concentration of free radicals [69,70].

Fructose and glucose levels both increase sharply when tomatoes ripen. The total soluble solids (TSS) concentration is influenced by the carbohydrate, organic acid, protein, fat, and mineral components. Our results suggest that shifts in the glucose/fructose ratio and organic acid levels may be responsible for the observed reduction in TSS in our investigation [66]. Compared with SS and SOS circumstances, the availability of silane, heptasiloxane, octasiloxane, and 2,4-Di-tert-butylphenol increases SSC during SOES application, which decreases ROS production by triggering antioxidant systems.

The results of the genotype effects revealed that tomato genotypes responded differently to SS, SOS, SOES, and SOBS applications under water stress. According to ASC, CAC, SSC, and TPC data, drought stress reduced the quality of tomato tolerant genotypes treated with SOS, SOES, and SOBS.

Different reactions were seen in terms of the leaf biochemical responses in plants treated with SS, SOS, SOES, and SOBS under stressful conditions. The highest levels of lipid peroxidation (LP), a metabolic process that results in the oxidative degradation of lipids by reactive oxygen species, were observed in the untreated and stressed genotype condition (SS). As a result of this process, the lipids in the cell membrane may break down, which can damage the cell and lead to its death. Low accumulations of biochemical compounds such as TPC, PC, SSC, AC, GPA, and CAT are responsible for this increase in LP. The genotypes treated under SOES and SOBS conditions, on the other hand, showed the highest levels of TPC, PC, SSC, AC, GPA, and CAT, which led to the reduction of LP. Furthermore, the SOES application may have induced the antioxidant systems, which may have contributed to the availability of silane, heptasiloxane, octasiloxane, and 2,4-Di-tert-butylphenol in the leaf extract.

Different responses were observed for the tolerant and sensitive genotypes during stress stages. Owing to the low accumulation of SSC, PC, TPC, AC, GPA, and CAT in sensitive geometries, the findings of leaf biochemical parameters showed the maximum

LP. Different response profiles between the tolerant genotypes were found. Under the first stages of stress, Raza Pashayi demonstrated the highest levels of TPC, GPA, and CAT, whereas Sandra's genotype had the highest levels of SSC, PC, and AC. Raza Pashyi recorded the highest values for GPA, AC, and SSC traits during the second stress stage, while Sandra had the maximum values for TPC, CAT, and AC.

5. Conclusions

According to our findings, the genotypes responded differently to the application of SS, SOS, SOES, and SOBS at various stress stages. In contrast to untreated and stressed plants, tomato plants treated with SOS, SOES, and SOBS showed a slight decrease in the morpho-physiological and fruit physicochemical attributes in response to drought stress. Additionally, a combination of the two stress stages resulted in a greater decrease in these features than either the first or second stress stage alone. All tomato genotypes exposed to SOES and SOBS exhibited significant levels of TPC, ASC, and SSC characteristics along with low amounts of TA in fruit. In fruit TPC, ASC, TSS, CAC, and SSC, the in vitro tolerant genotypes (Sandra and Raza Pashyi) outperformed the in vitro intolerant genotypes (Braw and Yadgar). In the leaf tolerant genotypes treated with SOES and SOBS, the lowest levels of lipid peroxidation and the highest levels of TPC, AC, SSC, PC, GPA, and CAT were found. Based on the findings of this study, Raza Pashyi and Sandra are ideal for growing in places with limited water availability. Furthermore, these genotypes are beneficial for breeding projects aimed at developing drought-tolerant tomato cultivars. Furthermore, the use of oak leaf powder, oak leaf extract, and biofertilizer reduced the effect of drought stress on tomato plants. However, the use of oak leaf powder and oak leaf extract can be described as novel agricultural practices because they are low-cost, simple to use, and time-consuming, and they can meet the growing demands of the agricultural sector by providing environmentally sustainable techniques for enhancing plant resistance to abiotic stress. The usage of the combination of leaf crude extract, oak leaf powder, and arbuscular mycorrhizal fungus should be investigated further under stress conditions. In order to determine the biostimulation effects of oak leaf powder and oak leaf extract, it is important to test their impacts on plant growth and production under normal conditions.

Supplementary Materials: The following supporting information can be downloaded at https://www.mdpi.com/article/10.3390/agriculture12122082/s1, Table S1. F and probability (P) values of different morpho-physiological traits of tomato genotypes treated with different treatments under the first, second, and their combinations of stress stages; Table S2. F and probability (P) values of different fruit physicochemical characters of tomato genotypes treated with various treatments under the first, second, and their combinations of stress stages; Table S3. F and probability (P) values of different leaf biochemical parameters of tomato genotypes treated with different treatments under the first, second, and their combinations of stress stages; Table S4. Interaction effects of tomato genotypes and treatments on the different morpho-physiological characters under the first, second, and their combinations of stress stages; Table S5. Interaction effects of tomato genotypes and treatments on the different physicochemical traits under the first, second, and their combinations of stress stages; Table S6. Interaction effects of tomato genotypes and treatments on the different biochemical traits under the first, second, and their combination of stress stages; Figure S1. Experimental design layout of the variables investigated in this study for each genotype; Figure S2. Effect of different treatments on the root morphology of the Yadgar genotype under different stress conditions; Figure S3. Effect of different treatments on the fruit morphology of the Raza Pashayi genotype under first stress conditions.

Author Contributions: Conceptualization, N.A.-r.T. and F.M.W.G.; methodology, K.S.R. and D.D.L.; data curation, D.D.L. and K.S.R.; formal analysis, N.A.-r.T.; writing—original draft preparation, N.A.-r.T. and K.S.R.; writing—review and editing, N.A.-r.T., K.S.R. and D.D.L. All authors have read and agreed to the published version of the manuscript.

Funding: This research received no external funding.

Institutional Review Board Statement: Not applicable.

Informed Consent Statement: Not applicable.

Data Availability Statement: The article and supplementary files contain all data.

Acknowledgments: The authors would like to thank the University of Sulaimani College of Agricultural Engineering Sciences staff for their help and assistance during this project.

Conflicts of Interest: The authors declare no conflict of interest.

References

1. Klunklin, W.; Savage, G. Effect on quality characteristics of tomatoes grown under well-watered and drought stress sonditions. *Foods* **2017**, *6*, 56. [CrossRef] [PubMed]
2. Lahoz, I.; Pérez-de-Castro, A.; Valcárcel, M.; Macua, J.I.; Beltrán, J.; Roselló, S.; Cebolla-Cornejo, J. Effect of water deficit on the agronomical performance and quality of processing tomato. *Sci. Hortic.* **2016**, *200*, 55–65. [CrossRef]
3. FAOSTAT. Food and Agriculture Organization of the United Nations, Statistics Division. Statistical Data on Crops, Tomatoes, World. 2020. Available online: http://www.fao.org/faostat/en/#data/QCL (accessed on 24 October 2022).
4. Aldrich, H.T.; Salandanan, K.; Kendall, P.; Bunning, M.; Stonaker, F.; Külen, O.; Stushnoff, C. Cultivar choice provides options for local production of organic and conventionally produced tomatoes with higher quality and antioxidant content. *J. Sci. Food Agric.* **2010**, *90*, 2548–2555. [CrossRef] [PubMed]
5. Bergougnoux, V. The history of tomato: From domestication to biopharming. *Biotechnol. Adv.* **2014**, *32*, 170–189. [CrossRef] [PubMed]
6. Abbasi, S.; Sadeghi, A.; Safaie, N. Streptomyces alleviate drought stress in tomato plants and modulate the expression of transcription factors ERF1 and WRKY70 genes. *Sci. Hortic.* **2020**, *265*, 109206. [CrossRef]
7. Srinivasa Rao, N.K.; Bhatt, R.M.; Sadashiva, A.T. Tolerance to water stress in tomato cultivars. *Photosynthetica* **2000**, *38*, 465–467.
8. Jangid, K.K.; Dwivedi, P. Physiological responses of drought stress in tomato: A review. *Int. J. Agric. Environ. Biotechnol.* **2016**, *9*, 53–61. [CrossRef]
9. Liu, Y.; Huang, W.; Xian, Z.; Hu, N.; Lin, D.; Ren, H.; Chen, J.; Su, D.; Li, Z. Overexpression of SLGRAS40 in tomato enhances tolerance to abiotic stresses and influences auxin and gibberellin signaling. *Front. Plant Sci.* **2017**, *8*, 1659. [CrossRef]
10. Tamburino, R.; Vitale, M.; Ruggiero, A.; Sassi, M.; Sannino, L.; Arena, S.; Costa, A.; Batelli, G.; Zambrano, N.; Scaloni, A.; et al. Chloroplast proteome response to drought stress and recovery in tomato (*Solanum lycopersicum* L.). *BMC Plant Biol.* **2017**, *17*, 40. [CrossRef]
11. Stajner, D.; Saponjac, V.T.; Orlovi, S. Water stress induces changes in polyphenol pro file and antioxidant capacity in poplar plants (*Populus* spp.). *Plant Physiol. Biochem.* **2016**, *105*, 242–250.
12. Janni, M.; Coppede, N.; Bettelli, M.; Briglia, N.; Petrozza, A.; Summerer, S.; Vurro, F.; Danzi, D.; Cellini, F.; Marmiroli, N.; et al. *In vivo* phenotyping for the early detection of drought stress in tomato. *Plant Phenomics* **2019**, *2019*, 6168209. [CrossRef] [PubMed]
13. Liang, G.; Liu, J.; Zhang, J. Effects of drought stress on photosynthetic and physiological parameters of tomato. *J. Am. Soc. Hortic.* **2020**, *145*, 12–17. [CrossRef]
14. Cherit-Hacid, F.; Derridij, A.; Moulti-Mati, F.; Mati, A. Drought stress effect on some biochemical and physiological parameters; accumulation on total polyphenols and flavonoids in leaves of two provenance seedling *Pistacia lentiscus*. *Int. J. Res. Appl. Nat. Soc. Sci* **2015**, *3*, 127–138.
15. Li, X.; Liu, F. Drought Stress Memory and Drought Stress Tolerance in Plants: Biochemical and Molecular Basis. In *Drought Stress Tolerance in Plants*; Springer: Cham, Germany, 2016; Volume 1, pp. 17–44.
16. Li, F.; Chen, X.; Zhou, S.; Xie, Q.; Wang, Y.; Xiang, X.; Hu, Z.; Chen, G. Overexpression of SlMBP22 in tomato affects plant growth and enhances tolerance to drought stress. *Plant Sci.* **2020**, *301*, 110672. [CrossRef]
17. Zhang, Y.; Wan, S.; Liu, X.; He, J.; Cheng, L.; Duan, M.; Liu, H.; Wang, W.; Yu, Y. Overexpression of CsSnRK2.5 increases tolerance to drought stress in transgenic Arabidopsis. *Plant Physiol. Biochem.* **2020**, *150*, 162–170. [CrossRef]
18. Carvalho, J.L.N.; Hudiburg, T.W.; Franco, H.C.J.; DeLucia, E.H. Contribution of above- and belowground bioenergy crop residues to soil carbon. *Glob. Change Biol. Bioenergy* **2017**, *9*, 1333–1343. [CrossRef]
19. Cherubin, M.R.; Oliveira, D.M.D.S.; Feigl, B.J.; Pimentel, L.G.; Lisboa, I.P.; Gmach, M.R.; Varanda, L.L.; Morais, M.C.; Satiro, L.S.; Popin, G.V.; et al. Crop residue harvest for bioenergy production and its implications on soil functioning and plant growth: A review. *Sci. Agric.* **2018**, *75*, 255–272. [CrossRef]
20. Armada, E.; Portela, G.; Roldán, A.; Azcón, R. Combined use of beneficial soil microorganism and agrowaste residue to cope with plant water limitation under semiarid conditions. *Geoderma* **2014**, *232–234*, 640–648. [CrossRef]
21. Coskun, D.; Britto, D.T.; Huynh, W.Q.; Kronzucker, H.J. The role of silicon in higher plants under salinity and drought stress. *Front. Plant Sci.* **2016**, *7*, 1072. [CrossRef]
22. Chakma, R.; Saekong, P.; Biswas, A.; Ullah, H.; Datta, A. Growth, fruit yield, quality, and water productivity of grape tomato as affected by seed priming and soil application of silicon under drought stress. *Agric. Water Manag.* **2021**, *256*, 107055. [CrossRef]
23. Meena, M.L.; Gehlot, V.S.; Meena, D.C.; Kishor, S.; Kumar, S. Impact of biofertilizers on growth, yield and quality of tomato (*Lycopersicon esculentum* Mill.) cv. Pusa Sheetal. *J. Pharmacogn. Phytochem.* **2017**, *6*, 1579–1583.

24. Gagné-Bourque, F.; Bertrand, A.; Claessens, A.; Aliferis, K.A.; Jabaji, S. Alleviation of drought stress and metabolic changes in timothy (*Phleum pratense* L.) colonized with *Bacillus subtilis* B26. *Front. Plant Sci.* **2016**, *7*, 584. [CrossRef] [PubMed]
25. Hashem, A.; Tabassum, B.; Fathi Abd_Allah, E. Bacillus subtilis: A plant-growth promoting rhizobacterium that also impacts biotic stress. *Saudi J. Biol. Sci.* **2019**, *26*, 1291–1297. [CrossRef] [PubMed]
26. Yin, P.; Yang, L.; Li, K.; Fan, H.; Xue, Q.; Li, X.; Sun, L.; Liu, Y. Bioactive components and antioxidant activities of oak cup crude extract and its four partially purified fractions by HPD-100 macroporous resin chromatography. *Arab. J. Chem.* **2019**, *12*, 249–261. [CrossRef]
27. Sánchez-Burgos, J.A.; Ramírez-Mares, M.V.; Larrosa, M.M.; Gallegos-Infante, J.A.; González-Laredo, R.F.; Medina-Torres, L.; Rocha-Guzmán, N.E. Antioxidant, antimicrobial, antitopoisomerase and gastroprotective effect of herbal infusions from four Quercus species. *Ind. Crops Prod.* **2013**, *42*, 57–62. [CrossRef]
28. Jong, J.K.; Bimal, K.G.; Hyeun, C.S.; Kyung, J.L.; Ki, S.S.; Young, S.C.; Taek, S.Y.; Ye-Ji, L.; Eun-Hye, K.; Ill-Min, C. Comparison of phenolic compounds content in indeciduous *Quercus* species. *J. Med. Plants Res.* **2012**, *6*, 5228–5239. [CrossRef]
29. Brossa, R.; Casals, I.; Pintó-Marijuan, M.; Fleck, I. Leaf flavonoid content in *Quercus ilex* L. resprouts and its seasonal variation. *Trees* **2009**, *23*, 401–408. [CrossRef]
30. Korbekandi, H.; Chitsazi, M.R.; Asghari, G.; Bahri Najafi, R.; Badii, A.; Iravani, S. Green biosynthesis of silver nanoparticles using *Quercus brantii* (oak) leaves hydroalcoholic extract. *Pharm. Biol.* **2015**, *53*, 807–812. [CrossRef]
31. Lavado, G.; Ladero, L.; Cava, R. Cork oak (*Quercus suber* L.) leaf extracts potential use as natural antioxidants in cooked meat. *Ind. Crops Prod.* **2021**, *160*, 113086. [CrossRef]
32. Himes, F.L.; Tejeira, R.; Hayes, M.H.B. The reactions of extracts from maple and oak leaves with iron and zinc compounds. *Soil Sci. Soc. Am. J.* **1963**, *27*, 516–519. [CrossRef]
33. Lateef, D.; Mustafa, K.; Tahir, N. Screening of Iraqi barley accessions under PEG-induced drought conditions. *All Life* **2021**, *14*, 308–332. [CrossRef]
34. Rahman, A.; Kandpal, L.M.; Lohumi, S.; Kim, M.S.; Lee, H.; Mo, C.; Cho, B.K. Nondestructive estimation of moisture content, pH and soluble solid contents in intact tomatoes using hyperspectral imaging. *Appl. Sci.* **2017**, *7*, 109. [CrossRef]
35. Rasul, K.S.; Grundler, F.M.W.; Abdul-razzak Tahir, N. Genetic diversity and population structure assessment of Iraqi tomato accessions using fruit characteristics and molecular markers. *Hortic. Environ. Biotechnol.* **2022**, *63*, 523–538. [CrossRef]
36. Pedro, A.M.K.; Ferreira, M.M.C. Simultaneously calibrating solids, sugars and acidity of tomato products using PLS2 and NIR spectroscopy. *Anal. Chim. Acta* **2007**, *595*, 221–227. [CrossRef]
37. Abbasi, N.A.; Ali, I.; Hafiz, I.A.; Alenazi, M.M.; Shafiq, M. Effects of putrescine application on peach fruit during Storage. *Sustainability* **2019**, *11*, 2013. [CrossRef]
38. Ferrante, A.; Spinardi, A.; Maggiore, T.; Testoni, A.; Gallina, P.M. Effect of nitrogen fertilisation levels on melon fruit quality at the harvest time and during storage. *J. Sci. Food Agric.* **2008**, *88*, 707–713. [CrossRef]
39. Juknys, R.; Vitkauskaitė, G.; Račaitė, M.; Venclovienė, J. The impacts of heavy metals on oxidative stress and growth of spring barley. *Cent. Eur. J. Biol.* **2012**, *7*, 299–306. [CrossRef]
40. XLSTAT. *Statistical and Data Analysis Solution*; Addinsoft: Boston, MA, USA, 2019.
41. Tahir, N.A.-R.; Lateef, D.D.; Mustafa, K.M.; Rasul, K.S. Under natural field conditions, exogenous application of moringa organ water extract enhanced the growth- and yield-related traits of barley accessions. *Agriculture* **2022**, *12*, 1502. [CrossRef]
42. Ullah, U.; Ashraf, M.; Shahzad, S.M.; Siddiqui, A.R.; Piracha, M.A.; Suleman, M. Growth behavior of tomato (*Solanum lycopersicum* L.) under drought stress in the presence of silicon and plant growth promoting rhizobacteria. *Soil Environ.* **2016**, *35*, 65–75.
43. Fathalipour, S.; Mardi, M. Synthesis of silane ligand-modified graphene oxide and antibacterial activity of modified graphene-silver nanocomposite. *Mater. Sci. Eng. C* **2017**, *79*, 55–65. [CrossRef]
44. Togashi, N.; Shiraishi, A.; Nishizaka, M.; Matsuoka, K.; Endo, K.; Hamashima, H.; Inoue, Y. Antibacterial activity of long-chain fatty alcohols against *Staphylococcus aureus*. *Molecules* **2007**, *12*, 139–148. [CrossRef]
45. Gugliotti, M.; Baptista, M.S.; Politi, M.J. Reduction of evaporation of natural water samples by monomolecular films. *J. Braz. Chem. Soc.* **2005**, *16*, 1186–1190. [CrossRef]
46. Kolyada, M.N.; Osipova, V.P.; Berberova, N.T.; Shpakovsky, D.B.; Milaeva, E.R. Antioxidant activity of 2,6-Di-tert-butylphenol derivatives in lipid peroxidation and hydrogen peroxide decomposition by human erythrocytes *in vitro*. *Russ. J. Gen. Chem.* **2018**, *88*, 2513–2517. [CrossRef]
47. Varsha, K.K.; Devendra, L.; Shilpa, G.; Priya, S.; Pandey, A.; Nampoothiri, K.M. 2,4-Di-tert-butyl phenol as the antifungal, antioxidant bioactive purified from a newly isolated *Lactococcus* sp. *Int. J. Food Microbiol.* **2015**, *211*, 44–50. [CrossRef] [PubMed]
48. Zhao, F.; Wang, P.; Lucardi, R.D.; Su, Z.; Li, S. Natural sources and bioactivities of 2,4-Di-tert-butylphenol and Its analogs. *Toxins* **2020**, *12*, 35. [CrossRef]
49. Jaradat, N.; Ghanim, M.; Abualhasan, M.N.; Rajab, A.; Kojok, B.; Abed, R.; Mousa, A.; Arar, M. Chemical compositions, antibacterial, antifungal and cytotoxic effects of *Alhagi mannifera* five extracts. *J. Complement. Integr. Med.* **2021**. [CrossRef]
50. Tonisi, S.; Okaiyeto, K.; Hoppe, H.; Mabinya, L.V.; Nwodo, U.U.; Okoh, A.I. Chemical constituents, antioxidant and cytotoxicity properties of *Leonotisleonurus* used in the folklore management of neurological disorders in the Eastern Cape, South Africa. *3 Biotech* **2020**, *10*, 141. [CrossRef]

51. Gautam, V.; Kohli, S.K.; Arora, S.; Bhardwaj, R.; Kazi, M.; Ahmad, A.; Raish, M.; Ganaie, M.A.; Ahmad, P. Antioxidant and antimutagenic activities of different fractions from the leaves of *Rhododendron arboreum* Sm. and their GC-MS profiling. *Molecules* **2018**, *23*, 2239. [CrossRef]
52. Bhat, M.Y.; Talie, M.D.; Wani, A.H.; Lone, B.A. Chemical composition and antifungal activity of essential oil of *Rhizopogon* species against fungal rot of apple. *J. Appl. Biol. Sci.* **2020**, *14*, 296–308.
53. Farag, S.M.; Essa, E.E.; Alharbi, S.A.; Alfarraj, S.; Abu El-Hassan, G.M.M. Agro-waste derived compounds (flax and black seed peels): Toxicological effect against the West Nile virus vector, Culex pipiens L. with special reference to GC–MS analysis. *Saudi J. Biol. Sci.* **2021**, *28*, 5261–5267. [CrossRef]
54. Al-Samman, A.M.; Siddique, N.A. Gas chromatography-mass spectrometry (GC-MS/MS) analysis, ultrasonic assisted extraction, antibacterial and antifungal activity of *Emblica officinalis* fruit extract. *Pharmacogn. J.* **2019**, *11*, 315–323. [CrossRef]
55. Ibrahim, M.F.M.; Abd Elbar, O.H.; Farag, R.; Hikal, M.; El-Kelish, A.; El-Yazied, A.A.; Alkahtani, J.; Abd El-Gawad, H.G. Melatonin counteracts drought induced oxidative damage and stimulates growth, productivity and fruit quality properties of tomato plants. *Plants* **2020**, *9*, 1276. [CrossRef] [PubMed]
56. Eziz, A.; Yan, Z.; Tian, D.; Han, W.; Tang, Z.; Fang, J. Drought effect on plant biomass allocation: A meta-analysis. *Ecol. Evol.* **2017**, *7*, 11002–11010. [CrossRef] [PubMed]
57. Wang, X.; Xing, Y. Evaluation of the effects of irrigation and fertilization on tomato fruit yield and quality: A principal component analysis. *Sci. Rep.* **2017**, *7*, 350. [CrossRef]
58. Zhou, R.; Yu, X.; Ottosen, C.O.; Rosenqvist, E.; Zhao, L.; Wang, Y.; Yu, W.; Zhao, T.; Wu, Z. Drought stress had a predominant effect over heat stress on three tomato cultivars subjected to combined stress. *BMC Plant Biol.* **2017**, *17*, 24. [CrossRef]
59. Khan, S.H.; Khan, A.; Litaf, U.; Shah, A.S.; Khan, M.A.; Bilal, M.; Ali, M.U. Effect of Drought Stress on Tomato cv. Bombino. *J. Food Process Technol.* **2015**, *6*, 465. [CrossRef]
60. Kafi, M.; Rahimi, Z. Effect of salinity and silicon on root characteristics, growth, water status, proline content and ion accumulation of purslane (*Portulaca oleracea* L.). *J. Soil Sci. Plant Nutr.* **2011**, *57*, 341–347. [CrossRef]
61. Sonobe, K.; Hattori, T.; An, P.; Tsuji, W.; Eneji, A.E.; Kobayashi, S.; Kawamura, Y.; Tanaka, K.; Inanaga, S. Effect of silicon application on sorghum root responses to water stress. *J. Plant Nutr.* **2010**, *34*, 71–82. [CrossRef]
62. Wang, M.; Wang, R.; Mur, L.A.J.; Ruan, J.; Shen, Q.; Guo, S. Functions of silicon in plant drought stress responses. *Hortic. Res.* **2021**, *8*, 254. [CrossRef]
63. Kim, H.-J.; Li, X. Effects of phosphorus on shoot and root growth, partitioning, and phosphorus utilization efficiency in Lantana. *HortScience* **2016**, *51*, 1001–1009. [CrossRef]
64. Bavaresco, L.G.; Osco, L.P.; Araujo, A.S.F.; Mendes, L.W.; Bonifacio, A.; Araújo, F.F. *Bacillus subtilis* can modulate the growth and root architecture in soybean through volatile organic compounds. *Theor. Exp. Plant Physiol.* **2020**, *32*, 99–108. [CrossRef]
65. Ortiz-Castro, R.; Campos-García, J.; López-Bucio, J. *Pseudomonas putida* and *Pseudomonas fluorescens* influence Arabidopsis root system architecture through an auxin response mediated by bioactive cyclodipeptides. *J. Plant Growth Regul.* **2020**, *39*, 254–265. [CrossRef]
66. Medyouni, I.; Zouaoui, R.; Rubio, E.; Serino, S.; Ahmed, H.B.; Bertin, N. Effects of water deficit on leaves and fruit quality during the development period in tomato plant. *Food Sci. Nutr.* **2021**, *9*, 1949–1960. [CrossRef] [PubMed]
67. Ma, W.-F.; Li, Y.-B.; Nai, G.-J.; Liang, G.-P.; Ma, Z.-H.; Chen, B.-H.; Mao, J. Changes and response mechanism of sugar and organic acids in fruits under water deficit stress. *Peer J.* **2022**, *10*, e13691. [CrossRef]
68. Menezes-Silva, P.E.; Sanglard, L.; Ávila, R.T.; Morais, L.; Martins, S.C.V.; Nobres, P.; Patreze, C.M.; Ferreira, M.A.; Araújo, W.L.; Fernie, A.R.; et al. Photosynthetic and metabolic acclimation to repeated drought events play key roles in drought tolerance in coffee. *J. Exp. Bot.* **2017**, *68*, 4309–4322. [CrossRef]
69. Jian-Kang, Z. Plant salt tolerance. *Trends Plant Sci.* **2001**, *6*, 66–71.
70. Wang, L.; Meng, X.; Yang, D.; Ma, N.; Wang, G.; Meng, Q. Overexpression of tomato GDP-L-galactose phosphorylase gene in tobacco improves tolerance to chilling stress. *Plant Cell Rep.* **2014**, *33*, 1441–1451. [CrossRef]

MDPI
St. Alban-Anlage 66
4052 Basel
Switzerland
Tel. +41 61 683 77 34
Fax +41 61 302 89 18
www.mdpi.com

Agriculture Editorial Office
E-mail: agriculture@mdpi.com
www.mdpi.com/journal/agriculture

www.ingramcontent.com/pod-product-compliance
Lightning Source LLC
LaVergne TN
LVHW071305170726
843515LV00006BA/1490

* 9 7 8 3 0 3 6 5 6 9 6 4 2 *